OPTRONIC TECHNIQUES IN DIAGNOSTIC AND THERAPEUTIC MEDICINE

OPTRONIC TECHNIQUES IN DIAGNOSTIC AND THERAPEUTIC MEDICINE

Edited by
Riccardo Pratesi

*Institute of Quantum Electronics
and Department of Physics
Florence, Italy*

Springer Science+Business Media New York

Library of Congress Cataloging-in-Publication Data

Workshop on Optronic Techniques in Diagnostic and Therapeutic Medicine
 (1990 : Florence, Italy)
   Optronic techniques in diagnostic and therapeutic medicine /
 edited by Riccardo Pratesi.
        p.   cm.
     "Based on the proceedings of a Workshop on Optronic Techniques in
 Diagnostic and Therapeutic Medicine, held March 26-27, 1990, in
 Florence, Italy"--T.p. verso.
   Includes bibliographical references and indexes.
   ISBN 978-0-306-43938-4   ISBN 978-1-4615-3766-3 (eBook)
   DOI 10.1007/978-1-4615-3766-3
   1. Lasers in medicine--Congresses. 2. Optoelectronics-
 -Congresses.   I. Pratesi, R. II. Title.
   [DNLM: 1. Laser Surgery--congresses. 2. Lasers--diagnostic use-
 -congresses. 3. Lasers--therapeutic use--congresses.   WB 117 W926o
 1990]
 R857.L37W67   1990
 617'.05--dc20
 DNLM/DLC
 for Library of Congress   91-24101
                                                                      CIP

Based on the proceedings of a Workshop on
Optronic Techniques in Diagnostic and Therapeutic Medicine,
held March 26–27, 1990, in Florence, Italy

ISBN 978-0-306-43938-4

© 1991 Springer Science+Business Media New York
Originally published by Plenum Press,New York in 1991
Softcover reprint of the hardcover 1st edition 1991

To Nadia
with love

PREFACE

The papers in this Volume were given at a two-day Conference on the subject of Optoelectronics in Medicine. The meeting was held in Florence, and promoted by the Consortium *Centro di Eccellenza Optronica* (C.E.O.). It represented the first of a series of Meetings on Optoelectronics that C.E.O. is organizing in order to stimulate new developments in this field and more efficient cooperation among local, national, and international research centers, industries, utilizers, etc..

Italian scientists have contributed consistently to the development of laser sources and to their applications to Medicine. A significant role has also been played by research institutes and industries in Florence. However, in this Conference, and in the Proceedings only a few Italian scientists were invited to present a lecture, thus offering the local and national communities as wide an international view as possible. Many more were present, however, as chairmen, and contributed successfully to making the discussions stimulating and fruitful. As Editor, I had to substitute last-minute missing manuscripts with papers of my own, in order to keep the scheduled index of papers.

The contributions presented at the Conference are written as extended, review-like papers to provide a broad and representative coverage of the fields of light sources, optoelectronic systems for medical diagnosis, and light and laser applications to Medicine.

The aim of the Conference was to bring together scientists and clinicians that belong to different disciplines and consequently do not usually meet, but who have as a common link the use of optoelectronic instrumentation, techniques and procedures.

The Conference was an important opportunity to discuss latest developments and emerging perspectives on the use of light sources and optoelectronic techniques for diagnostic and therapeutic purposes.

Progress in the medical uses of light can only be achieved with systematic research on the understanding of the basic processes that regulate the interactions of optical radiation with living matter and of the subsequent photobiological events. Optoelectronic systems should be carefully developed to ensure optimized performances, and cooperation among scientists of numerous disciplines is an essential step towards this goal. Moreover, the information on recent progress and potentials of specific discoveries and/or achievements must be at the disposal of this community if it is to produce promptly innovative systems and procedures of clinical utility.

The various Parts of this Volume offer a reasonably wide view of present uses of light in Medicine, with several tutorial papers on physical and biological processes of fundamental relevance for clinical applications. The part dedicated to laser sources has been limited to the very important class of miniaturized lasers, which are entering the bio-medical field and will represent a key component for the development of reliable and compact optoelectronic systems for diagnosis and therapy.

Riccardo Pratesi
Conference Chairman

Florence, March 1991

INTRODUCTORY REMARKS : The Consortium CEO

Many aspects of optics and microelectronics have today come together to create a new discipline, that of "Optoelectronics". An important factor in the development of this emerging field has been the progress made in communications and material physics. In the literature, the term "optronic" is already frequently used with reference to a system rather than a single device.

The role of optoelectronics in many areas of production is destined to be as fundamental and innovatory as that of microelectronics. It is already playing an important and growing role in industry. In the laboratory and in medicine, in environmental monitoring and in consumer production, it is fast becoming the keystone for further development.

As is well known, the city of Florence is home to innumerable priceless art treasures. Less well known is the fact that the city boasts many public institutions, University Faculties and Research Institutes which have an international reputation in basic and applied research in the fields of solid-state physics, optics, laser sources, optical fibres, photodetectors, signal and image processing, and artificial intelligence.

Florence is also home to industrial concerns of international standing such as Officine Galileo S.p.A., which employ technologies connected with the above-mentioned fields.

Such a background made Florence a very suitable location for the Consortium "Centro di Eccellenza Optronica" (CEO).

Associated with the CEO are:

- Florence University, through its Departments of Physics and Electronic Engineering;

- The National Research Council (CNR), through its Florentine Research Institutes, Institute of Quantum Electronics (IEQ) and Institute of Electromagnetic Waves (IROE);

- The Institute of Optics (INO), Florence;

- Officine Galileo S.p.A. (OG), Florence.

The CEO is currently engaged in important scientific projects both at national and international levels.

Florence OPTRONICS is a congress forum through which the CEO periodically provides the scientific community with an opportunity to present ideas and verify research results. *"Optronic Techniques in Diagnostic and Therapeutic Medicine"* is the first meeting organized by CEO - Florence OPTRONICS.

G. Longobardi

CEO Director

x

ACKNOWLEDGEMENTS

We thank the participants for making the first edition of C.E.O.'s meetings on Optoelectronics a stimulating conference. Special thanks are due to each of the presenters for the high quality of their contributions and for preparing the manuscripts which comprise this Volume. We are also indebted to R.L. Byer and F. Hillenkamp, who contributed very much with their lectures to make the conference a success. Particular thanks to R.R. Anderson, J.-L. Boulnois, and B. Wilson, who provide additional, excellent papers, making the Volume more complete.

We gratefully acknowledge Valdo Spini for his introductory speech as the representative of the City of Florence, Curzio Cipriani, Nicola Rubino, and Tito Arecchi for welcoming the participants on behalf of the University of Florence, C.N.R., and the Consortium C.E.O., respectively.

We are indebted to our colleagues for having accepted to participate in the Conference as chairmen of the various sessions, namely, A.Andreoni (University of Naples), R.Cubeddu (CEQSE-CNR, Milan), V.Degiorgio (University of Pavia), H.Emanuelli (National Cancer Institute, Milan), A.Marino (ENEA, Rome), S.Martellucci (University of Rome), S.Riva-San Severino (University of Palermo), M.Rubino (University of Bari), V.Russo (University of Salerno), P.Santoianni (University of Naples), A.M.Scheggi (IROE-CNR, Florence), S.Solimeno (University of Naples), A.Sona (CISE, Milan), F.Strumia (University of Pisa), O. Svelto (CEQSE-CNR, Milan).

The Conference was made possible by the generous and timely financial support of many public and private institutions. They provided funds which made the Conference possible, and enabled us to create a relaxed and intellectually stimulating atmosphere during the meeting.

Special thanks to the Italian National Research Council (C.N.R.), in particular to the Committees for Physical Sciences, Technology, and Bio-Medical Sciences. The sponsorship and cooperation from the CNR Special Project on Electro-optic Technologies is also acknowledged. The contribution of the European Community was particularly valuable.

The financial participation of Regione Toscana, Provincia di Firenze and CESVIT, Comune di Firenze, and Ente Provinciale del Turismo was very welcome, as it indicated the good degree of participation in the initiatives of our scientific community. ENEA also contributed financially to our Conference.

We gratefully acknowledge the financial support of several banks and industries, namely Cassa di Risparmio di Firenze, Banca Toscana, Monte dei Paschi di Siena, Banca Mercantile; Officine Galileo, El-En, S.M.A., O.T.E., Optikon, Sirio Panel, Space

Laser, Spectra Physics Italia, Valfivre Sistemi Laser, C.S.O., Elicam, DB-Electronic Instruments, New Tech, Perimed, Proel Tecnologie.

We tried to provide the Conference members with some of the most beautiful settings available in Florence. In this regard we are grateful to the Rector of the University of Florence for putting at our disposal the magnificent Aula Magna and the technical assistance of the university staff; and to the Mayor of Florence for the kind hospitality during the social events held in the Salone dei Dugento. We also thank Tibiarum Concentus for the delightful concert.

Many people worked to make this meeting a success. We thank our colleagues of the Scientific Committee of C.E.O.; special thanks to Antonella Somigli, Conference Secretary, for her expert and cheerful help, and to the secretarial staff, namely Carla Pardini and Umberto Vanni from I.E.Q.-C.N.R., Daniela Fubiani and Silvano Mascalchi from I.N.O., Sabrina Paoletti and Simone Boanini from O.G., for cheerfully working many long hours in the final days before the meeting, and for careful attention to many details.

The efficient and qualified assistance of Globus Travel Agency is also acknowledged.

Finally, we acknowledge Stephen J. Coffey of the Interdepartmental Language Centre, University of Pisa, for his accurate reviewing of the various manuscripts written by non-English authors in order to ensure a uniform language style in the present Volume.

The cheerful help of C.Pardini in preparing the final edition of this Volume is gratefully acknowledged.

Organizing Committee :

L. Landi
Officine Galileo

G. Longobardi
Istituto Nazionale di Ottica

R. Pratesi
Istituto di Elettronica Quantistica

OBITUARIES

We wish to remember in these pages our dear friends and colleagues Carlo A.Sacchi, Aldo Cingolani, and Vincenzo Gallucci, who died prematurely just before or soon after this Conference.

They were well known to the international scientific community for both their standard of research and their degree of humanity.

Carlo A.Sacchi dedicated a large part of his research activity to the studies of the interaction mechanisms of light with biological matter, and to the development of optical techniques for bio-medical applications. He was a pioneer in the promotion of laser techniques to Biology and Medicine in Italy, and his research and organizational work greatly contributed to the international dimension of our national efforts. He was invited to present a talk on laser-tissue interactions at this Conference. We are indebted to Franz Hillenkamp for having accepted to give this talk in Carlo's memory, which he did in a masterful fashion.

Aldo Cingolani had only a temporary encounter with laser biomedicine. Being a brilliant scientist, he was able to catch several basic aspects of the photophysics of biomolecules, and to leave his signature on numerous important papers of bio-medical interest. Like Carlo, Aldo played an important role in the promotion of laser medicine in Italy, cooperating with many clinicians and biologists. He was coming back to this field with renewed interest when death came.

Vincenzo Gallucci was a well known clinician, who joined the laser medical team only recently. As the pioneer of heart transplant in Italy and a skilled operator in coronary by-pass, he was invited to perform the first clinical laser angioplasty applications in coronary surgery in Italy. He achieved considerable success, and his experience was reported during the Conference. His death represents a great loss for the scientific medical community and for patients, and a severe blow to the development of laser coronary angioplasty both in Italy and at an international level in view of his ability and scientific rigour.

We are indebted to these friends for the example they have set us. They will live on in the work and in the memory of their friends and collaborators.

F.T. Arecchi

G. Longobardi

R. Pratesi

CONTENTS

PART 1. MINIATURE LASER SOURCE TECHNOLOGY

Properties of High Power Semiconductor Lasers 3
  D.F. Welch

Diode Laser Pumped Miniature Solid State Lasers 15
  R. Pratesi

PART 2. LIGHT-TISSUE INTERACTIONS AND OPTICS OF TISSUES

Biophysical Bases of Laser-Tissue Interactions 29
  J.L. Boulnois

Optical and Thermal Modeling of Tissues: Dosimetry 45
  M.J.C. van Gemert and A.J. Welch

PART 3. DIAGNOSTIC TECHNIQUES

Holography in Medical Diagnostics 61
  G. von Bally

Laser Reflectance Spectroscopy of Tissue 73
  B. Wilson and R.R. Anderson

Monitoring and Imaging of Tissue Blood Flow by Coherent
  Light Scattering 89
  G.E. Nilsson, A. Jakobsson, and K. Wårdell

Transillumination Imaging 101
  P.C. Jackson, H. Key, and P.N.T. Wells

Fluorescence Techniques in Diagnostic Medicine with
  Special Consideration of Optoelectronic Methods 117
  E. Unsöld

Survey of the UV and Visible Spectroscopic Properties of Normal
  and Atherosclerotic Human Artery Using Fluorewcence EEMS 129
  R. Richards-Kortum, R. Rava, J. Baraga, M. Fitzmaurice, J. Kramer,
  and M. Feld

Development of Near-IR Angiography 139
    R.W. Flower

PART 4. THERAPEUTIC TECHNIQUES

4.1 LAMPS

Ultraviolet Radiation Lamps for the Phototherapy of Psoriasis 151
    B.L. Diffey

Phototherapy and Photochemotherapy of Psoriasis 165
    T.B. Fitzpatrick

Light Therapy for Neonatal Jaundice 177
    J.F. Ennever

The Hazards of Cosmetic Tanning with UVA Radiation 191
    A.R. Young

4.2 LASERS

The Role of the Neodymium Yittrium Aluminum Garnet
        (Nd:YAG) Laser in Medicine 197
    H. Barr and S.G. Bown

Diode Laser Photocoagulation in Ophthalmology 207
    R. Brancato, G. Leoni, and G. Trabucchi

Lasers in Dermatology 213
    R.R. Anderson and J.A. Parrish

Photochemiotherapy of Tumours: Molecular and Biophysical Bases 227
    G. Jori and E. Reddi

Laserchemotherapy of Tumours: Clinical Aspects 237
    P. Spinelli, M. Dal Fante, and A. Mancini

Laser Angioplasty: State of the Art 249
    H.J. Geschwind

UV Laser Angioplasty: Clinical Aspects 259
    A. Fracasso and V. Gallucci

PART 5. FUTURE DIRECTIONS

Initial Applications and Potential of Miniature Lasers in Medicine 271
    R. Pratesi

Future Trends in Laser Medicine 287
    J.A. Parrish

Contributors **299**

Index **303**

PART 1

MINIATURE LASER SOURCE

TECHNOLOGY

PROPERTIES OF HIGH POWER SEMICONDUCTOR LASERS

David F. Welch

Spectra Diode Laboratories
80 Rose Orchard Way, San Jose, CA 95134-1356, USA

Recent advances in semiconductor lasers has resulted in their utilization in several surgical techniques including endoscopic photocoagulation, and tissue welding. Their utilization is a consequence of the high power, high efficiency, high reliability, low price, and compact size, of which the combination is unique to semiconductor lasers. In this article I will try to outline the advantages of semiconductor lasers in general and then deal with specific characteristics of several laser diodes.

BASIC THEORY

The semiconductor laser is different than other laser systems in that the population inversion required to reach the lasing threshold results from electrical charge injection into the active region, contrary to solid state lasers which are typically optically pumped. When the quasi-Fermi levels associated with the injected electron and holes of the forward biased p-n junction separate by an energy greater than the bandgap of the active region, the stimulated emission rate exceeds that of the spontaneous emission rate and the material lases.

To reduce the threshold current of the laser the active region is imbedded in a layer structure that not only supplies electrical confinement to the active layer but confines the optical mode to a region overlapping the active layer[1-4]. The waveguide is fabricated by the growth of lower refractive index materials on either side of the active region. The laser resonator is completed by the formation of partially reflecting mirrors by cleaving the semiconductor crystal perpendicular to the active region.

A schematic representation of a basic diode laser is presented in Figure 1. The GaAs active region is bounded by AlGaAs cladding layers. The higher bandgap AlGaAs results in electrical confinement of the injected carriers to the active region while the lower index of refraction of the AlGaAs forms a waveguide around the active region for optical confinement.   The current is confined to a narrow stripe of width S by incorporating a highly resistive layer either by the fabrication of a dielectric insulator or an isolation implantation layer. Current spreading in the y-direction results in a gain guided laser as discussed in more detail below. The crystal is cleaved to form mirrors at the output facets. The reflectivity of a cleaved GaAs surface is approximately 32%.

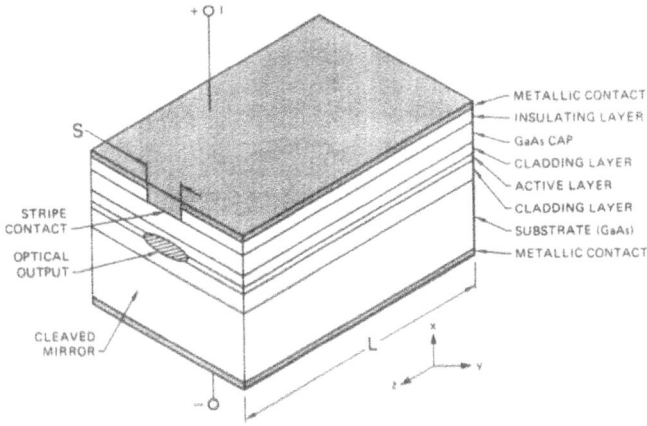

Fig.1. Schematic diagram of a semiconductor laser diode.

For most applications the facets are coated with a series of dielectric layers to increase the reflectivity of the back facet to greater than 95% while the front facet reflectivity is adjusted to either 5% or 30%, depending on the application.

The emission wavelength of the laser is a result of the bandgap of the material in the active region. There are four primary operating regions of semiconductor lasers as depicted in Figure 2. GaInAsP lasers, lattice matched to InP, emit in the 1.2 - 1.5 μm wavelength region[4]. Their primary use is for telecommunications where the dispersion and loss of the optical fibers are minimized. The high temperature sensitivity of the InGaAsP material system limits the output powers of these lasers to a few hundred milliwatts cw before the dissipated power quenches lasing. The temperature dependence of a laser is characterized by an increase in its threshold current. The threshold increase with temperature can be empirically determined by

$$I_{th1}/ I_{th2} = \exp \{(T_1 - T_2)/ T_o \}$$

where I_{th1} and I_{th2} are the threshold currents at temperature T_1 and T_2. T_o is a characteristic temperature of the material that is primarily dependent on the active layer material and secondarily dependent on material quality and laser design. Typical values of T_o for the InGaAsP material system are between 40°K and 90°K.

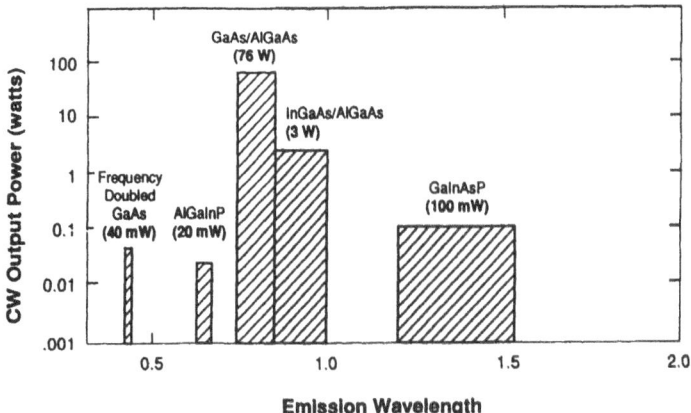

Fig. 2. Demonstrated output powers as a function of emission wavelength.

Emission wavelengths in the range from 0.78 µm to slightly greater than 1 µm have been demonstrated from compounds grown on GaAs substrates. The vast majority of the lasers consist of AlGaAs active regions, where a variety of wavelengths can be accessed by varying the concentration of Al in the active region. The lattice constant of AlGaAs is nearly identical to that of GaAs and, therefore, is easily grown on GaAs substrates. Wavelengths from 0.89 µm to 1.0 µm have been demonstrated by adding In to the GaAs active region[5-7]. Although InGaAs is not lattice matched to GaAs, the active region is thin enough to accommodate the strain associated with the mismatch[8,9]. GaAlAs lasers have demonstrated the highest efficiency and output power of any semiconductor laser system in excess of 60%[10,11]. Due to the low sensitivity of GaAs lasers to temperature, output powers as high as 76 W cw have been demonstrated from a monolithic array of lasers[12]. Typical T_0 values for GaAs lasers range between 150°K to 250°K.

AlGaInP lasers lattice matched to GaAs substrates are used for emission in the 0.63 µm to 0.68 µm wavelength region[13-15]. Laser demonstration in this material system as been achieved only recently, as a result the demonstrated output power is limited to a few tens of milliwatts. Preliminary temperature sensitivity measurements, however, indicate that T_0 is approximately 70°K. Consequently high power operation is highly dependent on the efficiency of the laser.

In order to access short emission wavelengths GaAs laser diodes have been coupled to nonlinear materials such as $KNbO_3$ for direct frequency doubling to wavelengths between 420 and 430 nm. The output powers in this range are limited by the input power from a single mode GaAs laser diode. Although GaAs has demonstrated very high output powers for multi-mode emitters, single mode lasers have been limited to 100 mW cw operation. As a result the emission in the blue has been limited to 40 mW cw.

As improvements in single mode lasers are demonstrated, the efficiency and total output powers of frequency doubled systems will improve dramatically.

Laser diodes can be classified as either gain guided lasers or index guided lasers. Gain guided lasers rely on current confinement to channel the charge to a narrow stripe of the active region. The optical mode that develops is a result of the preferential amplification along the active region. The output power from gain guided lasers, which can be fabricated with active region widths varying from a few microns to greater than 500 µm[18,19], scales with the extent of the aperture size.

Gain guided lasers, however, do not supply any optical confinement to the lasing mode in the direction parallel to the p-n junction. The output from a gain guided laser is, therefore, typically multi-spatial mode at large aperture sizes and is astigmatic[20,21]. The applications which these lasers are suited for are systems that do not require diffraction limited optics. Gain guided arrays are well suited for most medical applications where spots sizes greater than 50 µm are required.

Index guided lasers, on the other hand, augment the current confinement of gain guided with a real refractive index profile parallel to the p-n junction, resulting in an optical waveguide contiguous with the gain region in the direction transverse to the current flow. The index profile is produced by varying the composition of the material which bounds the active region. The output of an index guided laser is typically a single longitudinal and transverse mode. In addition, the astigmatism of these sources is less than a few microns. Single mode index guided lasers, however, can only be fabricated to approximately 4 µm in width prior to the onset of multi-transverse mode operation[11]. As a result these sources are limited in reliable output power to approximately 100 mW cw[22].

The differentiation between gain guided and index guided lasers is realized in the temporal and spatial coherence of the laser. Due to the negative index profile of a gain guided laser the laser operates in a multiple transverse modes. As a result the strehl ratio, or the percentage of the beam that can be represented by a gaussian distribution, is small. The positive index profile of the index guided laser, confines the optical mode, as a result the laser operates in a single transverse mode with strehl ratios in excess of 0.9. The following sections present the properties of both high power gain guided and index guided laser and laser arrays.

The thickness of the active region of high power lasers is approximately 10 nm. When the thickness of the active is less than the de Broglie wavelength of the electron, and the active region is bounded by a higher bandgap material a potential well is realized and the energy levels of the electron and hole are quantized. These structure, designated as quantum well lasers, have demonstrated very low threshold currents[23] on the order of 100 A/cm^2 and due to the low optical overlap of the optical mode with the active region the optical loss[10] is reduced to 3 cm^{-1} and the catastrophic power is increased to as high as 12 MW/cm$^{2(11)}$. All of the structures discussed below are fabricated with quantum well active regions.

GAIN GUIDED LASER ARRAYS

Large aperture gain guided laser arrays are fabricated by confining the current to a specific aperture size. Figure 3 presents a schematic diagram of a gain guide array. The AlGaAs quantum well active region is bounded by higher bandgap AlGaAs cladding and confining layers. The composition of Al in the active region ranges from $Al_{0.08}Ga_{0.92}As$ to GaAs for emission wavelengths between 0.78 µm and 0.88 µm. Selective proton implantation in the contact layer produces highly resistive regions, as a result the current is confined to a few hundred micron wide region.

The light output as a function of input current (L-I) of a gain guided array of aperture widths of 100 µm and 200 µm is presented in Figure 4. The maximum output power demonstrated from these arrays are 6 W cw and 8 W cw respectively. The output powers are limited by thermal dissipation in the laser and not catastrophic degradation of the facet. The threshold current of the high power lasers is on the order of 1 A. Similar designs that have been optimized for reliable emission at 1 W cw output and not maximum output power exhibit threshold currents in the range of 0.3 - 0.5 A.

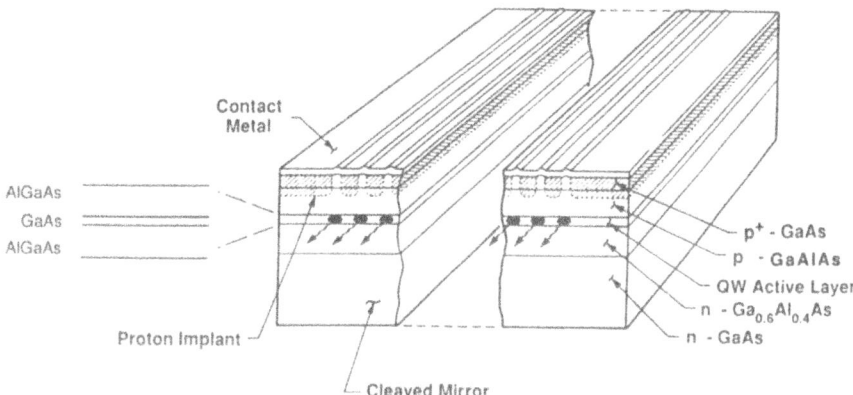

Fig. 3. Schematic diagram of a gain guided laser array.

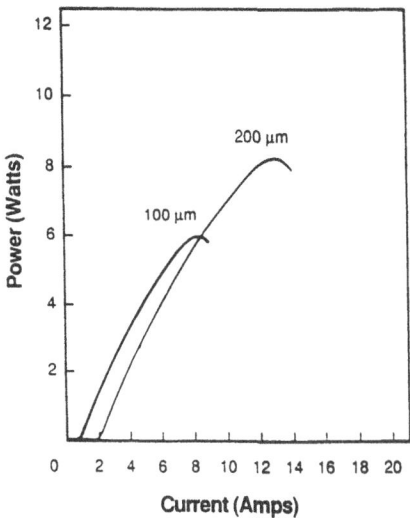

Fig.4.   Light output as a function of input current for a gain guided laser array of aperture width of 100 μm and 200 μm.

The radiation patterns of the 100 μm aperture array is presented in Figure 5. Although the radiation pattern is stable with increasing power to 3 W cw, the multi-lobe is a result of multiple transverse mode operation. The modes radiate within an envelope of approximately 10° in width. The radiation pattern perpendicular to the p-n junction is a gaussian profile with a divergence ranging from 20° to 45°, dependent on the particular epitaxial design of the laser. The spectral output is a series of modes associated with the multiple transverse and longitudinal mode operation. The spectral width is approximately 2 nm.

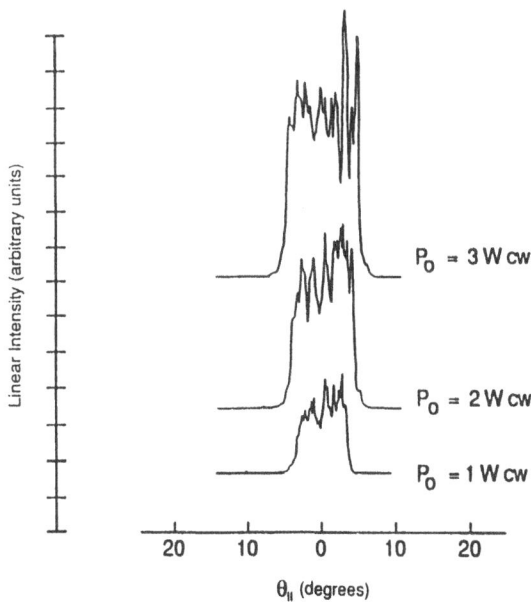

Fig. 5. Far field radiation patterns from a gain guided laser array.

To access wavelength between 0.88 μm and 1.0 μm, the laser structure is modified by incorporating an InGaAs quantum well active region. Although the InGaAs active region is not lattice matched to the GaAs substrate, if the active layer thickness is limited to less than approximately 10 nm the stress associated with the lattice mismatch is accommodated by the biaxial strain within the crystal and the device operates essentially identical to GaAs lasers. The emission wavelength can be extended to greater than 1.0 μm with the incorporation of as much as 30% In.

Figure 6 presents the L-I characteristic of a 100 μm aperture gain guided array with an InGaAs active region. The maximum output power is 3 W cw, comparable to GaAs lasers of similar waveguide design. The emission wavelength in this case is 960 nm.

Lasers designed to be reliable at greater than 3 W cw are fabricated by extending the source size to as large as 1 cm. Figure 7 presents the light output characteristics of a 1 cm monolithic bars of lasers, periodically spaced along the 1 cm length. The emitting aperture is 2 mm and the near field fill factor is 20 %. The maximum output power from this structure is 76 W cw while the total efficiency is greater than 45% at an output power of approximately 30 W cw (the heatsink temperature was 0°C). This is the highest cw output power of any semiconductor laser. Demonstration of 76 W cw from a monolithic laser array is indicative of the capabilities that semiconductor lasers offer. No other laser system can demonstrate this type of output power, efficiency, and reliability from a source that is 1 cm x 1 mm in size.

Reliable cw operation of 1 cm bars is limited to approximately 10 W cw. Higher peak powers, however, can be obtained by increasing the near field fill factor to 100% across the 1 cm bar and operating the laser in a quasi-cw condition. As a result the laser can be operated for pulse lengths that of 200 μs to 1 ms at a repetition rate of 100 Hz. A maximum output power of greater than 200 W has been demonstrated from these structures. Output power densities of greater than 3.5 kW/cm^2 have been demonstrated from 2-D stacks of 1 cm bars. The quasi-cw bars are operated reliably at powers of 60 W, or for a five bar stack 300 W.

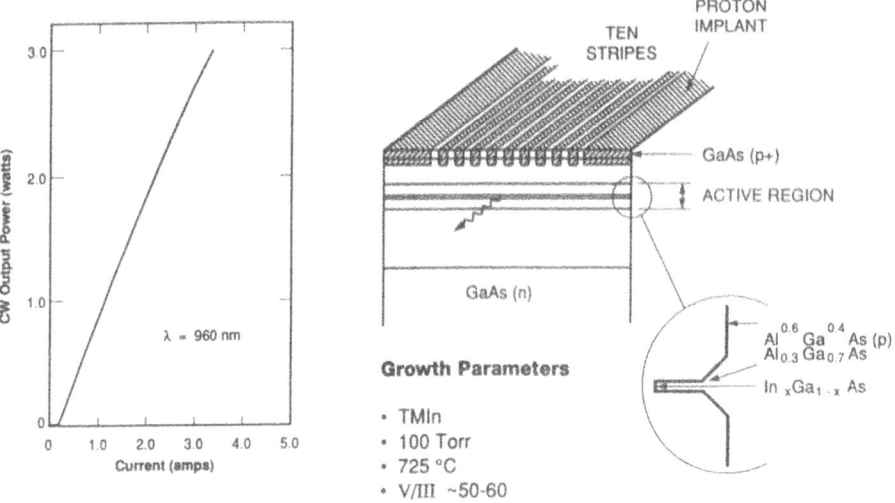

Fig.6.   Light output as a function of input current for an InGaAs gain guide laser array.

8

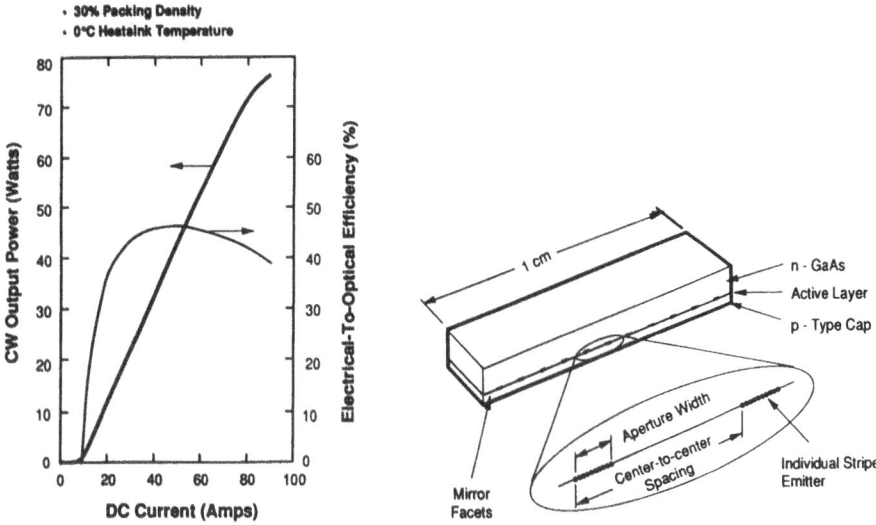

Fig.7.   Light output as a function of input current for a monolithic 1 cm wide laser array. The emitting aperture extends over 20% of the total aperture.

The reliability of semiconductor lasers is limited by either thermal dissipation, high current density, or catastrophic failure of the emitting facet. Catastrophic failure is a result of optical absorption at the emitting facet. The light is absorbed at the emitting facet by surface states and the associated depletion of the carriers at the surface. The absorption locally heats the material causing a reduction in the bandgap. The bandgap reduction further increases the absorption and the facet resulting in a positive feedback condition. At a critical temperature the material experiences thermal runaway and the emitting facet is destroyed by ablation of the facet.

Thermally induced degradation mechanisms are associated with the formation of point defects or darkline defects. In either case the defect is thermally activated and has a corresponding activation energy associated with its migration. Typical activation energies of semiconductor laser diodes are in the range of 0.2 - 0.7 eV.

TABLE I

RELIABILITY DATA OF GAIN GUIDED ARRAYS

APERTURE	OUTPUT POWER	LIFETIME
200 μm	1 watt cw	>25,000 hours
200 μm	2 watts cw	10 - 30,000 hours
500 μm	3 watts cw	10 - 20,000 hours
1 cm/30%	10 watts cw	>5,000 hours
1 cm	60 watts quasi-cw	>10^9 pulses, 200 μs

The reliability of the various gain guided arrays is summarized in Table 1. Lasers operating in the range of 1-3 W cw have demonstrated reliable operation on the order of 20,000 hours at a heatsink temperature of 25°C. Ten Watt cw bars exhibit reliable operation in the range of 5,000 hours at 25°C, while quasi-cw bars operating at an output power of 60 W are reliable over 10^9 pulses.

INDEX GUIDED LASER DIODES

In contrast to gain guided laser diodes, index guided lasers are fabricated for single mode operation. By incorporating a real refractive index step in the direction parallel to the p-n junction, the optical mode is confined to the active region and when the width of the stripe is limited to less than 4 μm, the waveguide will typically support only the lowest order transverse mode. Consequently the output is diffraction limited and the mode is stigmatic.

In Figure 7 the characteristics of an index guided laser diode similar to the SDL-5410 is presented. The threshold current of the laser is 16 mA and the differential efficiency is 87%. With a series resistance of 2.5 ohms, the total power conversion efficiency is greater than 60%. This is the highest reported conversion of electrical to optical power for any optical system. The maximum output power from the 5 μm aperture is 500 mW, the highest brightness for any laser diode reported.

The spatial and spectral characteristics of the index guided laser diode are shown in Figures 8 and 9. The diode radiates in the lowest transverse mode up to powers of 180 mW. The far field pattern is single lobed with divergence of 8° x 22° parallel and perpendicular to the p-n junction respectively. Single longitudinal behavior is also maintained over this operating range. A unique characteristic of this particular index guided laser is the large range over which longitudinal mode stability is maintained. The diode operates on a particular longitudinal mode over an operating range of 70 mW. As discussed below a further consequence of the longitudinal mode stability is the reduction in longitudinal mode hopping noise. This is shown in both the plots of spectral linewidth and relative intensity noise of the laser.

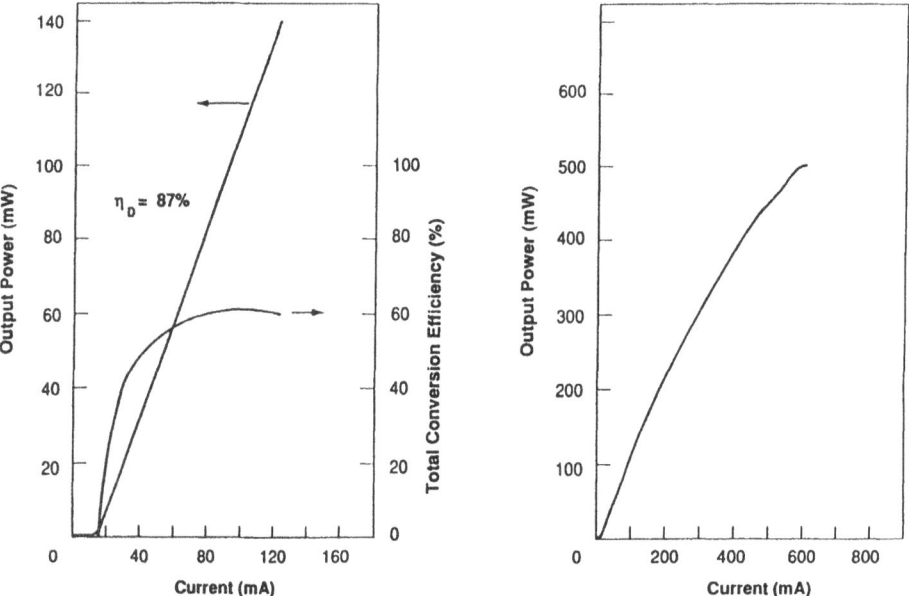

Fig.8. Light output as a function of input current for a 4 μm wide index guide laser diode.

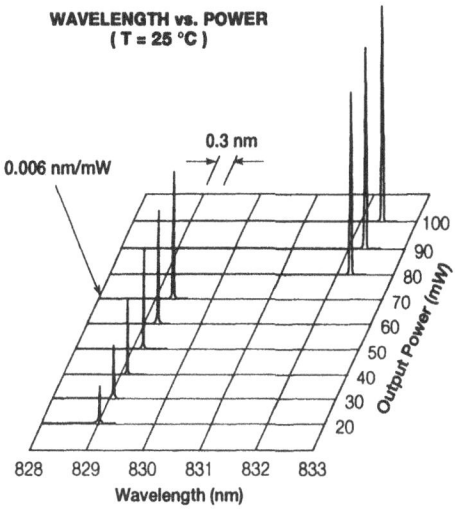

Fig.9. The spectral output of an index guided laser diode.

The linewidth of the longitudinal mode has been measured to be as narrow as 3 MHz at an output of 100 mW, as shown in Figure 10. The linewidth is linear with inverse power and does not show a linewidth floor, while the intercept at infinite power is 380 kHz. For this particular laser diode the longitudinal mode occurred at a power output of 30 mW cw. The measurement of the linewidth indicates no increase in the linewidth in the region of the longitudinal mode hop, the longitudinal mode hopping noise, therefore, is insignificant to the other noise contributions of the laser.

The relative intensity noise of the single mode laser is presented in Figure 11. The RIN is measured to be less than -145 dB/Hz at frequencies between 50 MHz and 1 GHz. The noise peak is associated with the relaxation oscillation frequency which occurs in the 3 - 4 GHZ range.

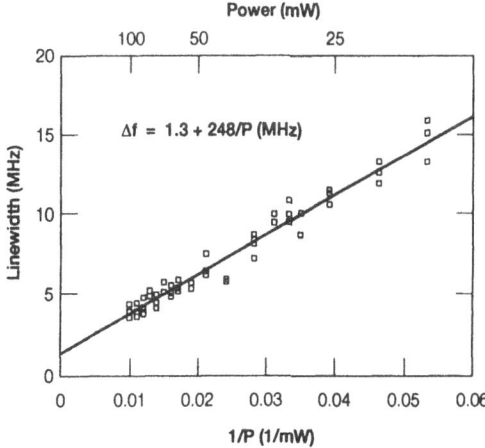

Fig.10. The spectral linewidth as a function of output power for a index guided laser diode.

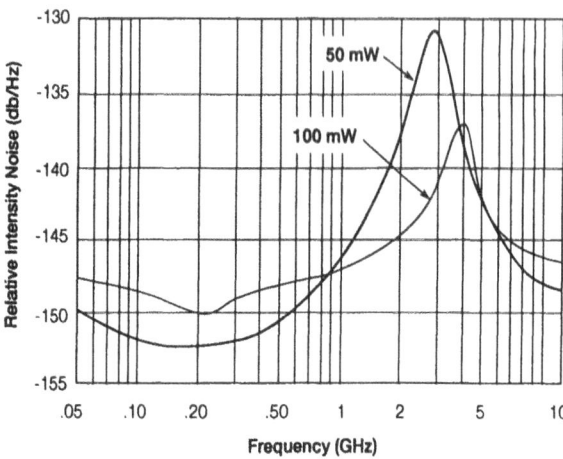

Fig.11. The relative intensity noise of an index guided laser diode.

As is the case of gain guide arrays the maximum output power out of a single mode laser diode is limited by either catastrophic degradation and/or thermal dissipation. Laser diodes fabricated from GaInAsP and emit in the range of 1.3 - 1.5 μm are limited by thermal dissipation when operated under cw conditions. GaAlAs diodes have historically been limited by catastrophic degradation of the mirror facet until recently where development of non-absorbing mirrors[24] has resulted in higher catastrophic power levels. By raising the catastrophic damage level of the facets the cw output power of the diodes has become limited by thermal dissipation at higher output powers. Further increases in the total efficiency of AlGaAs laser diodes has reduced the thermal dissipation, and as a result, even higher output powers have been achieved. The maximum output power from a GaInAsP index guided laser diode has been 200 mW while that of GaAlAs index guided diodes have reached 500 mW.

Two of the possible degradation modes of laser diodes result from either optically induced damage at the facet or thermally induced recombination centers in the bulk. Optical facet damage as a function of time is related to the catastrophic damage threshold of the laser. The higher maximum output power the laser can operate at, the less susceptible the laser is to degradation of the facet. Typically, facet damage results in sudden failure of the diode when operated under constant output power conditions. Thermally induced degradation is a result of the total efficiency and the output power of the diode. Thermally limited operation of AlGaAs diodes occurs above 500 mW output while that for GaInAsP diodes have been reported at 200 mW. From these considerations, we believe that the reliability of AlGaAs diodes should be as good or better than that of GaInAsP laser diodes.

The reliability of the single mode laser has been studied for approximately 8000 hrs. at an operating temperature of 50°C and an output power of 100 mW cw, Figure 12. The average degradation of the five diodes tested extrapolates to a lifetime of greater than 20,000 hrs. at 50°C. In this case the lifetime of the laser is determined by an increase in the operating current by greater than 50% to maintain the came output power. Assuming an activation energy of the thermally activated degradation mechanisms of 0.5 eV, the room temperature lifetime is extrapolated to be greater than 60,000 hrs.

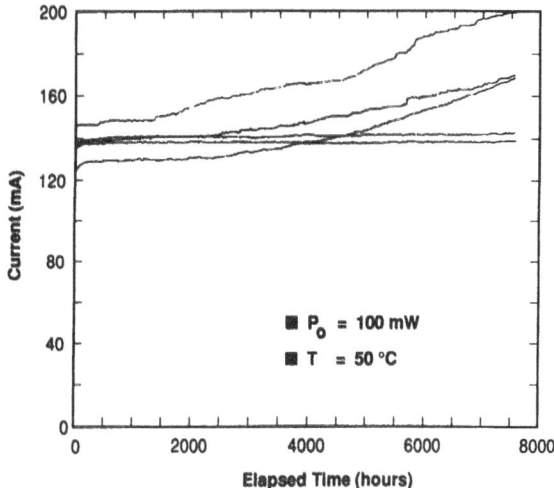

Fig.12. Reliability characteristics of an SDL-5410 operated at 100 mW cw at a heatsink temperature of 50°C.

CONCLUSIONS

Semiconductor lasers have demonstrated tremendous output powers and efficiencies. In addition, high reliability and compact size offer advantages that cannot be achieved in any other laser system. These systems are currently being demonstrated in many medical applications including endoscopic photocoagulation and tissue welding. Several other applications areas are being developed that utilize the high brightness of the laser source for tissue ablation and destruction. Photoactivated dyes are also being studied for applications that can utilize laser diodes.

The output powers available from laser diodes has increased superlinearly for the past five years, in addition the price per watt of optical power has decreased. Further improvement of device characteristics will result in a widespread applications base, replacing many of the existing gas and solid state laser systems as well as precipitating new applications which can utilize the unique properties of semiconductor lasers.

REFERENCES

1. G.H.B. Thompson, "Physics of Semiconductor Laser Devices", S. Wiley & Sons, 1980.
2. H.C. Casey, M. B. Pannish, "Heterostructure Lasers", Academic Press, 1978.
3. H. Kressel, J.K. Butler, "Semiconductor Lasers and Heterojunction LED's", Academic Press, 1977.
4. G.P. Agrawal, N.K. Dutta, "Long Wavelength Semiconductor Lasers", Van Nostrand Reinhold, 1986.
5. W.T. Tsang, "Extension of lasering wavelengths beyond 0.87 μm in GaAs/AlGaAs double heterostructure lasers by In incorporation in the GaAs active layers during molecular beam epitaxy", Appl. Phys. Lett., 38:661 (1981).
6. D.F. Welch, C.F. Schaus, S. Sun, P.L. Gourley, and W. Streifer, Gain characteristics of strained quantum well lasers, Appl. Phys. Lett., 56:10 (1990).
7. D.F. Welch, M. Cardinal, W. Streifer, and D. Scifres, High-power single mode InGaAs/AlGaAs laser diodes at 910 nm, Electron. Lett., 26:233 (1990).

8. J.W. Matthews, and A.E. Blakeslee, Defects in epitaxial multilayers, I: Misfit dislocation, J. Cryst. Growth, 27:118 (1974).

9. I.J. Fritz, S.T. Picraux, L.R. Dawson, T.J. Drummond, W.D. Laidig, and N.G. Anderson, Dependence of critical layer thickness on strain for $In_xGa_{1-x}As/GaAs$ strained-layer superlattices, Appl. Phys. Lett., 46:967 (1985).

10. D.F. Welch, M. Cardinal, W. Streifer, D.R.Scifres, High brightness, high efficiency, single quantum well laser diode array, Electron. Lett., 23:1240 (1987).

11. D.F. Welch, W. Streifer, D.R. Scifres, High power single mode laser diodes, in: "Laser Diode Technology and Applications", SPIE Proceedings vol.1043 (1989), pag. 54.

12. M. Sakamoto, D.F.Welch, J.G. Endriz, D.R. Scifres, W. Streifer, 76 Watt continuous-wave monolithic laser diode arrays, Appl. Phys. Lett. 54:2299 (1989).

13. Kazuhiko Itaya, Masayuki Ishikawa, Hazime Okuda, Yukio Watanabe, Koichi Nitta, Hideo Shiozawa, and Yutaka Uematsu, Effects of facet coating on the reliability of InGaAlP visible light laser diodes, Appl. Phys. Lett. 53:1363 (1988).

14. T. Tanaka, S. Minagawa, and T. Kajimura, Transverse-mode-stabilized ridge stripe AlGaInP semiconductor lasers incorporating a thin GaAs etch-stop layer, Appl. Phys. Lett. 54:1391 (1989).

15. T.H. Chong, K. Kishino, Remarkable reduction of threshold current density of 670 nm GaInAsP/AlGaAs visible lasers by increasing al content of AlGaAs cladding layers, Electron. Lett., 25:761 (1989).

16. W. Kozlovsky, C.D. Nabors, and R. Byer, Efficient second harmonic generation of a diode-laser-pumped CW Nd:YAG laser using monolithic $MgO:LiNbO_3$ external resonant cavities, IEEE J. Quantum Electron., 24:913 (1988).

17. L. Goldberg, M.K. Chun, Efficient generation at 421 nm by resonantly enhanced doubling of GaAlAs laser diode array emission, Appl. Phys. Lett., 55:218 (1989).

18. D.F. Welch, B. Chan, W. Streifer, D.S. Scifres, High power, 8w cw, single quantum well laser diode array, Electron. Lett., 24:113 (1988).

19. W. Streifer, D.R. Scifres, G.L. Harnagel, D.F. Welch, J. Berger, M. Sakamoto, Advances in solid state pumps, IEEE J. Quantum Electron., 24:883 (1988).

20. D.R. Scifres, D. Lindstrom, R.D. Burnham, W. Streifer, Phase-locked (GaAl)As laser diode emitting 2.6 W CW from a single mirror, Elect. Lett. 19:169 (1983).

21. T.R. Chen, D. Mehuys, Y.H. Zhung, M. Mittelstein, H. Wang, P.L. Derry, M. Kajanto, and A. Yariv, Broad-area tandem semiconductor laser, Appl. Phys. Lett., 53:1468 (1988).

22. D.F. Welch, R. Craig, W. Streifer, D. Scifres, High reliability, high power, single mode laser diodes, to be published in Electronic Letters (1990).

23. W.T. Tsang, Extremely low threshold (AlGa)As graded-index waveguide separate confinement heterostructure lasers grown by molecular beam epitaxy, Appl. Phys. Lett., 40:217 (1982).

24. H. Nakashima, S. Semura, T. Ohta, T. and T., AlGaAs window stripe buried multiquantum well lasers, Japan. J. Appl. Phys., Vol. 24, No. 8, August (1985).

DIODE-PUMPED MINIATURE SOLID STATE LASERS

R. Pratesi

Istituto di Elettronica Quantistica del CNR, and
Dipartimento di Fisica dell'Universita'
Firenze - Italia

Miniature Diode Laser Pumped Solid State Lasers, extended by nonlinear optical techniques, offer unprecedented capability in power, efficiency and wavelength diversity at peak and average powers that are useful for biological and medical applications (R.L.Byer, Stanford University).

INTRODUCTION

The most interesting and important developments in the field of laser sources during the last few years are represented by the new generation of high power semiconductor diode lasers and by the new class of all-solid-state lasers, in which diode lasers have replaced the traditional flashlamp pumping. The principles of operation of semiconductor diode lasers (or coherent light emitting diodes, CLEDs) have already been reviewed in this Volume and in References 1 and 2,3. In this paper an updated review of CLED-pumped solid-state lasers is presented. More detailed discussions on this fast developing topic can be found in References 4 and 5.

BASIC PRINCIPLES OF OPERATION

CLED-pumped solid-state lasers represent a new class of miniaturized, all-solid-state devices based on the emerging technology of diode-laser pumping of a solid-state crystal. This type of laser incorporates a high-power CLED in place of flashlamp to excite the laser crystal.

The advantages of CLED-pumped solid-state lasers range from improved performance that was unattainable with many earlier lasers, to greater ease of manufacturing the products. The all-solid-state structure of these lasers allows the improvement of the performances typical of solid-state lasers, namely: excellent beam profile, high stability, single frequency operation, low noise.

The replacement of flashlamps with CLEDs permits important improvements of solid-state laser performances. First of all, the CLED represents the most efficient device to convert electrical power into optical power (efficiencies in excess of 60% have been reported)[6] : therefore, the overall efficiency of conversion of the electrical power to

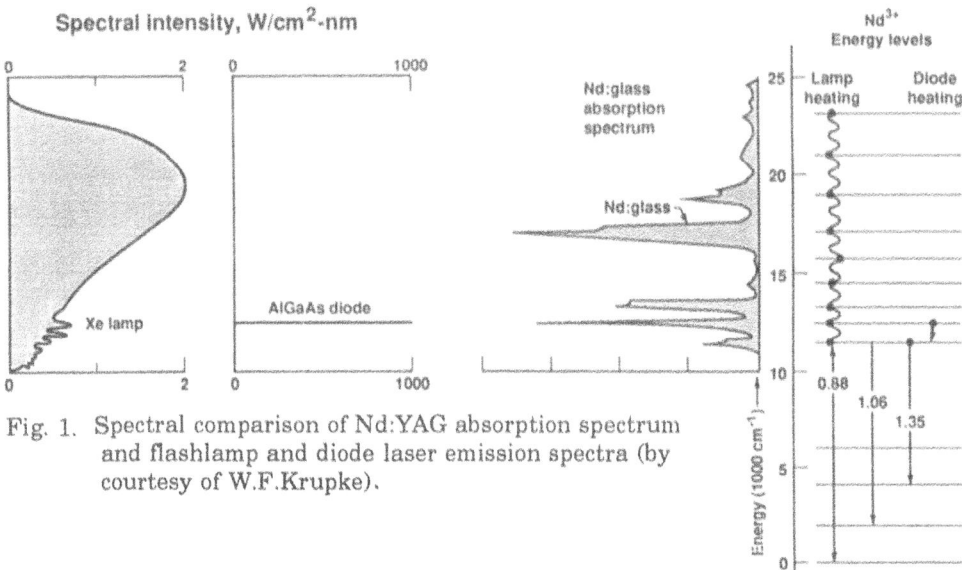

Spectral intensity, W/cm²-nm

Fig. 1. Spectral comparison of Nd:YAG absorption spectrum
and flashlamp and diode laser emission spectra (by
courtesy of W.F.Krupke).

the output power of the solid-state laser can be greatly increased. Moreover, the emission spectrum of the CLED is very narrow compared to the spectra of incoherent optical sources, and CLED pumping can, then, be very efficient.

Since the CLED output wavelength can be finely tuned to the absorption peak of the laser medium to excite a pump level that lies only slightly above the upper laser level, the density of waste heat generated is several times lower than that produced by flashlamps. This drastically reduces the amount of external cooling required and permits the design of lasers with diffraction-limited output beams of extremely high stability.

Because of their directional emission, the efficiency of transport to, and deposition of CLED pump radiation in the solid-state gain element is several times higher than can be achieved with flashlamps that emit broadband radiation into 4 π steradians.

Finally, another important merit of CLED pumping is represented by the possibility of remote pumping. CLED pump sources, in fact, can also be efficiently coupled into a solid-state laser gain element via multimode fibers, keeping the pump and power conditioning equipment remote from the laser head and work area. This permits improved laser stability and higher output power operation with still very compact laser heads.

CLED-Pumping Geometries

The key parameter for high electrical-to-optical efficiency of CLED-pumped solid-state lasers is represented by the matching between the laser mode and the gain (i.e. pump) distribution. In general, CLED pumping takes two forms: end pumping and surface (side) pumping, while the two most popular configurations of the solid-state medium are: cylindrical rod, and geometric slab.

i) End-Pumped Lasers. End-pumping of laser rods has produced the highest efficiency from any configuration. The optical mode from the pump CLED can be matched to the lasing mode of the active solid-state material, thus minimizing the

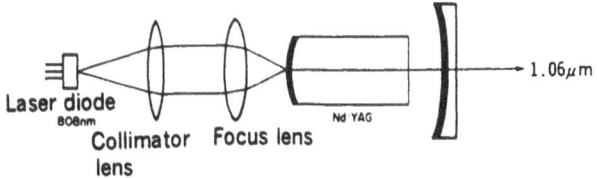

Fig. 2. Schematic diagram of end-pumped laser.

coupling efficiency between pump and laser. Figure 2 shows a typical schematic diagram of end-pumped solid-state laser. Typical overall conversion efficiency is 35% at power levels of 1 W in the fundamental spatial mode.[7] Optical-to-optical efficiency as high as 50% has been achieved, approaching the theoretical limit of 78% given by the pump/output photon energy defect.

The output power regime with end-pumping ranges from the milliwatt level to less than 10 W. The output power from this type of laser is, in general, limited by the power obtainable from the single CLED emitter. Although high power CLED devices are available, the output radiation pattern cannot be efficiently coupled to the active solid-state material. In any case the rod geometry suffers from thermal and stress induced birefringence and gain nonuniformity at high power levels, and it is not suitable for high power operation.

ii) <u>Side-Pumped Lasers</u>. Figure 3 shows a schematic diagram of a side-pumped laser. The lasing medium has usually the form of a geometric slab. The geometry of the slab results in the laser mode propagation through the material by total internal reflection. An advantage of this zig-zag configuration is an increased effective length compared with the physical length of the active material and scalability to high output power. "Wall-plug" efficiencies are generally lower than those obtained with end-pumping, while higher average output power levels are achieved.

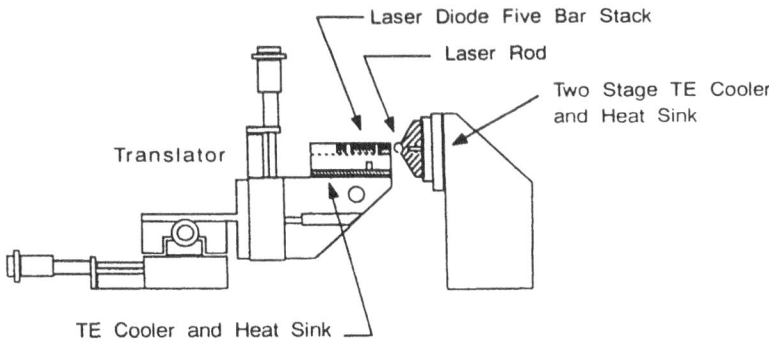

Fig. 3. Schematic drawing of a side-pumped laser. Both laser bars and laser crystal are temperature controlled by thermo-electric (TE) coolers [8] (by courtesy of A.D.Hays).

Scaling to high power regimes is obtained by increasing slab sizes and pumping diode number. Two dimensional (2D) integrated arrays of CLEDs are appropriate sources in side-pumped geometries. Technological development is in progress and power densities of ~ 1kW/cm^2 have been reported. Presently, higher pump power densities are commercially available with 2D stacked linear CLED arrays (commonly known as laser "bars"). Duty factors are still limited by the rate at which waste heat is removed from the densely packed arrays. Microchannel cooling structures integrated into the densely packed diode arrays have also been successfully introduced.[9] In future, this technology should lead to substantial improvements in the power radiated per unit area.

The power levels attainable with the side-pumping geometry range from about 10 W to a few kilowatts.

iii) <u>Multi-CLED Pumping</u>. With current technology, direct pumping of solid-state lasers with 2D CLEDs is limited in average output power by the difficulty of stacking CLED bars while maintaining efficient cooling. The commercial availability of CLED arrays and bars with 1-3 W cw and ~ 100 W/1cm cw output power, respectively, has suggested the use of Multi-CLED pumping geometry as an alternate approach to high average power solid-state lasers. It makes use of existing technology and offers the advantages of reduced cost and power scaling and ease of thermal and optical engineering. Large area fiber bundles can be used to deliver high power densities into a rod or slab solid-state laser from less-densely packed diode arrays. The CLEDs can be mounted remotely from the solid-state laser to facilitate electrical driver design and cooling.

Figure 4 shows the schematic of the fiber-coupled CLED pumped zig-zag slab laser.

For operation at multikilowatt average output power, geometries with moving laser medium have been projected in order to reduce thermal load of the lasing material. Moving slab and multidisk lasers have been evaluated for high average power operation.[10]

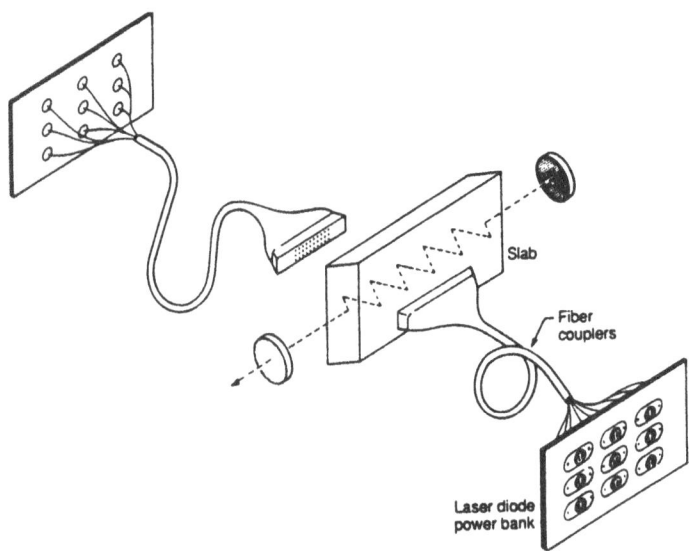

Fig. 4. Schematic of the fiber-coupled CLED pumped slab laser (adapted from Ref.5).

Several novel designs for CLED-pumped solid-state lasers with both cw high - output power and diffraction limited beam quality have recently been reported and introduced into the market.

i) Monolithic,Non-Planar Ring Oscillator (NPRO)

A miniature (12x9x3mm) monolithic Nd:YAG ring resonator with nonplanar ring light path (Fig.5) was successfully introduced by Byer at al. [11] in 1985 following the requirement for a stable single-frequency laser source for environmental applications. This device combines the characteristics of a CLED into a monolithic ring structure to achieve single axial mode, unidirectional oscillation, with extremely narrow bandwidth.[4]

Recent progress has led to output powers up to 53 mW cw in both fundamental spatial and axial modes, with a measurement limited bandwidth of less than 3 kHz over 100 ms measurement time.[12] Present optical-to-optical and overall electrical-to-optical efficiencies are ~ 60 % and ~ 20 %, respectively.

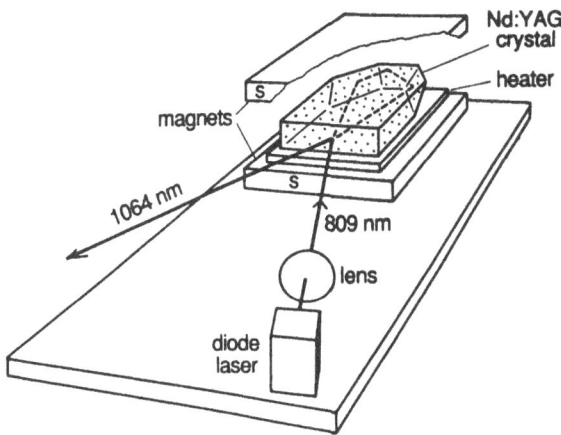

Fig. 5. Schematic of a Non-Planar Ring Oscillator [5].

ii) Side-Pumped, Tightly Folded Resonator Laser

Figure 6 shows the schematic drawing of a particularly efficient side-pumped Microlaser. The pumping light from the diode bar is coupled into a small, specially coated Nd:YLF block (5x5x20mm) via a small piece of optical fiber. The two cavity mirrors are aligned so that the intracavity laser beam bounces back and forth through the pumped region between opposing sides with vertices aligned at each diode location.

The output power is 3.9 W when pumped with 10 W from the 1 cm diode bar. The output beam is a diffraction limited Gaussian single-spatial mode. The conversion efficiency is 39 %, a high value for a side-pumped geometry, comparable to that typically achieved with end pumping [13].

The laser has also been efficiently operated in the giant pulse regime by using a high diffraction efficiency TeO_2 acousto-optic Q-switch (0.6 mJ/pulse at 1 kHz repetition rate; pulse width : < 10 ns; peak power : 70 mW)[13] .

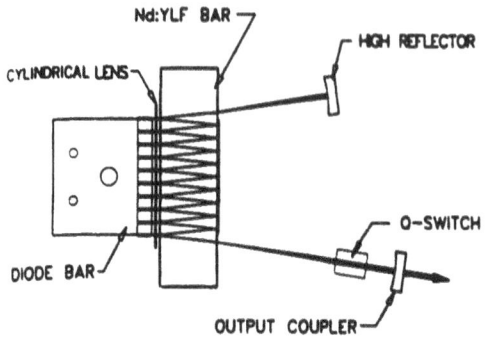

Fig. 6. Side-pumped, tightly-folded resonator Microlaser [13].

iii) Microchip lasers

The "microchip" laser concept has been developed to produce inherently single-frequency lasers [14]. The cavity length chosen is so short (typically from 150 to 750 μm) that only a single longitudinal cavity mode falls under the gain curve. Moreover, the geometry of the microchip cavity and the pumping configuration strongly favor oscillation in the TEM_{00} single transverse mode. Direct deposition of the dielectric laser mirrors onto the solid-state gain material ensures high mechanical stability to the laser.

Moreover, the very short cavity length allows: i) continuous tuning over the entire gain bandwidth of the laser by changing the optical length of the cavity, for example by applying a transverse stress to the laser crystal, as shown in Fig.7; ii) short cavity lifetime, which permits the generation of very short gain-switched pulses (much less than 100 ps).

CLED-pumping of Microlasers has been demonstrated with end-pumping geometry, using a 0.65 x 1.0 x 2.0 mm, 1.1 wt.% Nd:YAG crystal. The overall efficiency is greater than 4% for emission at 1.32 μm, with maximum cw power of 51 mW.

The low pump threshold, high efficiency, single-frequency operation, and the unique tuning capabilities of microchip lasers make CLED-pumped microchip lasers extremely promising.

Low cost mass production of microchip lasers is possible due to the simple fabrication of the chip and to the small amount of material used for each laser.

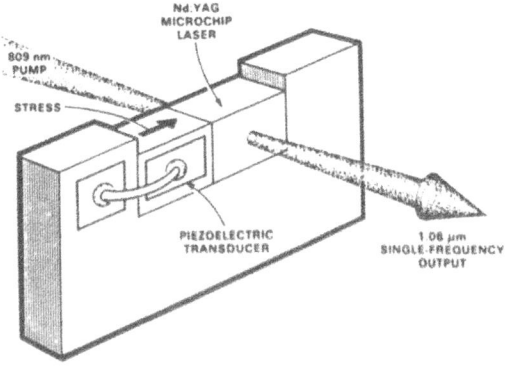

Fig. 7. Schematic of a piezoelectrically tunable Microchip laser [14].

As is known, solid-state lasers can be operated to generate i) high-peak power, short duration pulses with the technique of Q-switching; ii) ultrashort pulses by mode-locking a multimode cavity; iii) new frequencies by nonlinear harmonic generation and frequency mixing.

The miniaturization process involved in the construction of Microlasers has led to corresponding research to obtain small size, possibly monolithic components for actively pulsing and frequency shifting Microlasers.

i) Short Pulse Operation

Q-Switched Microlasers. For "giant pulse" (Q-switch) operation a shutter must be inserted into the laser cavity, but, even so, the overall length of the microlaser can be maintained below 50 mm. The shutter can be either an acousto-optical (AO) switch or an electro-optical (EO) switch. The electro-optical switch is particularly appealing due to its very small size, which permits a very short (25mm) cavity. An extremely compact (~ 20 mm in length) AO Q-switched laser has also recently been reported with pulse width as short as 3 ns.[15]

Several commercial models of Q-switched Nd:YAG Microlasers are suitable as replacements for standard Q-switched Nd:YAG lasers in eye microsurgery and genetic manipulations.

Mode-Locked Microlasers. Microlasers offer the possibility of generating ultra short pulses with very high peak powers. Recent work [16] using frequency modulation and amplitude modulation techniques has led to the generation of sub 10-ps pulses at high repetition rate. The soliton Raman pulse compression technique has then be applied to produce pulses as short as 80 fs.[16] These all-solid-state systems are particularly suitable for efficient harmonic generation, synchronous pumping of optical parametric oscillators, tunable solid-state-crystals, etc.

ii) Nonlinear Frequency Conversion

Harmonic Generation. In early experiments, second harmonic generation (SHG) of microlaser output was achieved by placing the nonlinear crystal within the laser resonator to take advantage of the high circulating power in the laser. SHG that produced milliwatts of green output using either KTP or $LiNbO_3$ as the nonlinear crystal was demonstrated. Recent advances in single-frequency laser operation and in nonlinear materials have also led to greatly improved conversion efficiencies using external resonant SHG. In a recent experiment about 500 and 200 mW average output powers have been reported for a frequency doubled and tripled Q-switched Nd:YLF laser with typical conversion efficiencies of 35 and 13 %, respectively.[13]

Self-Frequency Doubling. An interesting approach to generating in a very simple way second harmonic radiation from a laser oscillator is represented by the use of the so-called "self-doubling" crystals. Self-doubling lasers are capable of oscillating simultaneously at a given frequency and at its harmonic. Without additional (lossy) intracavity components these lasers are of simple design and the high intracavity power should, in principle, provide good conversion efficiencies.

Recently, cw CLED-pumped miniature self-doubled lasers have been realized by using a crystal of Neodymium doped yttrium aluminium borate (Nd:YAB). Output powers at 532 nm as high as 10 to 30 mW have been reported.[17]

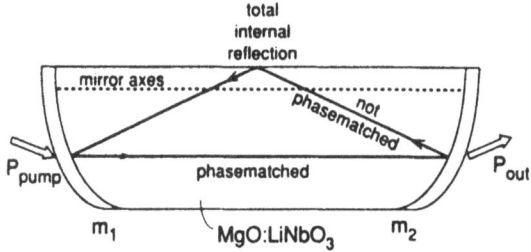

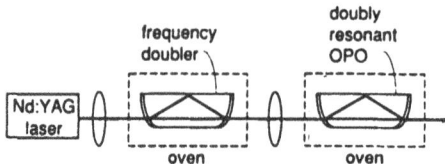

Fig. 8.   Schematic of an optical parametric oscillator, with resonantly doubled Nd:YAG Microlaser. The ring geometry of the OPO permits simultaneous resonant oscillation of both of the OPO-generated waves.

These initial experimental results indicate that self-frequency doubling diode pumped Nd:YAB lasers are simple devices, and compact laser systems for the generation of green cw laser radiation are feasible.

Optical Parametric Oscillator (OPO). The optical parametric oscillator has received consistent attention in view of the wide tuning range, good conversion efficiency, and high beam quality that make this nonlinear device a very promising solid-state source of tunable coherent radiation over the spectral range that extends from UV to IR.[12] The OPO requires high optical quality and high damage threshold crystals and a diffraction-limited, single-axial-mode pump source for stable operation.[18] Recent advances in high performance lasers and crystal growth technologies allow these requirements to be satisfied.

Tuning ranges from 0.42 to 2.5 μm, 0.6 to 4.2 μm, and 2.6 to 9 μm have been reported with barium borate (BBO), $LiNbO_3$ and $AgGaSe_2$ crystals, respectively. Microlaser pumping of OPOs represents a very interesting approach to the realization of an all-solid-state source of coherent, tunable radiation with remarkable stablity and reliability characteristics (Fig.8).[12,19]

RARE EARTH ION MICROLASERS

CLED pumping geometry can be successfully applied to produce Microlasers operating in a variety of wavelengths. Considerable effort is being devoted at present to obtain efficient laser emission from crystals doped with rare earth trivalent ions.

The number of possible laser transitions under CLED pumping is limited by the number of rare earth ions with absorption at CLED output wavelengths. With present high power CLEDs emitting in the 750-860 nm spectral region the rare earth ions are limited to Nd, Pm, Dy, Er, and Tm.[4] Several crystals have emission wavelengths in the NIR, of particular interest for medical applications. Shorter NIR transitions are also investigated in view of efficient frequency doubling for generation of violet-blue radiation.

The most promising rare earth Microlasers are now briefly reviewed.

Nd:YAG Microlaser

The Neodymium ion is an excellent dopant for Microlasers: it exhibits strong absorption in the emission bands of GaAs, GaAlAs, GaAsP LEDs and CLEDs.

i) The 946 nm transition. The 946 nm transition in Nd:YAG has recently attracted interest as a CLED-pumped source for frequency doubling to 473 nm, thus providing an efficient and compact blue laser source. The 946 nm transition has its lower laser level in the ground state manifold, and thus suffers from temperature dependent reabsorption losses.

Recently, the high efficiency of diode-laser pumping has allowed the exploitation of this transition for efficient cw operation at room temperature.[20] Moreover, it has been shown that only a small lowering of the laser operating temperature is sufficient to increase the system performance by a factor of 10 over room temperature operation.[8] Thus, the modest cooling requirements, easily achieved with a two stage thermoelectric cooler, make this a practical source for frequency doubling to blue region of the spectrum.

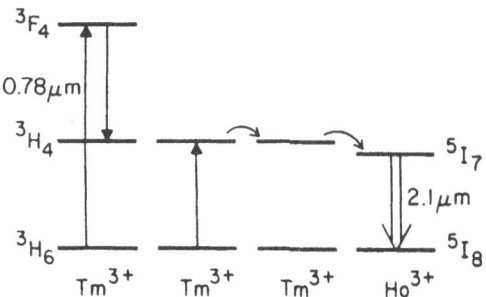

Fig. 9.   Energy-level scheme illustrating the mechanism of laser pumping in a Tm-sensitised Ho:YAG crystal [21].

ii) The 1.32 μm transition. Nd:YAG lasers operating at 1.32 μm are being introduced onto the medical market in view of the higher absorption of 1.32 μm radiation by biological tissues with respect to 1.06 μm radiation. This in part compensates for the reduction in output power at 1.32 μm.

Ho:YAG Microlaser

The medical interest in Holmium lasers relies on the value of the absorption coefficient of water at its wavelength, 2.1 μm, which is intermediate between those corresponding to CO_2 and Nd:YAG emission wavelengths.

Moreover, the 2.1 μm transition in Ho^{+3} is attractive because of its long upper laser level lifetime (~ 5 ms) and its eye-safe wavelength. Despite the three level nature of this transition, several studies of diode-laser pumping have been initiated because the long upper laser level lifetime may permit larger energy storage than Nd:YAG for the same cw pump power.

In order to improve the pumping efficiency of these three-level laser crystals and to achieve operation at room temperature the interesting technique of "sensitization" has been introduced. The crystals are codoped with suitable rare earth and/or other element trivalent ions. With flashlamp pumping the crystal is codoped with Er^{+3}, Tm^{+3}, and Cr^{+3} to absorb a larger fraction of the pump power and achieve higher pumping densities. Energy is transferred from Cr^{+3} absorption bands to Tm^{+3} and then to the Ho^{+3} upper laser level. In the case of NIR CLED pumping, the Ho:YAG crystal is codoped only with Tm^{+3} and pumped at 780 nm (Fig.9). Lowering of the threshold and room temperature emission with CLED pumping have been reported.[21]

Er:YAG/Er:YLF Microlaser

Highly efficient CLED-pumped Microlasers at room temperature operating near the peak of the water absorption near 3 μm have many potential medical applications. CLED-pumping could provide substantial advantages in efficiency, compactness, and compatibility with fiber/integrated optical systems.

An excellent candidate for this type of laser is the 2.94 μm Er:YAG laser or the Er:LiYF$_4$ (YLF) laser emitting at 2.80 μm. Recently, successful operation of an Er-Microlaser has been obtained by CLED-pumping at the 796 nm Er:YLF absorption peak a 30% erbium-doped LiYF$_4$, 8 mm long, with high reflectivity coatings on the end faces to form a monolithic laser cavity.[22]

With an absorption length of only 1 μm in water, precise cutting of biological materials is in principle possible. However, tissue damage can occur from the irradiated zone because of the possibility of cavitation and shock wave formation as well as thermal diffusion.

Cr;Tm:YAG Microlaser

NIR laser ablation is most effective at the wavelengths of 1.93 and 2.94 μm, which correspond to tissue absorption peaks. In addition to operating near an absorption peak of tissue, it is also important to consider the availability of a suitable fiber optic delivery system at the output wavelength of the laser system. Silica fibers retain good transmission characteristics in the NIR up to 2.1 μm.

A 1.96 μm Cr;Tm:YAG Microlaser has been developed with the aims of producing a suitable source of radiation for surgical applications and of utilizing the optical fiber delivery systems based on the well developed silica technology.[23]

TUNABLE MICROLASERS

The search for new, more efficient laser crystals for high power, tunable emission is continuing with particular effort.

Chromium-LiCAF laser

The development of a Microlaser operating at 790 nm has been reported [24] with a crystal of LiCAF doped with chromium pumped by 670 nm CLEDs. The Cr:LiCAF laser is a potential substitute for the tunable alexandrite laser: with flashlamp pumping, emission in the 720-840 nm interval has been observed.

Cobalt-MF laser

Cobalt-doped magnesium fluoride (Co^{+2}:MgF$_2$) lasers have tunable emission in the spectral range from 1.5 to 2.5 µm when optically pumped at 1.3 µm. Diode lasers operating at 1.3 µm are presently being developed at increasing output power levels for applications in optical fiber communications. Their use as pumping sources for $Co:MgF_2$ lasers is another potential application for generating tunable 1.5 - 2.5 µm radiation.

NARROW LINEWIDTH OPERATION

Frequency-stable lasers are required in a variety of applications, including spectroscopy of biomolecules and diagnostic techniques.

Microlasers with monolithic resonators (i.e. resonators fabricated by applying appropriate coatings to the ends of the laser crystal) are unsuitable for longitudinal mode control, tuning, and single-frequency operation, unless special geometries, such as the NPRO, are introduced. Recently, stable, single-mode operation of a Nd:glass Microlaser with an external resonator, including frequency tuning and line narrowing etalons, has been achieved.[25] The laser is tunable over 12 nm and capable of being frequency stabilized to better than 10 kHz.

Short-term free-running linewidths of 3-10 kHz have been reported for diode-laser-pumped Nd:YAG ring oscillators with non-planar geometry.[26-27] These impressive frequency properties can be attributed to the CLED pumping scheme together with the monolithic nonplanar ring geometry. CLED pumping avoids the frequency noise associated with flash lamps, and the unidirectional oscillation made possible by the non-planar ring geometry provides feedback isolation that eliminates spatial hole burning and therefore forces single-mode operation. Furthermore, the monolithic construction requires no external elements and is therefore less sensitive to ambient acoustics.

Further work has led to the stabilization of these lasers to 30 Hz.[28] More recently, Byer et al.[29] have demonstrated that the all-solid-state, CLED-pumped non-planar ring oscillator (NPRO) can be stabilized to the system noise limit of 3 Hz. Sub-hertz (0.33 Hz) operation was finally reported at this Workshop.

ACKNOWLEDGMENTS

Work supported by the Special CNR-Project "Tecnologie Elettroottiche".

REFERENCES

1. W. Streifer, D.R. Scifres, G.L. Harnagel, D.F. Welch, J. Berger, and M. Sakamoto, Advances in diode laser pumps, IEEE J.Quant.Electr. 24:883 (1988).
2. R. Pratesi, Diode lasers in photomedicine, IEEE J.Quant.Electr. 20:1433 (1984).
3. R. Pratesi, Semiconductor (diode) lasers: basic principles and potential applications in the biomedical field, Photobiochem.Photobiophys., Suppl.,57-75, (1987).
4. T.S. Fan and R.L. Byer, Diode laser-pumped solid-state lasers, IEEE J.Quant. Electr. 24:895 (1988).
5. R.L. Byer, Diode laser-pumped solid-state lasers, Science 239:742 (1988).
6. M. Sakamoto, D.F. Welch, J.G. Endriz, D.R. Scifres, and W. Streifer, 76 W cw monolithic laser diode arrays, Appl.Phys.Lett. 54:2299 (1989).
7. J. Berger, D.F. Welch, D.R. Scifres, W. Streifer, and P.S. Cross, High power, high efficient Nd:YAG laser end pumped by a laser diode array, Appl.Phys.Lett. 51:1212 (1987).

8. A.D. Hays and R. Burnham, Quasi-cw diode array side-pumped 946 nm Neodymium laser, Digest of Technical Papers, CLEO 90, paper CMA1 (May 1990), p.4.

9. D. Mindinger, R. Beach, W. Benneth, R. Solarz, and V. Sperry, Laser diode cooling for high average power applications, SPIE vol.1043: 351-358 (1989).

10. S. Basu, and R.L. Byer, Average power limits of diode-laser-pumped solid state lasers, Appl.Opt. 29:1765-1771 (1990).

11. T.J. Kane, and R.L. Byer, Monolithic unidirectional single-mode Nd:YAG ring laser Opt.Lett. 10:65-67 (1985).

12. R.L. Byer, Nonlinear frequency conversion enhances diode-pumped lasers, Laser Focus World p.77 (March 1989).

13. T.M. Baer and D.F. Head, High peak power Q-switched Nd:YLF laser using a tightly folded resonator, Digest of Technical Papers, CLEO 90, paper CMF2 (May 1990), p.24.

14. J.J. Zayhowski, and A. Mooradian, Single-frequency microchip Nd lasers, Opt.Lett. 14:24-26 (1989).

15. W.M. Grossman, M. Gifford, and R.W. Wallace, Short-pulse Q-switched 1.3- and 1-µm diode-pumped lasers, Opt.Lett. 15:622-624 (1990).

16. A.I. Ferguson, Mode-locked diode pumped solid state lasers, Digest of Technical Papers, CLEO 90, paper CMF1, p.24,(May 1990).

17. I. Schutz, I. Freitag, and R. Wallenstein, Miniature self-frequency-doubling cw Nd:YAB laser pumped by a diode-laser, Opt.Commun. 77:221-225 (1990).

18. Y.X. Fan, R.C. Eckardt, R.L. Byer, C. Chen, and D. Jiang, Barium borate optical parametric oscillator, IEEE J.Quant.Electr. 25:1196-1199 (1989).

19. C.D. Nabors, R.C. Eckardt, W.J. Kozlovsky, and R.L. Byer, Monolithic MgO:LiNbO$_3$ optical parametric oscillator pumped by a frequency doubled diode laser pumped Nd:YAG laser, Digest of Technical Papers, CLEO 90, paper FK2, pp.416-418,(May 1990).

20. J. Hong, B.D. Sinclair, W. Sibbett, M.H. Dunn, A diode laser pumped and Q-switched 946-nm Nd:YAG laser with frequency doubling to 473 nm, unpublished report (1990).

21. T.Y. Fan, G. Huber, R.L. Byer, and P. Mitzscherlich, Continuous-wave operation at 2.1 µm of a diode-laser-pumped, Tm-sensitized Ho:Y$_2$Al$_5$O$_{12}$ laser at 300 K, Opt.Lett. 12:678-680 (1987).

22. L. Esterowitz, and R. Stoneman, Diode pumped Er : LiYF$_4$ cw laser at 2.8 µm with 10% slope efficiency , SPIE Vol.1040:99-102 (1989).

23. J.F. Pinto, and L. Esterowitz, A 1.96 µm Cr; Tm:YAG medical laser, Digest of Technical Papers, CLEO 90, paper CTUK4 (May 1990), p.190.

24. R. Sheps, SPIE OE/LASE'91, paper 1410-19 (1991).

25. P. Nachman, J. Munch, and R. Yee, Diode-pumped, frequency-stable, tunable, continuous-wave Nd:glass laser, IEEE J Quant.Electron. 26:317-322 (1990).

26. T.J. Kane, A.C. Nilsson, and R.L. Byer, Frequency stability and offset locking of a laser-diode-pumped Nd:YAG monolithic nonplanar ring oscillator, Opt.Lett. 12:175-177 (1987).

27. S.P. Bush, A. Gungor, and C.C. Davis, Studies of the coherence properties of a diode-pumped Nd:YAG ring laser, Appl.Phys.Lett. 53:646 (1988).

28. T. Day, A.C. Nilsson, M.M. Fejer, A.D. Farinas, E.K. Gustafson, C.D. Nabors, R.L. Byer, 30-Hz-linewidth, diode-laser-pumped, Nd:GGG nonplanar ring oscillators by active frequency stabilization, Electron.Lett. 25:810 (1989).

29. T. Day, E.K. Gustafson, and R.L. Byer, Active frequency stabilization of a 1062 nm, Nd:GGG, diode-laser-pumped nonplanar ring oscillator to less than 3 Hz of relative linewidth, Opt.Lett. 15:221-223 (1990).

PART 2

LIGHT-TISSUE INTERACTIONS AND

OPTICS OF TISSUES

BIOPHYSICAL BASES OF LASER-TISSUE INTERACTIONS

Jean-Luc Boulnois

Technomed International
North Woods Business Park, 100 Rosenwood Drive, Suite 140
Danvers, Mass. 01923 USA

INTRODUCTION

Numerous reviews of the principal photobiologic interactions between radiation and living tissues have been presented for the optical part of the electromagnetic radiation spectrum.[1-7] In the specific case of laser radiation, the unique characteristic of monochromaticity of the incident field and the spatial and temporal coherence of the emission, find increasing use in medical applications, in both diagnosis and therapy. Together with the resulting biologic response of the irradiated tissues, these characteristics determine the **specific modes of interaction**, namely, the various processes of conversion of the incident electromagnetic energy within biomolecules.

At present, several uncorrelated mechanisms are used in laser photomedicine. The high power densities reached on submillimeter spot sizes under quasicontinuous irradiation provide spatially localized heating used in the **thermal** mode of interaction, forming the well known basis of surgical applications.[3-5,7-9] The so-called **photochemical** interaction mode, corresponding to the matching of laser frequencies with specific excitation bands of chromophore molecules or photosensitizers of precise cellular structures, has recently opened a spectacular field of applications in photodynamic therapy.[10-12]

The recently introduced techniques of photodecomposition with pulsed ultraviolet lasers[13-15] constitute yet another photochemical mode of tissue action that has been labeled **photoablative interaction**. The particular combination of temporal coherence in the generation of ultrashort pulses and high peak powers, together with spatial coherence that provides the ability to focus laser radiation, constitutes the essence of the photodisruptive mechanism used in the **electromechanical** mode of interaction.[16]

This chapter will review the photophysical principles governing these applications. The presentation relies on the observation that these seemingly different interaction modes appear to share an intrinsic unity: hence this provides a natural and convenient structure for the chapter.

The relationship between the duration of laser exposure and photophysical processes induced in tissue chromophores is the basis of the observed unity. Optimal laser parameters associated with a given photomedical process can be arranged into three distinct families. A single parameter can then be shown to distinguish these processes, namely, the exposure time sufficient for delivery of the foregoing energy dose. Consequently, a simplified classification of laser-tissue interaction modes is proposed, and three groups are distinguished by order of increasing time-scale:

1. electromechanical interaction (10 psec to 10 nsec pulses)
2. photochemical interaction (10 nsec to 100 μsec pulses)
3. thermal interaction (1 msec to 10 sec exposure, quasicontinuous wave)

LASER PHOTOMEDICAL CHART

In an incident radiant flux the energy dose supplied per unit area (the **energy fluence,** measured in J/cm^2) is a possible measure of the macroscopic transformation or the 'biologic damage' in the general sense (thermal, chemical, mechanical, or electrical) caused to the exposed and reacting tissues.

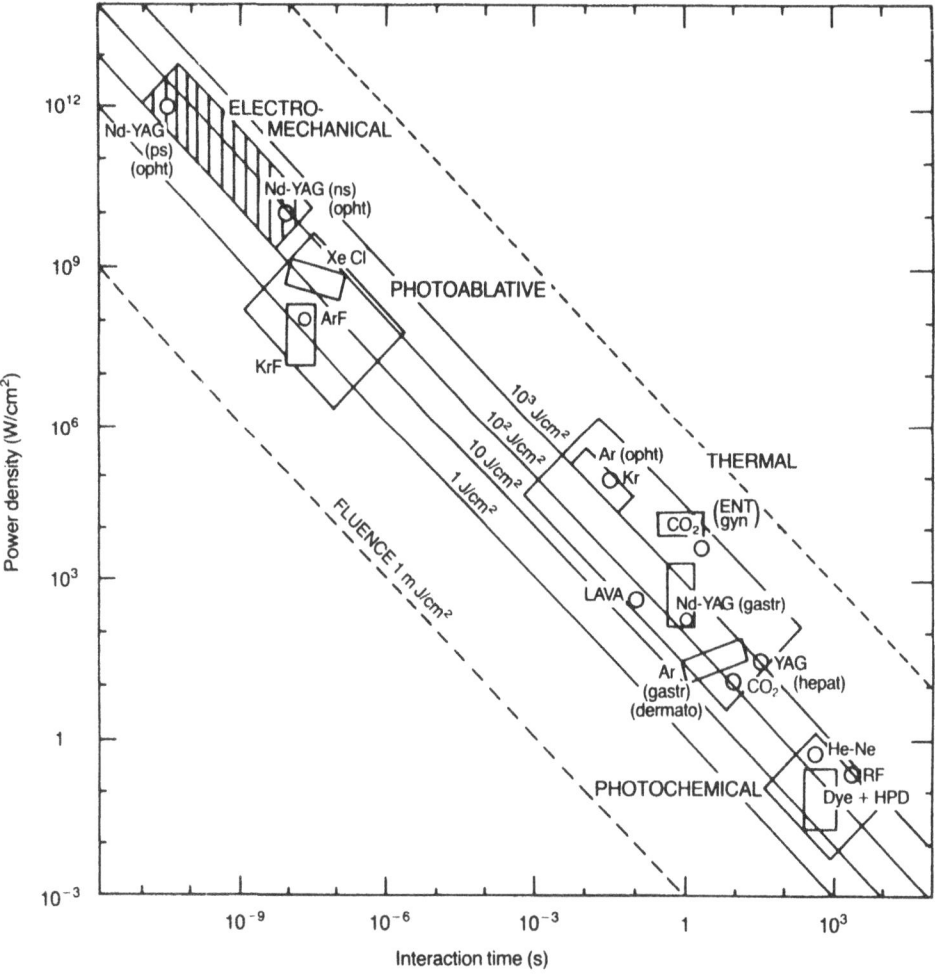

Fig.1 Medical lasers interaction map. The ordinate axis corresponds to the irradiance (in watts/cm^2 on a logarithmic scale) and is commonly labeled "power density"; the abscissa corresponds to the interaction time (in seconds, on a logarithmic scale); drawn diagonally are the lines of a constant fluence (in J/cm^2) by increasing magnitude.

30

Damage processes in pulsed systems differ from those associated with continuous waves (cw) lasers in that time-constants are substantially different. For cw operation, or when the pulse time-constant is of the order of the thermal diffusion time or the scattering lifetime, damage phenomena are controlled in depth by irreversible thermal effects of thermo-chemical transformation in the bulk. In the case of pulsed operation, from picosecond to microsecond irradiations, molecular time-constants are so short that radiant electric field effects predominate in a zone of extremely small extent.

Along these guidelines, a chart gathering most published photomedical laser data can be constructed,[8] as shown on Figure 1. It plots more than 50 experimentally determined optimal values of irradiance vs time, corresponding to most clinical and experimental photomedical applications, as given in the literature. A large variety of widely used lasers, such as Nd:YAG, Ar, Kr, CO_2 excimer, dye, or He-Ne lasers is also presented.

Two major features are displayed on this chart:

1. Contrary to what might be expected, the data are not scattered more or less randomly over the entire diagram. Rather, the experimental points are approximately clustered over 12 orders of magnitude around a straight diagonal line contained within a 1-1,000 J/cm^2 fluence band. Consequently, **time**, precisely the time of exposure during which this energy dose is to be delivered, appears to be the single parameter controlling the transformation process entirely;

2. Three separate groups of transformations clustered within this common fluence appear logically organized along a diagonal on this chart, according to the duration of interaction: they correspond precisely to the characteristic time-scales of the respective photobiologic damages involved.

ANALYSIS OF LASER-TISSUE INTERACTION PROCESSES

Thermal Interaction

All laser surgical applications, whether in the cutting or hemostatic mode, rely upon the conversion of electromagnetic energy to thermal energy. This is achieved by focusing a beam onto spot sizes a few micrometers or millimeters wide; such collimation is possible because of the spatial coherence of lasers which can supply high energy densities providing spatially confined heating of target tissues, resulting in thermal injury, tissue removal, or control of bleeding. The choice of wavelength determines the **depth of penetration** and thus influences the interplay between tissue removal and hemostasis.

In fact, the vast majority of therapeutic applications of lasers takes advantage of their capability for some spatial control over the degree and extent of tissue injury. The characterization of the photothermal biologic response following laser irradiation depends, however, on the structural level that is targeted.

At the microscopic level, the photothermal process originates from bulk absorption occurring in molecular vibration-rotation bands or, perhaps, in the vibrational manifold of the lowest electronic excited state, instantaneously followed by subsequent rapid thermalization through nonradiative decay. Since tissue structures may be considered as complex condensed phase media, rotation is hindered and vibration amplitudes are more or less damped; consequently, energy levels are not sharp but instead are broadened. It is then more appropriate to describe vibrationally excited electronic states in terms of **vibronic** states.[17] The reaction with a target molecule A proceeds in two nearly simultaneous steps: first, the absorption of a photon of energy hv, promoting A to a vibronic state A*; second, an inelastic scattering occurring on a 1-100 psec time-scale with a collisional partner M belonging to the surrounding medium. On colliding with A*, M instantaneously increases its kinetic energy by carrying away the internal energy released by A*. The microscopic origin of the temperature rise results from the amount of energy released to M.

In contrast to other photobiologic laser processes in which the choice of photon energy usually is selected to access a specific reaction channel, the biologic effects of heating (to first order) are nonspecific. The scattering properties of the medium may influence wavelength selection and, to some extent, the depth of penetration. However, the characteristic heating effects are largely controlled by molecular target absorption, usually from free water, hemoproteins, pigments (e.g., melanin), and other macromolecules such as nucleic acids and aromatics.

In normal photochemistry, as in photobiology, the energy range of interest lies between 1 and 10 eV, the latter value corresponding to the first ionization potential of most organic molecules: in fact the relevant spectral band extends from 1,000 nm in the infrared (IR) to 190 nm in the ultraviolet (UV). The respective coefficients of water, oxyhemoglobin (HbO_2), adenine, and melanin are plotted in Figure 2 over a wide spectral range. Absorption of tissue in the UV varies drastically depending on the concentration of DNA and aromatic residues of proteins, but in general most organic molecules absorb very strongly in this range.

The water absorption coefficient, which typically reaches 10^6 cm^{-1} in the vacuum UV at 100 nm, exhibits a dramatic cutoff around 190 nm and has no significant absorption throughout the entire UV range.

Nucleic acids, which constitute about 10% to 15% of a cell's dry weight, are the most widespread absorbers in the 190-300 nm spectral region. The resonance structure of the pyrimidine and purine bases of these acids are responsible for a strong absorption with a maxima ($\varepsilon \sim 10^4$ 1/mol-cm) in the region of 260-280 nm (e.g., adenine in Fig. 2). Consequently, penetration depths in the UV are extremely shallow (fractions of micrometers).

Hemoglobin (Hb), in its oxygenated structure HbO_2, presents two absorption maxima in the visible at 540 nm and 580 nm followed by a marked cutoff at about 600 nm.[18] For a physiologic solution of 150 g/L, an absorption coefficient of 107 cm^{-1} at 500 nm is readily obtained. The extinction spectrum cannot, however, be directly used to infer a penetration depth in full blood because light scattering by the densely packed erythrocytes will change light penetration by a factor of 2 or more, depending on vessel geometry.

By contrast, the absorption spectrum of melanin, does not exhibit marked absorption bands in the UV, visible, or near IR ranges.[19]

Infrared radiation, on the other hand, is absorbed mainly by water with increasingly stronger bands toward longer wavelengths (300 cm^{-1} at 3 μm), with typical absorption depths as small as 10 μm in the far IR. Clearly, in Figure 2 a spectral therapeutic window[20] is delineated between 600 nm and 1,200 nm. In this range, radiation penetrates tissues with fewer losses because of weaker scattering and absorption, thereby offering the possibility of reaching deep targets.

The wavelengths corresponding to the most widely used photomedical lasers are also shown in Figure 2.

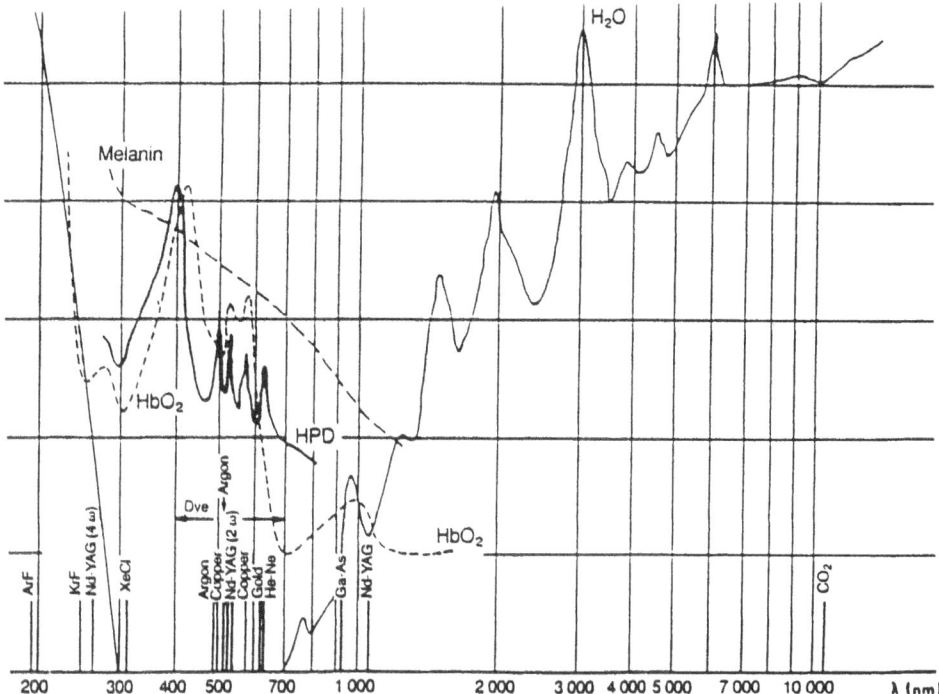

Fig.2 Molar extinction coefficient spectra of oxyhemoglobin, adenine, DOPA-melanin, hematoporphyrin derivatives, and water together with the wavelength positions of the most widely used medical lasers; energy and cross-section conversion scales.

In the IR, CO_2 lasers at 10.6 μm yield immediate vaporization in tissues with heavy water content and, hence they have an excellent cutting effect, especially at high power densities (~ 10 kwatts/cm^2), producing minimal necrosis but presenting poor hemostatic capabilities. Similar results are achieved with the 2.1 μm Holmium: YAG laser. By contrast, Nd:YAG laser radiation (1.06 μm) is capable of reaching deeper targets, especially in connective tissue where optical scattering by collagen fibers embedded in the ground substance determine the depth of penetration. Thermal exchanges take place in the bulk, most of the absorption occurring in hypervascularized and compactly connected tissue (liver tissue, for example, absorbs much more than stomach tissue). The cutting effect is less marked, but the hemostatic properties are widely recognized. In the visible, a host of lasers such as diode, argon, copper, Nd:YAG (2nd harmonic), tunable solid-state, or dye lasers are present, showing simultaneous interactions with hemoglobin, melanin, and other organic compounds, whereas gold lasers fall noticeably outside the HbO_2 absorption cutoff. Since it is strongly absorbed by hemoglobin, red globules, pigments, and melanin, argon laser radiation is used in subcutaneous vascular coagulations. Dye lasers are attractive since their tunability can be advantageously used to match particular absorption bands of specific chromophores. The UV spectral region is fairly well covered by excimer lasers (e.g., XeCl, KrF, ArF), which are powerful pulsed sources, but the Nd:YAG laser operating in the 4th or possibly 5th harmonic may perhaps become a serious competitor.

The first mechanism by which tissue is thermally affected is molecular denaturation (of, e.g., proteins, collagen, lipids, hemoglobin). For completeness, Table 1 summarizes the temperature ranges of successive transformations.[21]

Table 1 Histological Changes in Photothermal Processes

Conversion of Electromagnetic Radiation into Heat ↓ Elevation of Tissue Temperature	

Temperature	Effect
43°C - 45°C	Conformational changes Retraction Hyperthermia (cell mortality)
50°C	Reduction of enzyme activity
60°C	Protein denaturation Coagulation
80°C	Collagen denaturation Membrane permeabilization Carbonization
100°C	Vaporization and ablation

In the neighborhood of $T \sim 45°C$ (hyperthermic range) one observes a tissue retraction related to macromolecular conformational changes, bond destructions and membrane alterations. The range of protein denaturation is between 50°C and 60°C. As the molecule reaches its "melting temperature," the originally densely packed polynucleotide chain unfolds and a process called "chain melting" occurs, associated with a marked increase in light absorption around 260 nm. This denaturation results in tissue coagulation, which is exploited either to destroy small tumors or to stop hemorrhage (hemostasis). In the latter case the combination of thermal shrinkage together with hemostasis induces closing of vessel lumens which could subsequently be obstructed by a blood clot (thrombosis). Above the protein denaturation temperature, coagulation necrosis and vacuolation are produced. The temperature limit at which tissues become carbonized is about 80°C. Vaporization occurs beyond 100°C, predominantly from heated free water. The high vaporization heat of water (2,530 J/g) is advantageous, since the steam generated carries away excess heat, thereby preventing further temperature increase of adjacent tissue. Vaporization together with carbonization yield decomposition of tissue constituents. Laser ablation uses these properties to make precise incisions or resections, and the technique serves as the basis of all photosurgical or photocoagulative applications.

These irreversible structural changes reflect tissue thermogenesis caused by deep thermal conduction of the absorbed incident power. The major problem with material removal is to adjust the duration of laser exposure in order to minimize tissue injury and thermal damage to adjacent zones so as to obtain little necrosis. The scaling parameter for this time-dependent problem is the so-called **thermal relaxation time,** τ, associated with a characteristic diffusion length L. From heat diffusion theory, this latter quantity can be shown to be proportional to the square root of time together with a lumped physical parameter, the tissue diffusivity $K(cm^2/sec)$, characterizing the material thermal response (thermal conductivity, specific heat, and density). The relaxation time if then related to L through a relationship of the form:

$$L^2 = 4 \, K\tau \qquad\qquad (1)$$

For example, since the diffusivity of liquid water is $K = 1.43 \times 10^{-3}$ cm^2/sec, heat diffuses approximately to 0.8 mm in 1 sec in aqueous media. Similarly, typical thermal relaxation times associated with 10-μm vessels are on the order of 2×10^{-4} sec, whereas for microvasculature of 100 μm size they reach approximately 1.8×10^{-2} sec.

This relationship serves as the theoretical basis of a scheme, called **selective photothermolysis**, making use of pulsed irradiation to confine thermally mediated radiation damage to choose pigmented targets at the ultrastructural, cellular, or tissue structural level.[5,7,20] Selective absorption of a short laser pulse is converted into heat and transferred into cooler surroundings by thermal diffusion. According to Equation 1, since heat is essentially confined within a target of size L for a time τ, by choosing the laser pulse duration to be of the order of τ, it is possible to have the absorber temperature exceed the threshold required for given physical or chemical changes of the target, while maintaining the surrounding temperature below its threshold value.

Obviously, for this technique to apply, targets must have greater absorption than the surrounding medium. The requirement can be met by selecting the laser wavelength to coincide with absorption bands of endogenous chromophores in spectral regions where there is minimal competition from surrounding absorbers.[7] Selective photothermolysis effectively controls the spatial extend of damage by monitoring the light dose and the duration of laser exposure.

The water relaxation time at 10.6 μm suggests an interesting operation mode of CO_2 lasers in microsurgery which is called **superpulse mode**. The high temperatures needed for phase change (steam formation without appreciable heating of adjacent tissues, are reached only when the exposure duration is shorter than τ. Consequently, by pulsing the laser with **pulses shorter than 100 μsec**, it is possible to selectively vaporize specific aqueous tissues and still obtain extremely small necrosis. Typical 100 mJ/50 μsec pulses from a lower power CO_2 laser (10 watts) operating at 100 Hz vaporize 300 μm spots and cut with a velocity larger than 3 cm/sec, the peak power being approximately 2 kWatt.

In fact, the idea of delivering energy in pulses of high peak power is gaining currency; matching the pulse width and recursion with the relaxation time is likely to lead to less subjacent tissue damage than seen with cw lasers.

For completeness, well-established applications based on photothermal interactions are now briefly reviewed. In ophthalmology, short interaction times (10-100 msec) are commonly used with argon lasers (1-4 watts) or krypton lasers (1-3 watts) focalized on spot sizes of 30-100 μm. This results in power densities up to 10^5 watts/cm^2 and brings the local tissue temperature to around 60°C. Thermal photocoagulation lesions are initiated by radiant heating at specific chromophore sites and pigmented structures, e.g., melanin in the retinal pigment, epithelium, and choroid, hemoglobin in retinal and choroidal vessels, and macular xanthochromes.[22,23]

Power densities in the range of 10^4 watts/cm^2 are routinely employed with CO_2 lasers for consecutive irradiation pulses of, at most, 1 sec each, in ear, nose, throat, and laryngeal microsurgery[24,25] or gynecologic treatments.[26]

The management of various gastrointestinal bleeding lesions with Nd:YAG lasers operating in the hemostatic mode,[27] the resection of various tumors in gastroenterology, urology, tracheobronchial endoscopy, and in general surgery, are mostly performed with Nd:YAG, CO_2, or argon lasers at typical influences of 100-1,000 J/cm^2, with corresponding irradiation times on the order of seconds.[28]

Dermatology is another field in which lasers such as argon or copper prove advantageous, particularly on cutaneous colored lesions or exophytic lesions.[29] Selective thermal damage to endogenous tissue chromophores (e.g., hemoglobin, melanin) may induce minimal necrosis of healthy adjacent dermis and epidermis. Subcutaneous vascular coagulation techniques, employed for thrombosis of superficial dysplasia such as portwine stains, require careful control of the energy dose with cw lasers, in order to minimize scarring. In a recent development, pulsed copper vapor lasers at 578 nm have been investigated in portwine stain therapy.[30] The operating mode is similar to the foregoing "superpulse mode;" pulsed exposure durations are shorter than the time required for cooling the pigmented absorbers, thereby inducing selective transient heating of these targets. In the experiment, pulse duration was about 60 nsec, repetition rate 5 kHz, and energy around 3 mJ. The process was presumably thermal, via selective heating of blood vessels, owing to the fast deactivation of HbO_2 vibronic state and the high average power delivered by such lasers. Whereas τ for hemoglobin is about 10 msec, portwine stained blood vessels have a relaxation time of the order of a few msec,[31] due to erythrocyte scattering. This time is optimal for therapy, since it produces little heat loss during a laser pulse so that the temperature difference between the vessel wall and the dermis is maximized. The choice of wavelength is particularly judicious because if perfectly matches an absorption peak of hemoglobin.

Photochemical Interaction

Selective targeting in tissues can be obtained using either endogenous tissue chromophores, such as hemoglobin or melanin, or by introducing exogenous chromophores. If pulsed exposures of a duration shorter than the time required for cooling the target chromophore are used, it is possible to maximize the transient photoresponse occurring at absorption sites. For most chromophores, the corresponding time-scales rage from 1-20 nsec to about 10 µsec, and when the radiation-induced response involved an absorber's electronic or oscillatory motion , it is defined as a **photochemical transformation**. In most instances, the basic physical channels of photochemical interactions between laser radiation and cellular structures are only partially elucidated.

Nevertheless, it is possible to schematically distinguish two subfamilies:

1. Reactions in which molecules are involved as energy carriers or as catalytic regulators after experiencing a photoexcitation. In this case the chromphore receptors are said to be **photoactivated.** One of the most attractive applications of this type is laser spectral sensitization in which photodynamic therapy (PDT) of malignant tumors appears to be a highly promising technique.

2. Reactions in which the chromophore molecules are modified and converted into photoproducts. An example of general importance is the photoinactivation caused by short-wavelength ultraviolet light used in the recently introduced technique of **photoblative** microsurgery.

Photoactive Interaction. Within the "therapeutic window', radiation penetrates rather deeply into most tissue. If spectrally adapted chromophores are introduced and selectively retrained in specific cellular sites, narrowband irradiation can trigger selective photochemical reactions in vivo, inducing subsequent photobiologic transformations. Hence, energy can be selectively delivered to deep target cells. Normal processes such as melanogenesis can thus be photoactivated in vitiligo Psoralen Ultraviolet A (PUVA) therapy. Photodamage to abnormal cells can also be induced, as in photochemical modification of nucleic acids in psoriasis PUVA therapy by furocoumarins or in cytotoxic photosensitization therapy of tumoral cells by hematoporphyrin or its derivatives.

Table 2 Physical Principle of Photodynamic Therapy

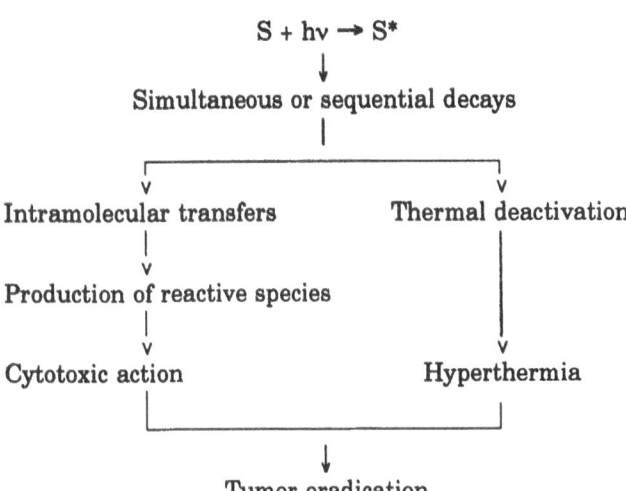

* Administration of photosensitizer S

* Selective tumor retention of photosensitizer

* Irradiation by monochromatic source (laser)

* <u>Resonant excitation</u> of Photosensitizer

$$S + h\nu \rightarrow S^*$$

Simultaneous or sequential decays

Intramolecular transfers   Thermal deactivation

Production of reactive species

Cytotoxic action   Hyperthermia

Tumor eradication

A chromophore compound capable of causing light-induced reactions in molecules that do not absorb light in the same wavelength range may be called a **photosensitizers**.[10,32] Following resonant excitation by a monochromatic source, the photosensitizer undergoes a series of simultaneous or sequential decays which result in intramolecular transfer reactions. The decays may ultimately culminate in the release of highly reactive cytotoxic species that cause, for example, irreversible exudation of some essential cellular component and destroy affected host tissues.[33] Most photosensitizers currently used are organic dyes.

The sequence of molecular reactions undergone by a given photosensitizer can be separated into several stages: resonant excitation, decay, substrate reaction, or reactant formation and oxidation. These kinetics can be further separated schematically into two types of 'mechanisms', depending on the substrate involved in the crucial stage of reactant formation.[34] Table 2 summarizes these different pathways, as they are encountered in a new therapeutic treatment of tumors based on selective photosensitized oxidation, and called photodynamic therapy (PDT).

<u>Photoablative Interaction</u>. As already stated, UV radiation is very strongly absorbed by most biomolecules (fig.2), specifically in a band extending from 200 nm to 320 nm. Absorption coefficients as large as 10^4-10^5 cm^{-1} are common, and absorption depths are consequently very small, a few micrometers at most. Such short-wavelength radiation hardly occurs in any natural biotopes at present, and organisms are poorly adapted to

it. Because of the high quantum energy of this radiation, it can modify and convert chromophores into photoproducts by photodissociation of functionally important compounds (proteins, nucleic acids, enzymes, pigments).

This feature has recently been exploited experimentally to produce well-defined, nonnecrotic, photoablative cuts of very small width (~50 μm) by exposure to excimer lasers at several UV wavelengths (ArF: 193 nm; KrF: 248 nm; XeCl: 308 nm), with short pulses (15 nsec) focused on various tissue (cornea,[13] skin,[14] plaques.[15] Typical irradiances are 10^8-10^9 watts/cm^{2}s. Similar sharp cuts (2-3 μm) with minimal thermal damage are also obtained with the 4th harmonic of the Nd:YAG laser at 266 nm, with excellent cutting aspect owing to the high spatial quality of the beam.[35] Control of thermal damage is clinically important since it generally produces undesirable biologic effects.

The photoablative process consists of a photodissociation, i.e., the direct breaking of intramolecular bonds in polymeric changes, caused by absorption of nearly 5 eV incoming photons followed by effluent ejection. In the photon energy range of 5-7 eV, biopolymers such as collagen may dissociate by absorption of a single photon. The microscopic mechanisms correspond to transitions of a macromolecule AB, which is promoted to a repulsive electronic state and thereby yields photoproducts A and B. The first step in the process is controlled by optical absorption, whereas the second step, dissociation, is controlled by molecular motion.

The mechanisms controlling the third step, ablation, are much more complex, and the fate of molecular photofragments is not easily predicted. A current assumption is that photoejection, which carries away most of the excess energy, is the primary process; but desorption may be an important secondary mechanism. At the present time the dynamics of nuclear motion are not clearly ascertained; but evidently the power density significantly determines the rate of momentum transfer. Radiative dissociation in repulsive electronic excited states yields products whose kinetic energy is necessarily the difference between the absorbed energy, hν, and the bound state molecular energy. As a result, not every excited state may achieve photodissociation, and a quantum yield specific to the photofragments involved is associated with the process.

Laser-induced photodissociation is an extremely rich field. A comprehensive photophysical analysis of polyatomic molecules would have to consider how structured their vibrational manifold is, whether the process is single photon or multiphoton, whether it is resonant or nonresonant, whether the incoming field is weak or strong, or single mode or multimode. One can thus anticipate that UV photodissociation, when applied to biomolecular models, will certainly prove to be even more complex.

It is interesting to notice that UV photoablation is inherently pulse-width limited. The limitation is determined by characteristic de-excitation times, on the order of 100 nsec at most, arising from recombination, attachment, or other quenching mechanisms, such as collisional decay of vibronic states. A significant thermal component resulting from these inefficiencies might then be associated with photoablative processes. This slightly impairs the special feature of UV photoablation, which makes it different from any other type of laser-processing. Theoretically, all the energy difference between absorption and bound-state energy ends up in the ablated particles (dissociation plus kinetic energy), leaving the unirradiated portions of the sample unaffected. In practice, thermal diffusion is sometimes observed on the edges of the etched zone, and long UV exposures would be associated with a large thermal component (30 μsec time-scale). Hence, short laser pulses and adequate repetition rates need to be chosen and this explains the position of the photoablative interaction family in Figure 1.

When validated, this experimental surgical technique may prove to be unique in its ability to produce sharp incisions with minimal thermal damage to adjacent normal structures. Table 3 summarizes the pathways of processes involved.

Table 3 Physical Principles of Laser Photoablation

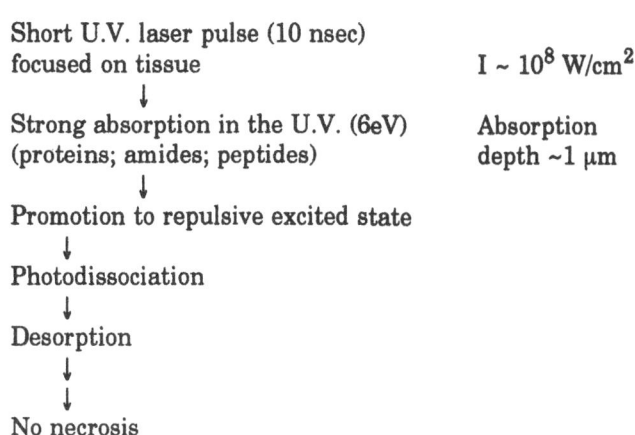

Short U.V. laser pulse (10 nsec)
focused on tissue $I \sim 10^8$ W/cm^2

↓

Strong absorption in the U.V. (6eV)   Absorption
(proteins; amides; peptides)   depth ~1 μm

↓

Promotion to repulsive excited state

↓

Photodissociation

↓

Desorption

↓
↓

No necrosis

Electromechanical Interaction

In electromechanical interaction, a fluence of about 100 J/cm^2 is delivered to possibly transparent tissues by, for example, a Nd:YAG laser with extremely short time exposures by means of either mode-locked 30-psec pulses or 10-nsec Q-switch pulses. The process is not maintained by linear absorption and is consequently not thermal. Rather, when focused at a target the high peak power laser pulsed (approximately 10^{10} watts/cm^2 for nanosecond pulses and 10^{12} watts/cm^2 for picosecond pulses) locally generates high electric fields (10^6-10^7 V/cm), comparable to average atomic or intramolecular Coulomb electric fields. Such large fields induce a dielectric breakdown of the target material, resulting in the formation of a microplasma, i.e., an ionized volume with a very large free-electron density. The shock wave associated with the plasma expansion generates a localized mechanical rupture over dimensions where the rise in pressure exceeds the yield strength of encountered tissues.[35] Table 4 summarizes the global sequence of the physical processes involved.

At the microscopic level, the mechanism responsible for optical breakdown is the massive generation of free electrons. The initial phase, which corresponds to a localized electron 'seeding' by ionization involving a few electrons only, should be distinguished from the subsequent massive electron photoproduction.

For ionization mechanisms, energies of about 7 to 10 eV must be supplied to bound electrons in order to separate them from individual molecules. In this initiation phase, ionization mechanisms differ, depending on pulse duration.[35,36] For Q-switched pulses (3-10 nsec duration), ionization is caused by thermionic emission resulting from focal heating of the target, where local 'temperatures' (i.e. specific molecular energies) exceeding several 10,000°C are reached all the more rapidly when impurities are present. For very short, mode-locked 20-psec pulse trains, molecular ionization is obtained by a non-linear process called **multiple photon absorption**. Because of the extremely high irradiance at the focal spot, photons can add up their energies coherently and provide the ionization energy, thus producing free electrons.

Table 4 Physical Principles of Laser-Induced Breakdown

Short laser pulse focused at target	$I \sim \dfrac{(E_p/\tau_p)}{W^2}$
$\downarrow$	
High power density	$I \sim 10^{12}$ W/cm^2
$\downarrow$	
High electric field	$E \sim 10^6$ V/cm
$\downarrow$	
Dielectric breakdown (multiphoton process)	$E_{laser} \sim E_{ionization\ molecules}$
$\downarrow$	
Plasma formation free electrons	$N_E \sim 10^{21}$/cm^3 $T > 20000°C$
$\downarrow$	
Spherical shock wave propagating at sound velocity	P (bars) $\sim 13\ E_p/R^3$ E_p(mJ) R(mm)
$\downarrow$	
Localized mechanical rupture for small radius	$P >$ Yield strength of tissues

The threshold for optical breakdown is higher for modelocked picosecond pulses than for Q-switched nanosecond pulses. In air, the threshold for a single 25 psec pulsed is about 10^{14} watts/cm^2, whereas it is only 10^{11} watts/cm^2 for a 10 nsec pulse. In biologic solutions, these numbers are reduced by a factor of 100. The total energy density delivered in achieving optical breakdown (~ 10 J/cm^2) is ultimately the same for a modelocked pulse train and a Q-switched pulse, even though peak power density is, on the average, 100 times higher for picosecond pulses.[37]

The key physical elements of the electromechanical mode of interaction is rooted in the subsequent massive photoproduction of extraneous free electrons. The process is termed **electron avalanche growth**. In the electric field of molecules or ions, already seeded free electrons absorb incoming photons and convert this energy into kinetic energy, thereby increasing their velocity. On colliding with neighboring molecules, rapidly moving free electrons ionize collision partners when their relative kinetic energy is sufficient and, hence, create extraneous free electrons.

The particularly important feature of this process is that there is no restriction on the photon energy. In fact, the cross section is largest for small photon energy. Each electron in turn absorbs more photons, accelerates, strikes more atoms or molecules, and ionizes them in an avalanche process with exponential growth.

In a few hundred picoseconds a very large free electron density (typically 10^{21} cm^{-3}) is thus created in the focal volume of the pulsed laser beam. This is called laser-induced breakdown of the dielectric medium.

The conditions required for plasma growth and sustainment are that losses such as inelastic collisions or free electron diffusion should not suppress the avalanche. Irrespective of pulse duration, if the rise time is sufficiently short, a permanent state is reach in about 100 psec or less. The origin of this steady state is the so-called 'plasma shielding effect.' As the plasma number density N_e increases, the characteristic "plasma frequency" w_p increases, as does photon scattering where w_p is representative of collective free-electron motion and is proportional to $N_e^{1/2}$. As a result, progressively less energy is coupled from the laser field to free electrons and a quasi-equilibrium is reached. The critical density, N_e^*, at which incident energy is not converted any further, is reached when the plasma frequency becomes equal to the pulsation of the incident electromagnetic wave.[37] For Nd:YAG laser radiation, the upper bound electron density is:

$$Ne^* < 10^{21}cm^{-3}$$

When this density is reached in less than 100 psec after initiation of a 25-psec pulse, the plasma radius extends to several tens of micrometers and pressure as well as temperature have reached their maximum values. The pressure typically reaches several hundred kilobars locally. The plasma-shielding steady state has then been reached and radiation cannot penetrate the medium any further. As the plasma shock wave expands, it cools and the pressure falls accordingly. For a mode-locked picosecond pulse train with pulses typically separated by 7-nsec intervals, the cooling rate between pulses is faster than with Q-switched nanosecond pulses, but the process is qualitatively similar.

Clearly, inside the spherical volume, where the pressure exceeds the yield strength of the biologic structures encountered, a localized mechanical rupture will take place, resulting in an 'opening or disruption' of these structures. This is the physical basis of new ophthalmic surgical modalities,[16] and the forementioned plasma shielding effect is of major importance in retinal protection during pulsed Nd:YAG laser ophthalmic treatments in the anterior segment of the eye.[38]

This photodisruption process is of particular interest for the noninvasive treatment of several ocular pathologies such as capsulotomies,[16] certain peripheral iridotomies, the removal of vitreous strands, or the dissociation of opacified membranes which frequently develop following cataract surgery. Since its inception, considerable interest has been generated by this technique which, owing to its simplicity, can be implemented in ambulatory treatments.

Other medical specialties are developing interest in this technique as well. Recently, optical breakdown techniques have been used[39] in cardiovascular models to investigate possible disintegration of atheromatous plaques or stenosis in small arteries. Experiments performed in dry conditions with Q-switched Nd:YAG pulses have been carried out at various wavelengths (1,064 nm, 532 nm, 266 nm), since spectroscopic study did not show any specific strong absorption band in the atheromatous plaque spectrum. As stated previously, longer wavelengths would probably be more favorable. Histologic examination showed an absence of thermal damage in the edges of the lumen supporting the assumption of a photodisruptive mechanism. Although preliminary results in vitro suggest feasibility of the recanalization technique, a main difficulty remaining is the transport within fibers of the necessary high peak powers.

A spectacular development of this photomechanical effect arises in the photoacoustic effect used in intracorporeal laser lithotripsy, particularly for the fragmentation of ureteric calculi. Whereas initially Q switched Nd:YAG lasers were employed,[40] in further developments, flashlamp-pumped pulsed dye lasers emitting in the visible spectrum between 450 nm and 600 nm and operating at 5-20 Hz repetition rates are now being used in conjunction with a fiber delivery system for the fragmentation of stones in the ureteral tract, or common bile duct and other stones within the biliary tree. Short pulses lasting 1-4 μsec, with energies ranging between 40 and 200 mJ and fluence levels 20-200 J/cm^2, locally heat an extremely small volume of the kidney stone porous matrix. This heat is rapidly conducted to the interstitial water confined inside the matrix microcavities, bringing it to its boiling point. A 1,670-fold volume expansion occurs when water is vaporized isobarically; the resulting mechanical strain creates the desired localized shattering of the stone. Since the water heat of vaporization is about 2,530 J/g, one establishes that 250 mJ pulsed heat up a focal volume equal to, at most, 0.1 mm^3 which, for a typical small kidney stone (a few millimeters), required an exposure of 10 seconds before completing the fragmentation. Besides their cost effectiveness and their possibility of operating at higher repetition rates, pulsed dye lasers offer a major advantage because their broad emission spectrum covers a large number of absorption bands of calculi constituents, which are essentially calcium or magnesium salts of phosphates or oxalates. This minimally invasive kidney stone laser shattering procedure is well suited to remove stones lodged in the lower part of the ureter, a problem traditionally corrected by surgery; it is complementary to the lithotripter noninvasive shockwave therapy which cannot reach this region because of anatomic configuration.

CONCLUSION

Basic analysis and comparison of different biomedical laser applications, emphasizing the sequences of photophysical steps involved have been presented in this chapter. It has been proposed that three groups of interactions may be distinguished according to their radically different tissue reactions, which depend on the duration of irradiation. Attenuation characteristics of primary absorbers, kinetics, and associated relaxation times have been analyzed for each group in support of the proposal. The review of well-established applications in surgery, photocoagulation, and tumor removal has been given in the perspective of understanding the primary laser-tissue photoprocesses. The review of such young fields as laser PDT, the use of pulsed lasers in the breakdown mode, or UV photoablative techniques has attempted to demonstrate the potential that laser may offer photomedicine in the near future.

More innovations will most likely appear, but unbridled optimism cannot be offered regarding their immediate applications. The complexity of the problems encountered underlines the necessity for careful analysis of the results along with the need for new theories of molecular dynamics. Laser photomolecular biology is a growing, exciting field. It clearly expands continuously its applications in photomedicine. We are witnessing the emergence of a new era in medicine which could be called **minimally invasive photomedical therapy.**

REFERENCES

1. L. Goldman and R. J. Rockwell, "Lasers in Medicine," Gordon and Breach, New York (1971).
2. W. R. Adey, Tissue interactions with nonionizing electromagnetic radiation, Physiol. Rev. 61:435-514 (1981).

3. J. R. Hayes and W. L. Wolbharsht, Models in pathology: Mechanisms of action of laser energy with biological tissues, in: " Laser Applications in Medicine and Biology," vol 1., W. L. Wolbharsht, ed., Plenum Publishing Co, New York (1975) pp 255-274.
4. R. Pratesi and C. A. Sacchi, eds., " Lasers in Photomedicine and Photobiology," Springer, New York (1980).
5. J. D. Regan and J. A. Parrish, " Science of Photomedicine," Plenum Publishing Co, New York (1982).
6. M. Grandolfo , S. M. Michaelson, and R. Rindi, eds., "Biological Effects and Dosimetry of Nonionizing Radiation," Plenum Publishing Co, New York (1983).
7. J. A. Parrish and T. F. Deutsch, Laser photomedicine, IEEE J. Quantum Electron. QE-20: 1386-1396 (1984).
8. J. L. Boulnois, Photophysical processes in recent medical laser developments: A review, Lasers Med. Sci. 1:47-66 (1986).
9. I. Kaplan, ed., " Laser Surgery," Academic Press, Jerusalem (1976).
10. J. D. Spikes and R. Straight, Sensitized photochemical processes in biological systems, Ann. Rev. Phys. Chem. 18:409-415 (1967).
11. I. Diamond, S. Granelli, and A. F. McDonagh, Photodynamic therapy of malignant tumors, Lancet 2:1177 (1973).
12. T. J. Dougherty, Photoradiation therapy, (abstract), Am. Chem. Soc. Mtg. N.014, Chicago, IL, Sept. 1973.
13. S. R. Trokel, R. Srinivasan, and B. Braren, Excimer laser surgery of the cornea, Am. J. Ophthalmol. 96:710-715 (1983).
14. R. J. Lane, R. Linsker, J. J. Wynne, et al., Ultraviolet-laser ablation of the skin, Lasers Surg. Med. 4:201-206 (1984).
15. W. S. Grundfest, P. Litvack, J. S. Forrester, et al., Laser ablation of human atherosclerotic plaque without adjacent tissue injury, J. Am. Coll. Cardiol. 5:929-937 (1985).
16. D. Aron-Rosa, J. Aron, J. C. Griesmann, et al., Use of the Nd:YAG laser to open the posterior capsule after lens implant surgery, J. Am. Intraoc. Implant. Soc. 6:352-354 (1980).
17. F. Dorr, Mechanisms of energy transfer, in: "Biophysics," W. Hoppe, W. Lohman, Z. Markl, et al., eds., Springer Verlag, New York (1983), pp 266-288.
18. E. J. van Kampen and W. G. Zijlstra, in: "Atlas of Protein: Spectra in the Ultraviolet and Visible Regions," vol II, D. M. Kirschenbaum, ed., Adam Hilger, London (1974), p.262.
19. M. L. Wolbarsht, A. W. Walsh, and G. George, Melanin, a unique biological absorber, Appl. Opt. 13:2184-2185 (1981).
20. J. A. Parrish, New Concepts in the therapeutic photomedicine: Photochemistry, optical targeting and the therapeutic window, J. Invest. Dermatol. 77:45-50 (1981).
21. J. M. Brunetaud, S. Mordon, J. Bourez, et al., Therapeutic applications of lasers, in: "Optical Fibers in the Biomedical Field," D. Boucher, ed., Proc. SPIE 405: 2-4 (1983).
22. F. A. L'Esperance, "Ocular Photocoagulation," C. V. Mosby, St. Louis, MO, (1975).
23. H. L. Little, H. C. Zweng, and R. R. Peabody, Argon laser slit lamp retinal photocoagulation, Trans. Am. Acad. Opthalmol. 74:85-90 (1970).
24. B. Karduck, M. G. Richter, and M. Blank, Laserchirurgie des Stimmbandes, Laryngol. Rhinol. Otol. 57:419-424 (1978).
25. C. Freche , J. Lotteau, and J. Abitbol, Le laser en ORL, in: "Concours Medicale," Masson, Paris (1979), pp 2607-2611.
26. R. Taoff, Use of the carbon dioxide laser in gynecological surgery, in: " Laser Surgery", vol III, I. Kaplan, ed., Academic Press, Jerusalem (1976), pp 235-238.
27. P. Kiefhaber, G. Nath, and K. Moritz, Endoscopical control of massive gastrointestinal hemorrhage by irradiation with a high-power neodynium-YAG laser, Prog. Surg. 5:140-155 (1977).

28. N. B. Groteluschen, M. Reilmann, V. Bodecker, et al., A high power Nd:YAG laser as a cutting tool in experimental surgery, in: " Laser Surgery," I. Kaplan, ed., Academic Press, Jerusalem (1976), pp 167-173.

29. T. Oshiro, " Laser Treatments for Nevi," Fukiun Printing Co, Ltd, Tokyo (1980).

30. T. F. Deutsch and A. R. Oseroff, New medical uses of lasers: A survey, Technical Digest, abstract WF3, CLEO, Baltimore (1985), p 84.

31. T. Lahaye and M. van Gemert, Optimal laser parameters for portwine stain therapy: A theoretical approach, Phys. Med. Biol. 30:573-576 (1985).

32. T. J. Dougherty, Hematoporphyrin as a photosensitizer of tumors, Photochem. Photobiol. 38:377-385 (1983).

33. K. R. Weishaupt, C. J. Gomer, and T. J. Dougherty, Identificant of singlet oxygen as the cytotoxic agent in photoinactivation of a murine tumor, Cancer Res. 35:2316-2329 (1976).

34. R. Bensasson, La photochimiotherapie par l'hematophorphyrine: Introduction, mecanismes moleculaires, in: " Laser Medical," 81-82, Masson, Paris (1984), pp 29-31.

35. M. W. Berns and R. N. Gaster, Corneal incisions produced with the 4th harmonic (266 nm) of the Nd:YAG laser, Lasers Surg. Med. 5:371-375 (1985).

36. C. A. Puliafito and R. F. Steinert, Short-pulsed Nd:YAG laser microsurgery of the eye: Biophysical considerations, IEEE J. Quantum Electron. 20:1442-1448 (1984).

37. D. W. Fradin, N. Bloembergen, and J.P. Letellier, Dependence of laser-induced breakdown of field strength on plasma duration, Appl. Phys. Lett. 22:635-637 (1973).

38. R. F. Steinert, C. A. Puliafito, and S. Trokel, Plasma formation and shielding by three ophthalmic Nd:YAG lasers, Am. J. Ophthalmol. 96:427-434 (1983).

39. G. Mayer, R. Astier, J. Englender, et al., Recanalization in vitro of atheromatous coronary arteries by short laser pulses, Technical Digest abstract from the Second International Congress of the ELA, Brussels, Jan 27-28 (1985).

40. N. S. Nishioka et al., Mechanisms of laser induced fragmentation of urinary and biliary calculi, Lasers in Life Sciences 3:231-245 (1987).

OPTICAL AND THERMAL MODELING OF TISSUES: DOSIMETRY

M.J.C. van Gemert[1,2], A.J. Welch[2,3]

[1] Laser Center, Academic Medical Center
Amsterdam, The Netherlands

[2] Biomedical Engineering Program
The University of Texas at Austin
Austin, Texas, USA

[3] Marion E. Forsman Centennial Professor
of Electrical and Computer Engineering and
Biomedical Engineering

I. INTRODUCTION

Medical therapies using lasers typically involve photochemical, thermal or mechanical reactions created through the absorption of light in the treated tissue. Absorption varies with tissue type and is a function of wavelength. During irradiation, absorption may change owing to dehydration, coagulation, carbonization, or ablation which ironically are the desired results of many accepted surgical procedures. The complex interaction of thermal and optical events during irradiation increases the difficulty of achieving successful clinical results.

Diagnostic devices use some form of energy to interrogate the tissue being examined, e.g. sound, heat, microwaves, light, X-rays. There are well-developed theoretical backgrounds for propagation of all these energy forms in biological tissue except light. Propagation of light in an absorbing and scattering medium is described by the equation of radiative transfer (Chandrasekhar, 1960; Ishimaru, 1978). New or improved diagnostic modalities that are possible with a better light propagation theory include blood oximetry, tumor localization using fluorescence, atheromatous plaque identification during laser angioplasty, and laser doppler perfusion devices.

The events depicted in Figure 1 comprise a coupled optical, thermal and damage model. Optical and thermal parameters are not constants, but depend upon the condition of the tissue which is a function of e.g. temperature, hydration and thermal damage. The distribution of light (fluence rate, W/m^2) within laser-irradiated tissue is a function of (1) the laser beam power, wavelength, and spatial irradiance profile, (2) the local optical properties (absorption, scattering and phase function) of the tissue, and (3) the index of refraction of the media (i.e. air, quartz fiber, etc.) delivering the laser beam and indices of refraction of the tissue. Often an approximation of the transport equation is used to estimate the fluence rate within tissue (Ishimaru, 1989). The rate of heat generation at a particular location is governed by absorption of photons by local chromophores and is numerically represented as the product of the fluence rate and the tissue absorption coefficient at that location (Ishimaru, 1978, p.158).

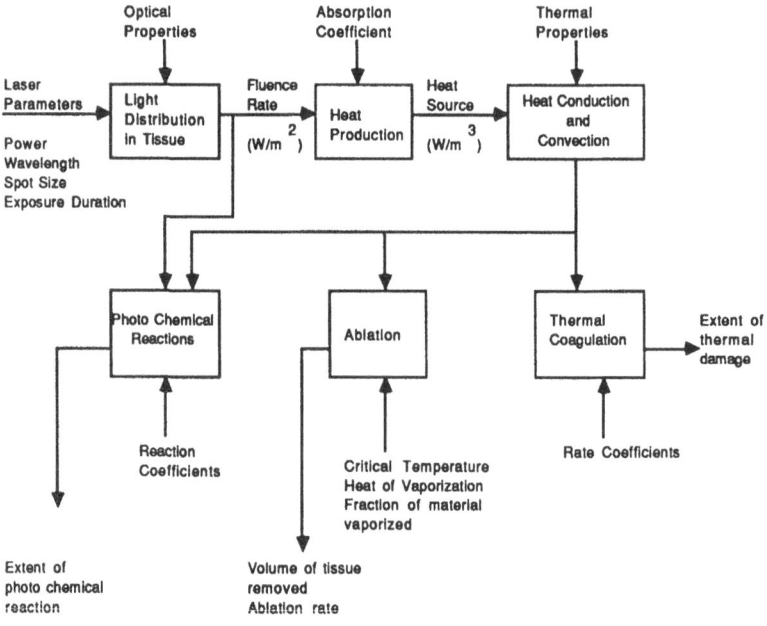

Fig.1. Diagram of Optical/Thermal Interaction

The rate of heat transfer is dependent upon thermal conductivity, thermal diffusivity and heat transfer coefficients. Tissue temperature is determined by the rate of heat generation and rate of heat transfer to cooler regions by conduction and convection. Sufficient temperature increases cause coagulation and/or ablation. The mathematical description of the optical-thermal response of tissue to laser irradiation requires a reasonable geometrical representation of the tissue. Mathematical models of tissue may require multiple layers and a realistic distribution of absorbing elements representing tissue chromophores.

II. OPTICAL MODELING

If tissues were light absorbing, non-scattering media, the spatial light distribution following laser irradiation could be described by simple exponential attenuation. We know, however, that tissues are turbid materials and do scatter light. Actual light distributions in tissue can be substantially different from those estimated when scattering is neglected. For example, scattering can cause the light energy available for absorption just below the surface of the tissue to exceed the incident light energy. At wavelengths where absorption is relatively low (in many tissues between 400 nm and 1500 nm), this increase in available light may be a factor of two or three. Also, scattering extends the available light beyond the lateral dimension of the incident beam. The magnitude of these scattering effects and their importance depends strongly on the scattering and absorbing abilities of the tissue, the refractive index of the tissue, and the diameter of the laser spot.

The absorption (μ_a) and scattering (μ_s) coefficients are defined such that the probability for absorption, or scattering as a photon travels an infinitesimal distance ds is $\mu_a ds$, or $\mu_s ds$, respectively. The reciprocal of μ_a or μ_s represents the average distance that a photon travels *without* being absorbed or scattered (mean free path of absorption or scattering). The absorption and scattering coefficients are expressed in units of reciprocal length (e.g. mm^{-1}).

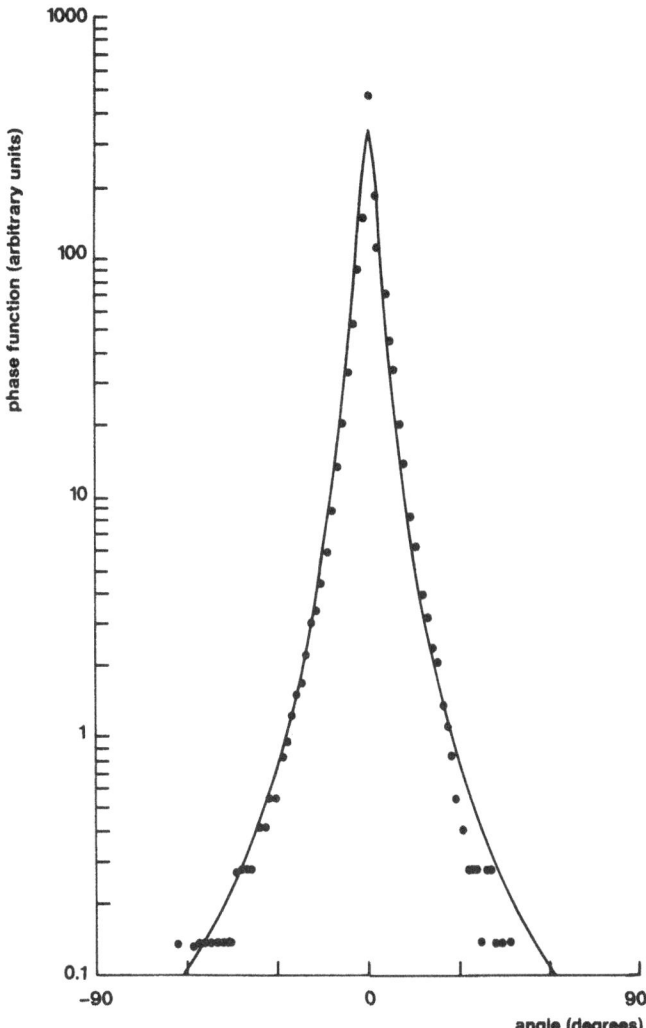

Fig.2. The Henyey-Greenstein phase function, Equation (2), as a fit to goniophotometer measurements of 0.02 mm thick human dermis (from Prahl, 1988)

Scattering is generally assumed to depend only on the angle θ between incoming and outgoing directions of the scattered photon. Such an assumption neglects the possibility that some tissues have a preferred scattering axis (such as striated muscle). The probability of scattering at an angle θ can either be constant (isotropic scattering) or depend on θ (anisotropic scattering). Human and animal tissues are now known to be strongly *forward* scattering (e.g. Wilson et al., 1987; Jacques et al., 1987; Parsa et al., 1989; van Gemert et al., 1989). A measure of the degree of anisotropy in scattering is the anisotropy factor g, where g = 1 means totally forward scattering and g = 0 means isotropic scattering. Mathematically, g is defined as the average cosine of the scattering angle θ over all possible angles,

$$g = \int_{0}^{\pi} \int_{0}^{2\pi} p(\theta) \cos\theta \, \sin\theta \, d\theta \, d\phi \qquad (1)$$

where p(θ) is the normalized scattering probability function (called phase function). For *in vitro* tissues, g turns out to be between 0.7 and 0.99. Experimentally, it has been found that in *vitro* tissues have a scattering probability that is close to the Henyey-Greenstein (1941) formula

$$p\ (\theta) = (1-g^2)\ [1 + g^2 - 2g\ \cos\ \theta]^{-1.5} \tag{2}$$

An experimentally obtained example is shown in Figure 2.

The equation of radiative transfer (e.g. Ishimaru, 1978, Chapter 7; Ishimaru, 1989) is generally assumed to be the "gold standard" for light propagation in turbid media. The equation originates from an energy balance of the radiance, incident upon a cylindrical volume with infinitesimal cross section and infinitesimal length. The radiance represents the (local) power density, flowing in a certain direction, confined within an infinitesimal solid angle. The transport equation is an integro-differential equation, and analytical solutions for situations of interest in clinical laser medicine are generally not available. Recently, numerical Monte Carlo computations have become available that produce solutions to the transport equation under clinical laser conditions (e.g. Wilson and Adam, 1983; et al., 1989a). The fluence rate ψ(r) at a given point r in the tissue is the total radiant power that passes through the surface of an infinitesimally small sphere divided by the cross sectional area of that sphere; fluence rate is expressed in power per area. Mathematically, the local fluence rate follows from the radiance at that point by integrating over all possible directions in space. Defining L(r,ŝ) as the radiance at r, pointing in the direction of the unit vector ŝ, the mathematical relation between ψ(r) and L(r,ŝ) is

$$\psi\ (r) = \int_{4\pi} L\ (r,\ \hat{s})\ dw \tag{3}$$

where dw denotes an infinitesimal solid angle pointing in the direction ŝ. The fluence rate has more practical significance than the radiance because absorption of photons by an absorbing chromophore, at location r inside the tissue, is independent of the direction of propagation of a photon. The fluence rate follows from the radiances by Equation (3), and the radiance follows from the equation of radiative transfer,

$$\frac{dL\ (r,\ \hat{s})}{ds} = -\ (\mu_a + \mu_s)\ L\ (r,\ \hat{s}) + \mu_s \int_{4\pi} p\ (\hat{s},\ \hat{s}')\ L\ (r,\ \hat{s}')\ dw' \tag{4}$$

The left hand side term represents the rate of change of the radiance at r and in direction ŝ, after traversing a cylindrical volume of infinitesimal length ds (in the direction of ŝ). The decrease in the radiance transversing the length ds due to absorption and scattering is

$$dL\ (r,\ \hat{s}) = -\ ds\ (\mu_a + \mu_s)\ L\ (r,\ \hat{s}) \tag{5}$$

whereas the increase in L(r, ŝ) due to scattering from the radiance from all other directions ŝ' into ŝ at position r is

$$dL\ (r,\ \hat{s}) = ds \int_{4\pi} \mu_s\ p\ (\hat{s},\ \hat{s}')\ L\ (r,\ \hat{s}')\ dw' \tag{6}$$

where the scattering angle θ is the angle between directions ŝ and ŝ'. Adding the contributions of (5) and (6), dividing by ds and letting ds go to zero yields the net rate of change of radiance L(r, ŝ) and hence the transport Equation (4).

Figure 3 shows an example of Monte Carlo computations of the center line fluence rate distribution for various laser beam diameters in a skin model consisting of an epidermis and a dermis. These results, representative for 577 nm irradiation, show that fluence rate divided by irradiance is substantially greater at large laser beam diameters (≥ 4 mm) than at small diameters (< 0.2 mm). The cause is that for larger diameter beams, scattering occurs from the rest of the beam onto the center line. As the diameter increases, more light is available for scattering. However, once the beam reaches a diameter of about 4 mm, further increases in beam diameter cannot affect the center line fluence rate any further because light is absorbed before reaching the center line. Monte Carlo computations are currently considered as the method of choice for estimation of the spatial fluence rate distribution in an absorbing and scattering tissue. Keijzer et al. (1989b) have recently used this for the analysis of autofluorescence of human aorta. Monte Carlo computations require data for μ_a, μ_s and g, and so far assume a Henyey-Greenstein scattering probability distribution. Recently, a comparison was made between fluence rate distribution measurements in tissue phantoms and Monte Carlo computations, showing excellent agreement (Flock, et al., 1989).

Experimental determination of the optical properties of (*in vitro*) tissues can be performed in an integrating sphere arrangement by measuring the percentages of light reflected and transmitted as a function of wavelength. An iterative solution of Equation (4) is used to obtain the best estimates of μ_a, μ_s and g for matching the measurement of transmission and reflection (Prahl, 1988; Wilson et al. 1987). A summary of optical properties for tissue and methods for measuring optical properties may be found in (Cheong et al., 1990). Figure 4 shows an example of optical properties of human aorta tissues as a function of wavelength (from Keijzer et al., 1989b). These results and others (van Gemert et al., 1989; Parsa et al., 1989) show that tissues are strongly forward scattering materials.

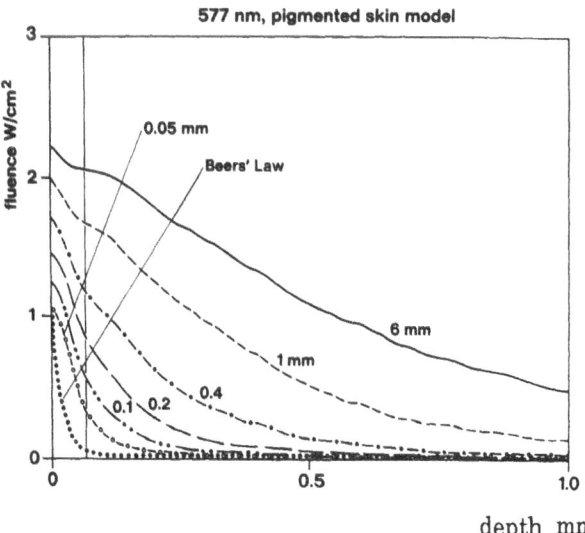

Fig.3.   Monte Carlo computations of fluence rate versus skin depth, for various spot sizes (from S.L. Jacques, Houston, unpublished, 1989).

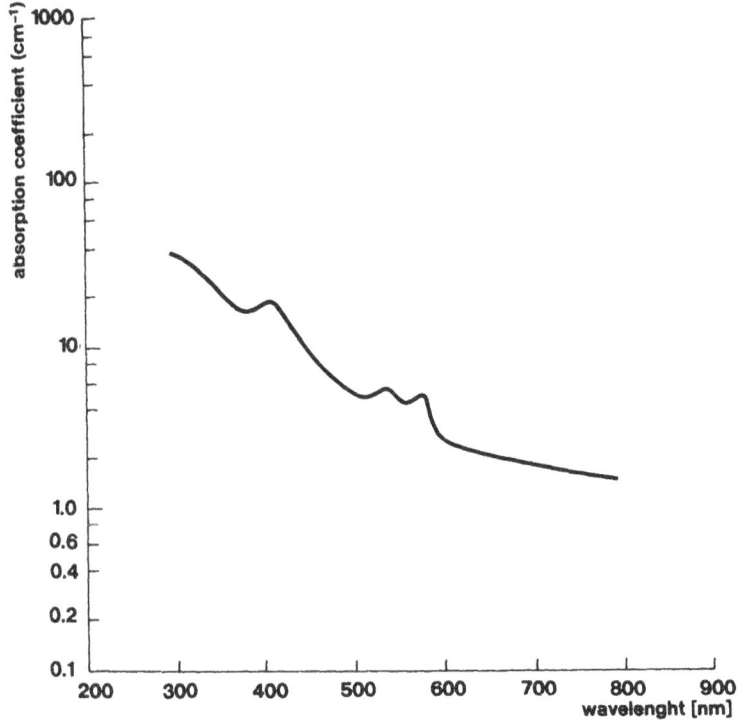

Fig. 4a. Absorption coefficient of human aortic media tissue as a function of wavelength (from Keyzer et al., 1989b)

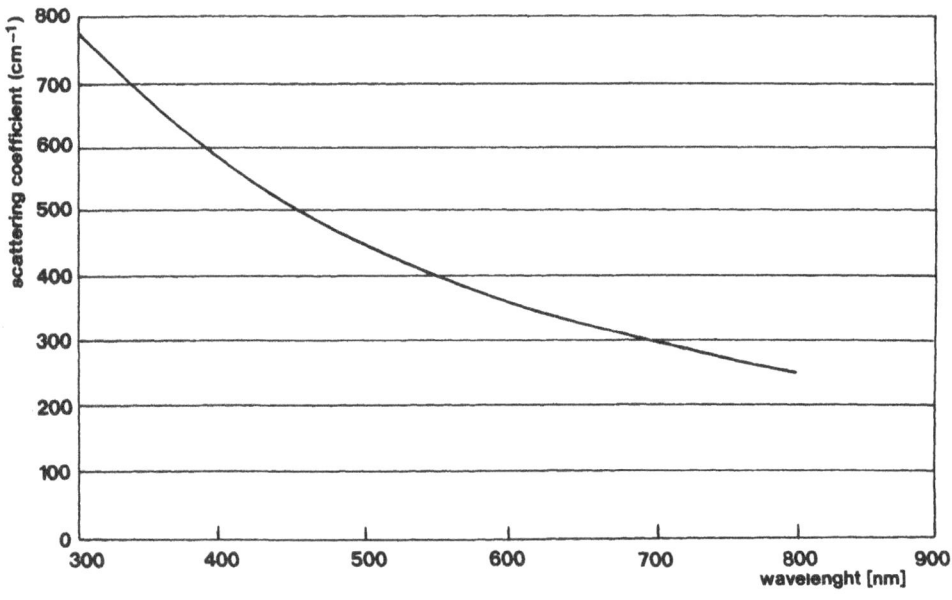

Fig. 4b. Scattering coefficient (see Fig. 4a)

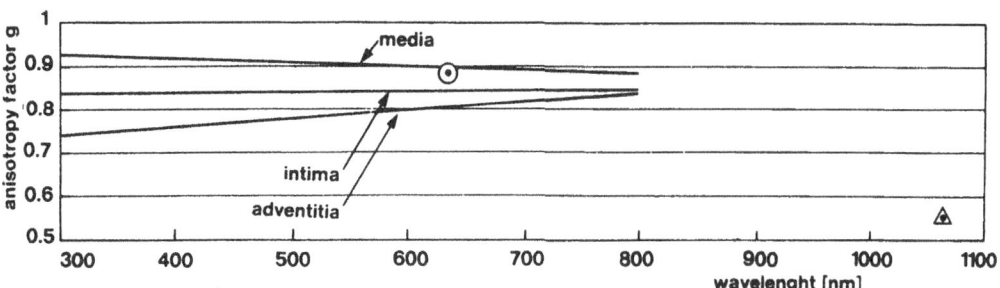

Fig. 4c. Anisotropy g factor (see Fig. 4a; the measurement point at 633 nm is for dog aorta using diffusion theory: G. Yoon et al., IEEE J. Quantum Electr., QE-23:1721 (1987))

In summary, optical modeling has reached the stage that Monte Carlo computations are considered the method of choice for computing fluence rate distributions in absorbing and (strongly) scattering tissues. The method requires data for the absorption and scattering coefficients and the anisotropy g-factor for scattering, assuming a Henyey-Greenstein scattering phase function. Approximate methods like diffusion theory are inappropriate because tissues are strongly forward scatterers and diffusion theory can only handle moderately non-isotropic scattering. Determination of these optical properties has recently been worked out from an exact solution of the transport equation for a slice of material fitting μ_a, μ_s and g to measured total back-scattering and total (non-collimated) transmission and collimated transmission. An integrating sphere set-up is used for these measurements. Future developments are: (a) to develop a tissue phantom that consists of a scattering and absorbing mixture, determination of the optical properties of that phantom, measuring fluence rate contributions and comparing them with computed distributions (Moes et al., 1989; Flock et al., 1989); (b) to perform integrating sphere measurements to in vitro tissues and characterizing μ_a, μ_s and g as a function of wavelength for several tissues at room temperature; (c) to perform these measurements for other temperatures.

III. THERMAL MODELING

The spatial distribution of the fluence rate in tissue is the key to the thermal use of lasers in medicine because the local, volumetric heat production (Q) is equal to the product of (local) absorption coefficient and (local) fluence rate

$$Q\ (r)\ ) - \mu_a\ (r)\ \psi\ (r) \tag{7}$$

Q is expressed in power per volume, and r denotes the location in the tissue. Equation (7) is the source term for heat production in the bio-heat equation for calculation of temperature distributions in laser irradiated tissue (Welch, 1984). This equation reads

$$\rho c\ \frac{\partial \Delta T}{\partial t} - Q\ (r)\ +\ k\nabla^2\ \Delta T \tag{8}$$

where ρ, c, k are the tissue density (kg m^{-3}), heat capacity (J kg^{-1} °C^{-1}) and thermal conductivity (Wm^{-1} °C^{-1}), $\Delta T(t, z, r)$ is the temperature rise at time t and location r(z,r) and ∇^2 is the Laplace differential operator. The left hand side of Equation (8) represents the local rate of storage of heat production per volume (W m^{-3}), the first term on the right hand side (rhs) is the local volumetric rate of heat production due to direct absorption of laser light energy, and the second term on the rhs is the volumetric rate of heat production losses due to heat conduction. Equation (8) assumes that laser irradiation times are sufficiently short (less than a few seconds) that convective heat losses such as blood perfusion (Welch et al., 1980) are insignificant. In this chapter we assume that both the thermal (ρ, c, k) and optical μ_a, μ_s, g) properties are independent of temperature, in concordance with the available literature. We acknowledge evidence that the optical properties (S.L. Jacques, private communication, 1989) and thermal properties (Valvano and Chitsabesan, 1987) are temperature dependent. Modeling which takes these effects into account has not yet been performed. Equation (8) is a diffusion equation that can be solved with proper initial and boundary conditions. An example of an initial condition is that $\Delta T(t, z, r)$ is zero for all z, r before the laser is switched on (at t = 0). An example of two boundary conditions is that, at r = 0, the radial behavior of ΔT is symmetrical while, at z = 0, the air-tissue interface is insulating, implying that the z-derivative of ΔT is zero at z = 0. For laser irradiation times t_L much shorter than the characteristic time for heat conduction, the solution to Equation (8) is

$$\Delta T\ (t,\ z,\ r)\ =\ \frac{\mu_a\ (z,\ r)\ \psi\ (z,\ r)}{\rho c}\ t \qquad\qquad (t \leq t_L) \qquad\qquad (9)$$

This is called: adiabatic heating. For longer irradiation times but still short relative to the time it takes the temperature to reach 63% of its steady state value, the solution of Equation (8) can be approximated as

$$\Delta T\ (t,\ z,\ r)\ =\ \frac{\tau \mu_a\ (z,\ r)\ \psi\ (z,\ r)}{\rho c}\ \left[1 - e^{-t/\tau}\right] \qquad\qquad (10)$$

where the time constant for heat conduction (τ) is defined as (van Gemert and Welch, 1989)

$$\tau\ =\ \frac{\rho c}{k\ (2.40)^2}\ \left(\frac{w_L^2}{1 + w_L\ \mu_{tr}/3)^2}\right) \qquad\qquad (11)$$

The parameters w_L, μ_{tr} are the $1/e^2$ laser beam radius and the transport optical attenuation coefficient

$$\mu_{tr}\ =\ \mu_a\ +\ \mu_s\ (1 - g) \qquad\qquad (12)$$

and the factor 2.40 is the first zero of the zeroth order Bessel function. Equation (10) turns out to be a reasonable solution to the bio-heat Equation (8) for t < 3τ, provided that the laser beam spot size is not too large for strongly scattering, low absorption media (van Gemert and Welch, 1989).

Exact solutions to Equation (8) are required for longer irradiation times, that is t >> τ. Such solutions have been obtained analytically (Wissler, 1976; Birngruber, 1980), and numerically (Takata, 1974; Priebe and Welch, 1979; van Gemert et al., 1982).

Thermal damage of the tissue at r,z depends on the temperature-time history at r,z. A first order Arrhenius rate theory model has been proposed in the past to model thermal damage (Henriques, 1947; Takata, 1974). The local damage integral (Ω) is defined as

$$\Omega\ (z,\ r) = C \int_0^\infty dr\ \exp\left[\ - E_{exp}\ /\ \big(RT\ (t,\ z,\ r)\big)\right] \qquad (13)$$

Various values for the frequency factor C and the experimental activation energy E_{exp} have been proposed for skin along with criteria for the damage constant Ω (see Table 1), where R is the molar gas constant (2 cal/mol/K). Typically, the threshold for irreversible damage is defined as $\Omega = 1$.

TABLE 1

Tissue	C(s^{-1})	E$_{exp}$(cal mol^{-1})
Skin (1) for 44° <T<50° C	1.3×10^{99}	1.5×10^5
Skin (5) for T>50° C	4.3×10^{64}	10^5
	9.4×10^{101}	1.6×10^5
Retina (2)	1.3×10^{99}	1.5×10^5
Retina (3)	10^{44}	0.7×10^5
Artery (4)	5.6×10^{63}	1.58×10^5

Experimental values proposed for the frequency factor (C) and the experimental activation energy (Eexp) occurring in the Arrhenius damage Equation (13).

(1) Henriques (1947)
(2) Welch, Polhamus (1984)
(3) Weinberg et al. (1984)
(4) Cheong, Welch (1989)
(5) Takata (1974)

Values for C and E_{exp} can be determined experimentally by using long irradiation times that permit the temperature to remain at a steady state for a significant portion of the irradiation time (Welch and Polhamus, 1984). If the tissue temperature can be approximated by a constant, the Equation (13) reduced to

$$\Omega\ (t,\ z,\ r) = t\ C\ \exp\left[\ - E_{exp}\ /\ \big(RT_{ss}\ (z,\ r)\big)\right] \qquad (14)$$

where $T_{ss}(z,r)$ is the local steady state temperature. For threshold damage ($\Omega = 1$) and a constant temperature of T_{ss} for t_0 seconds, this equation can be rearranged to produce

$$\ln t_0 = -\frac{E_{exp}}{R}\left(\frac{1}{T_{ss}}\right) + \ln C \tag{15}$$

By plotting experimentally determined threshold exposure durations and associated temperatures on a $\ln t_0$ versus $1/T_{ss}$ graph, it is possible to use a least squares fit to calculate $\ln C$ and E_{exp}/R. If after an irradiation of t_0 seconds threshold damage ($\Omega = 1$) occurs at z,r, at the boundary of the lesion within normal tissue, then the steady state threshold temperature is $T_{ss}(z,r)$.

Once rate constants have been established, it is possible to estimate threshold temperatures for damage no matter the irradiation time. Consider the temperature response in two time regions: very short exposures (no heat conduction during laser irradiation) and very long exposures (steady state temperatures are achieved). These regions represent the impulse response region and steady state response region.

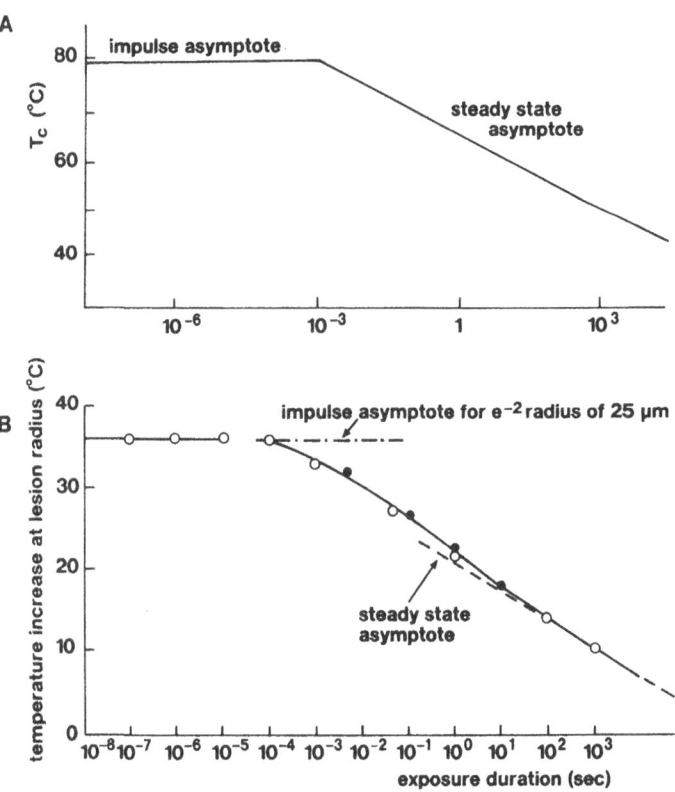

Fig. 5. Critical temperature versus irradiation time.
A. According to Lahaije and van Gemert (1985)
B. According to Welch and Polhamus (1984)

For very short exposure durations, the temperature at the end of irradiation is given by Equation (9). This time interval is so short that there is little or no accumulation of damage. That is $\Omega(t_0) = 0$. Damage occurs during the temperature relaxation response. The relaxation response depends only on the temperature at t_0, but not upon t_0 itself. Thus, a critical temperature $T_c(t_0)$, can be established for threshold damage for 'impulse' duration irradiations. If the exposure is long (steady state temperature response), then solving Equation (15) for T_{ss} yields

$$T_{ss} = \frac{E_{exp}/R}{\ln C + \ln t} \tag{16}$$

Curves for the critical impulse temperatures and steady state temperatures for $E_{exp}/R = 75,000$ K and $C = 8.3 \times 10^{95}$ s^{-1}, or $\ln C = 221$ have been published (Priebe and Welch, 1978; Welch and Polhamus, 1984; Lahaye and van Gemert, 1985). Curves of critical temperatures for threshold damage of skin are illustrated in Figure 5. The line in Fig. 5B represents a computer solution to the full heat conduction and damage integral Equations (8) and (13).

IV. DISCUSSION AND CONCLUSION

In summary, thermal modeling has reached the stage that temperature and damage distributions can be computed. So far, Monte Carlo computed light fluence rate distributions have not been used to estimate the rate of heat generation on thermal modeling. Although none of the thermal models include temperature dependence of the optical properties, the finite diffuse solution of the heat conduction equation by Takata (1974) includes thermal properties that are dependent upon computed values of local water content. In addition, rate coefficients in the Arrhenius model are not precisely known for the various tissues of interest. At best, therefore, thermal modeling of tissues can provide trends. Accurate dosimetry on the basis of thermal modeling is thus hardly available although exceptions have been reported (Mordon et al., 1987; Welch and Polhamus, 1984).

In conclusion, further developments in dosimetry during laser irradiation of tissue have to come from other disciplines like monitoring with MRI and monitoring with ultra-sound. The latter type of dosimetry has been used for the interstitial laser treatment of liver metastases in humans.

Acknowledgement

The authors acknowledge the processing of the manuscript by Chris Humphrey.

REFERENCES

Birngruber, R., 1980, Thermal modeling in biological tissues. in: "Lasers in Biology and Medicine," F. Hillenkamp, R. Pratesi, C.A. Sacchi, eds., Plenum Publishing Co., pp. 77-97.

Chandrasekhar, S., 1960, "Radiative Transfer", Dover Publications, New York.

Cheong, W.F., Prahl, S.A., and Welch, A.J., 1990, A review of optical properties of biological tissue, Submitted.

Flock, S.T., Wilson, B.C., and Patterson, M.S., 1989, Monte Carlo modeling of light propagation in highly scattering tissues II: comparison with measurements in phantoms, IEEE Trans. Biomed. Eng., BME36:1169.

van Gemert, M.J.C., de Kleijn, W.J.A., and Hulsbergen Henning, J.P., 1982, Temperature behaviour of a model port-wine stain during argon laser coagulation, Phys. Med. Biol., 27: 1089.

van Gemert, M.J.C., and Welch, A.J., 1989, Time constants in thermal laser medicine, Lasers Surg. Med., 9:405.

van Gemert, M.J.C., Jacques, S.L., Sterenborg, H.J.C.M., and Star, W.M., 1989, Skin optics, IEEE Trans. Biom. Eng., BME36:1146.

Henriques, F.C., 1947, Studies of thermal injury, Arch. Pathol., 43:4899.

Henyey, L.G., and Greenstein, J.L., 1941, Diffuse radiation in the galaxy, Astrophys. J., 93:70.

Ishimaru, A., 1978, "Wave propagation and scattering in random media", Vol. 1. Academic Press, New York.

Ishimaru, A., 1989, Diffusion of light in turbid material, Appl. Opt., 28:2210.

Jacques, S.L., Alter, C.A., and Prahl, S.A., 1987, Angular dependence of He-He laser light scattering by human dermis, Lasers Life Sci., 1:309.

Keijzer, M., Jacques, S.L., Prahl, S.A., and Welch, A.J., 1989a, Light distributions in artery tissue: Monte Carlo simulations for finite-diameter laser beams, Lasers Surg. Med., 9:148.

Keijzer, M., Richards-Kortum, R. R., Jacques,, S.L., and Feld, M.S., 1989b, Fluorescence spectroscopy of turbid media: autofluorescence of human aorta, Appl. Opt., 28:4286.

Lahaije, C.T.W., and van Gemert, M.J.C., 1985, Optical laser parameters for port wine stain therapy: a theoretical approach, Phys. Med. Biol., 30:573.

Moes, C.J.M., van Gemert, M.J.C., Star, W.M., Marijnissen, J.P.A. and Prahl, S.A., 1989, Measurements and calculations of the energy fluence rate in a scattering and absorbing phantom at 633 nm, Appl. Opt., 28:2292.

Mordon, S.R., Cornil, A.H., Brunetaud, J.M., Gosselin, B., and Moschetto, Y., 1987. Nd-YAG laser thermal effect: comparative study of rat liver in vivo by continuous wave and high power pulsed lasers, Lasers Med. Sci., 2:285.

Parsa, P., Jacques, S.L., and Nishioka, N.S., 1989, Optical properties of rat liver between 350 and 2500 nm, Appl. Opt., 28:2325.

Prahl, S.A., 1988, Light transport in tissue, 1988, Ph.D. dissertation, University of Texas at Austin.

Priebe, L.A., and Welch, A.J., 1978, Asymptotic rate process calculations of thermal injury to the retina following laser irradiation, J. Biomed. Eng., 100:49.

Priebe, L.A. and Welch, A.J., 1979, A dimensionless model for the calculation of temperature increase in biologic tissues exposed to nonionizing radiation, IEEE Trans. Biomed. Eng., BME26:244.

Takata, A.N., 1974, Development of criterion for skin burns, Aerospace Med., 45:634.

Valvano, J.W., and Chitsabesan, B., 1987, Thermal conductivity and diffusivity of arterial wall and atherosclerotic plaque, Lasers Life Sci., 1:219.

Weinberg, W.S., Birngruber, R., and Lorenz, B., 1984, The change in light reflection of the retina during therapeutic laser photocoagulation, IEEE J. Quant. Electr., QE-20:1481.

Welch, A.J., Wissler, E.H., and Priebe, L.A., 1980, Significance of blood flow in calculations of temperature in laser irradiated tissue, Biomed. Eng., BME27:164.

Welch, A.J., 1984, The thermal response of laser irradiated tissues, IEEE J. Quant. Elect., QE-20:1471.

Welch, A.J., and Polhamus, G.D., 1984, Measurement and prediction of thermal damage in the retina of the rhesus monkey, IEEE Trans. Biomed. Eng. BME31:633.

Wilson, B., and Adam, G., 1983, A Monte Carlo model for the absorption and flux distributions of light in tissue, Med. Phys., 10:824.

Wilson, B.C., Patterson, M.S., and Flock, S.T., 1987, Indirect versus direct techniques for the measurement of the topical properties of tissues, <u>Photochem. Photobiol.</u>, 46:601.

Wissler, E.H., 1976, An analysis of choroid-retinal thermal response to intense light exposure, <u>IEEE Trans. Biomed. Eng.</u>, BME23:207.

PART 3

DIAGNOSTIC TECHNIQUES

HOLOGRAPHY IN MEDICAL DIAGNOSTICS

G. von Bally

Laboratory of Biophysics
Institute of Experimental Audiology
University of Muenster
Kardinal-von-Galen-Ring 10
D-4400 Muenster
Federal Republic of Germany

INTRODUCTION

The current situation in medical optoelectronics is characterized by the fact that new results in today's optoelectronic research and development are at least as strong a motor for progress in this field of medical applications as the demands by medical experts for new diagnostic techniques. On the other hand, at a typical laser medicine meeting by far most of the contributions deal with surgical (high-power) applications and only few present new diagnostic possibilities of optoelectronic metrology. In addition, if one considers that today table-top surgical lasers are commercially available, it is obvious that optoelectronic metrology must play an increasingly predominant part in this medical research field.

One of these metrological methods is holography, which is known mainly as a photographic technique with the capability of storage and three-dimensional optical reconstruction of objects using a laser as light source. This does not at all cover the full variety of applications of laser optical methods based on the holographic principle. There are numerous applications in the fields of image processing and metrology, which have attracted the interest of research groups to evaluate the potentials of these methods in biomedical research.

From the various capabilities of holography for image processing and measuring purposes mainly holographic interferometric techniques have found more extended application in biological and medical research. Due to their special properties the different methods of holographic interferometry are applied to characteristic fields of biomedical investigations, where - similar to non-destructive testing - vibration and deformation analysis is of interest. Here features of holographic interferometry such as the possibility of non-contactive, three-dimensional investigations with a large field of depth are used with advantage. The main applications can be found in basic research in, for example audiology, dentistry, ophthalmology, and experimental orthopedics. Because of the great number of investigations and the variety of medical domains, in which these investigations were performed, this survey has to be confined to some characteristic examples.[1-4]

Like in all fields of optics and laser metrology a review on biomedical applications of holography would be incomplete if military developments and utilization were not mentioned. As will be demonstrated by selected examples, the increasing interlacing of science with the military does not stop at domains that traditionally are regarded as exclusively oriented to human welfare like biomedical research. The term "Star Wars Medicine", which is becoming an increasingly popular expression for laser applications-including holography - in medicine, characterizes the consequences of this development.[5]

BASICS OF HOLOGRAPHY

In order to record a hologram a laser beam is split into two sections - one part, the so called object beam, is used to illuminate the object, from where it is reflected onto a high resolving photographic material - usually a (hologram) plate (ref. Fig. 1a). Simultaneously, the other part of the laser beam is guided directly onto the hologram plate, where it acts as a wavefront reference (reference beam). Because of the coherence properties of laser light, object and reference beam interfere and form a microscopically fine pattern - the so called hologram. If - after development of the hologram plate - this interference pattern is illuminated only by the reference beam, it acts like a diffraction grating and generates - among others - the original wavefront as reflected from the object, correct in amplitude and phase. This is called holographic reconstruction. The holographically reconstructed image of the object is three-dimensional, which means one can observe this image from different perspectives as far as it is possible within the geometrical dimensions of the holographic plate. Since there are no imaging systems necessary for holographic recording and all parts of the reflected object wavefront are reconstructed, all (optically accessible) planes of the (spatial) object are simultaneously displayed sharply. Details on theory and practice can be found in Ref. 6.

THREE-DIMENSIONAL IMAGES WITH LARGE FOCAL DEPTH

The capability of three-dimensional imaging with the additonal advantage of a large field of depth is a basic feature of holography. Thus, - contrary to conventional two-dimensional photography - the focal plane of interest can be selected during reconstruction, i.e. after holographic recording and even after removal of the object. In this way, for example, the strongly curved fundus of the human eye can be recorded totally by a single hologram (fundus holography) so that any layer of interest can be investigated in the reconstructed image. The possibility of white light reconstruction does not only provide a three-dimensional display in biological or medical teaching, but in combination with image plane holography and by using limiting apertures inherent in the investigated medical object a considerable speckle reduction is achievable.

As an interferometric technique holography requires "interferometric" stability of the experimental set-up and the object during exposure, which means that the maximum allowable motion amplitude is in the order of one tenth of the used laser wavelength (some 1/10,000 mm). In order to meet this requirement in patients without anesthesia, (Q-switched) pulsed lasers such as ruby lasers with exposure times in the range of some 10 ns have to be used. Due to the short exposure times the use of Q-switched ruby lasers allows holographic imaging of fast moving objects. Therefore, it can be applied, for example, to investigations of chemical reactions, the distribution and size of droplets or jets, or the function of spray nozzles. Thus, in-line holographic arrangements have been used for the assessment of aerosols and pollution. Such investigations are especially important, for example, the development of sprays which should optimally penetrate to the lung, such as anti-asthmatic sprays, or to optimize the distribution of insecticides and pesticides in agriculture. The use of the same technique was proposed for application in the defoliation actions in Vietnam.[7]

HOLOGRAPHY WITH NON-VISIBLE WAVES

The use of non-visible waves like micro-, infrared-, ultrasonic- and X-rays is common and widely spread for diagnostic and therapeutical applications in medicine. Since today these waves can be generated coherently, holographic recording is possible, in principle, but with obstacles in detection and transformation in the visible spectrum at reconstruction.

While recording of holograms using micro- and infrared-waves in biological and medical research has up to now only been proposed, for example, for early skin and breast cancer diagnosis, extensive investigations by ultrasonic (acoustical) holography have been performed especially in gynecology and orthopedics.

Two-dimensional X-ray imaging is the most commonly used non-visible wave technique in medical diagnostics of internal structures within the human body. Thus, it has for a long time been the hope of researchers in the biomedical field to develop a three-dimensional X-ray imaging technique using holography. Such a method is expected to render possible recording of 3-d images of molecules, pinpointing cancer cells, and developing a safer and more effective X-ray therapy.

For quite a long time basic ideas have been proposed to overcome problems of finding an appropriate high resolving recording medium and a coherent X-ray source. However, an (expensive) technical realization was seriously investigated only after military interest arose especially within the frame work of the SDI program.

Thus, for recording of three-dimensional X-ray images a different holographic method has to be used, the so called multiplex holography. This technique can be a solution to the problem of three-dimensional imaging without a coherent X-ray source by combining holographically several two-dimensional X-ray pictures recorded from different views. The advantage compared to conventional tomography is the possibility of analysing any layer of interest in the image reconstructed from the synthetic hologram, although it has been generated only by a limited number of radiographs. The type of recording process of the two-dimensional images used for holographic multiplexing is obviously of no importance. It could be an ultrasonic-B-scan record, electron micrograph, or simply two-dimensional photographs. A semiautomatic combination of series recordings of computer tomograms (CT) or nuclear magnetic resonance (NMR) images and multiplex hologram display with the possibility of white light reconstruction is presently under study.[8]

HOLOGRAPHIC IMAGE PROCESSING

Holographic spatial filtering techniques can be applied not only to coherent wave fronts but also to non-coherent waves e.g. X- and Gamma-rays as used in radiology and nuclear medicine. Instead of taking serial two-dimensional radiographs for three-dimensional imaging by holographic multiplexing, the object can be projected simultaneously from different views, thus producing a coded image e.g. on a X-ray film. In these methods - known as coded source and coded aperture techniques - on-axis and off-axis Fresnel-zone plates, or discrete point distributions may be used for the coding process. Such coded aperture devices can be produced in form of holographic optical elements (HOE), the first well known practical application of which was the use as "Head-Up Display" for supersonic fighter aircraft pilots. If discrete point distributions are used, these may be an array of holes (Gamma-ray imaging), or distribution of radiation sources (X-ray imaging). Using a non-redundant distribution of the sources to optimize the signal-to-noise ratio, decoding can be provided during reconstruction in laser light e.g. by means of a Fourier transform hologram of this distribution. This latter technique ("flashing tomosynthesis") is being studied with the aim of displaying layers of moving objects like the pulsating heart or fast flowing contrast media, among other objectives.

The capability to "recognize" wavefronts using holographic spatial filtering techniques gives rise to a variety of suggestions for biomedical applications of this feature of holography. Cell identification - especially differentiation between normal and cancerous cells - may be an important example for clinical use of holographic pattern recognition techniques. Coherent optical filtering by means of light optical diffraction is used for pattern recognition and image evaluation of electromicrographs of biological specimens. Because of the great interindividual variety of the shape of biological specimens as well as the problem of orientation and size variance the generation of appropriate filters is sophisticated. New optical transforms particulary suited for scale, positional, and rotational invariant correlations without loss in signal-to-noise ratio can be applied advantageously to holographic pattern recognition.

HOLOGRAPHIC ENDOSCOPY

Although today endoscopy has established its place in medical diagnostics, it has not yet exceeded the function of a qualitative, subjective observation method. Yet, modern diagnostic techniques increasingly require an objective, quantitative determination of form, structure and (micro-)movement of the object under study.

Optical techniques, which can meet these requirements, are those of laser holographic interferometry known from industrial non-destructive testing. The introduction of holographic metrology into endoscopy is the decisive step towards a metrological basis for a quantitative diagnostic within body cavities, with the advantages of holographic interferometry such as large focal depth imaging, non-destructive, non-contactive, and high-resolving analysis of structure, form, deformation and vibration (the latter amplitude-, frequency-, and phase-selective).

For clinical diagnostic purposes this opens - among others - the following possibilities:

o quantitative determination of micro-movements of the (endoscopic) object, which are not detectable by mere visual inspection;
o quantitative analysis of the local elasticity of the object area under study;
o analysis of structure changes even underneath the (endoscopically) visible object surface by holographic measurement of local elasticity differences.

In principle, there are two approaches in developing a holographic endoscope:

o the hologram can be recorded inside the instrument or
o by using an external recording device - a holographic camera.

In addition, modern developments in micro-optics provide essential miniaturization of endoscopic devices. The introduction of micro-optics into holographic endoscopy raises problems due to limitations in optical signal-to-noise ratio (SNR) caused by small apertures of the image transmitting systems as well as due to restrictions of common holographic recording materials for in-situ recording and reconstruction with fast-repetition rate and high (line) resolution. Optical fibers and electro-optical crystals can help to overcome these obstacles.

Recording a hologram inside the endoscopic device requires a small holographic set-up to be inserted into the tip of the instrument. The advantages of this arrangement are full three-dimensionality of the reconstructed image with a large focal depth as well as maintaining of parallax. In this case a large aperture is achieved, since here the entrance pupil corresponds to the hologram area. This results in low speckle-noise and high lateral resolution, which allows high magnification a posteriori. Drawbacks are a large outer diameter of the endoscope needle (typically in the order of centimeters) and the necessity to develop a complete new type of endoscope. Using a transmission on-axis, nearly in-plane fiber optical recording and a white light reconstruction arrangement it could be proved, that a posteriori magnification allows cell identification and analysis without excision of tissue samples.[10]

External holographic recording allows the use of conventional endoscopes (Fig. 1). This means that also modern developments of endoscopes with extremely small outer diameter can be applied. These advantages have to be paid for by loss of parallax and a small entrance pupil, which due to coherence properties of laser light and the diffraction limited imaging characteristics of the optical system causes gross speckles in the reconstructed holographic image, as well as restrictions by the given optical performance data of the endoscopic optic like focal depth, spherical aberrations etc.

Especially the very small endoscopes as used in certain medical fields, e.g. otoscopes for outer and middle ear inspection or salpingoscopes for fallopian tube investigations, show small limiting apertures of the imaging system and therefore a reduction in signal-to-noise ratio (SNR) by speckles, which can reduce the holographic image information considerably.

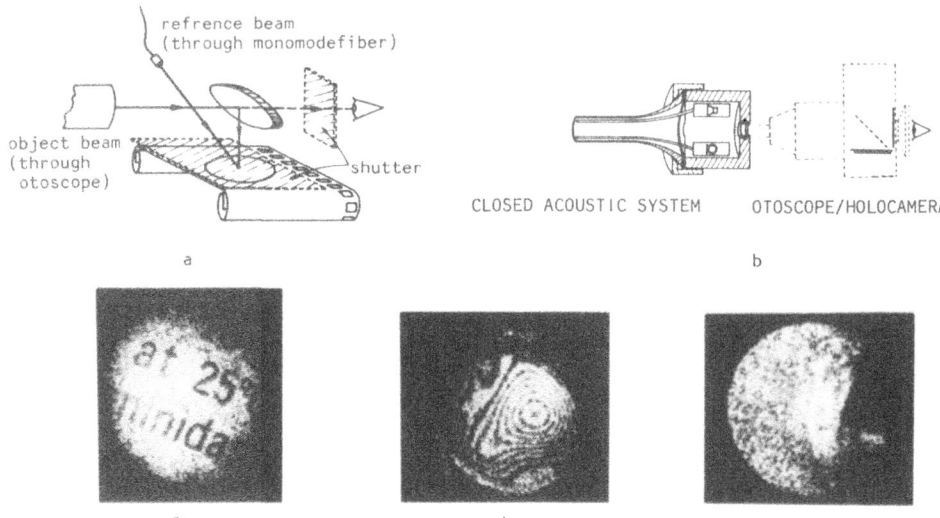

Fig. 1. Holographic endoscopic camera as developed at the Laboratory of Biophysics:

a) scheme of ray tracing for holographic recording and visual inspection (The illuminating object beam, not shown in the drawing, is guided via a multi-mode fiber bundle.);

b) in combination with an otoscope for middle ear inspection and a specially designed closed acoustic system to elicite tympanic membrane vibrations;

holograms recorded with the described holographic endoscopic system using a Q-switched ruby laser:

c) single exposure hologram (original letters 1mm high);

d) double-exposure holographic interferogram of a plane scale model of the human tympanic membrane;

e) in-vivo double-exposure holographic interferogram of a vibrating human tympanic membrane.

Thus, in holographic endoscopy it is especially important to use imaging systems with a high numerical aperture and a low f/No. to gain a high optical SNR.

Contrast enhancement with the additional advantage of suppression of unwanted reflections can be achieved by holographic subtraction.[11-13]

Holographic subtraction is realized by an optical phase shift of the laser light of 180 degrees between the two exposures of a double-exposure holographic record in either the object beam or reference beam. This phase shift results in a blackout of immobile parts of the object, while in moving areas interference lines change from bright to dark and vice versa in the holographic reconstruction. Such a holographic subtraction in combination with stroboscopic double-exposure holography is technically arranged by using an argon laser modulated by an acoustooptical modulator and a piezocrystal driven mirror, which generates an oscillating (optical) phase shift of 180 degrees in the reference beam between each exposure of the stroboscopic double-exposure recording.

For the development of a holographic endoscope, easy handling and flexibility are essential, which require the use of optical fibers.

Three major rays necessary for holographic recording can be guided through fibers: the illuminating object beam, the reflected object beam (in this case the endoscopic imaging system), and the reference beam. The requirements of maintaining mode structure and coherence of the laser light are different for each of these optical paths. While a multimode fiber bundle can be used for object illumination, the conditions for a defined wavefront and avoiding of mode interaction require a single mode fiber for reference beam guidance. Although launching of a beam from a Q-switched pulsed laser is critical because of the limited power transmission capability of single mode fibers, the use of this type of laser is possible if certain intensity limits of object and reference beam are obeyed.[14]

One important advantage of laser beam delivery via optical fibers is the fixed path length difference of object and reference beam in spite of changing position of the holographic set-up.

Using a coherent multimode fiber bundle for imaging, the quality of the reconstructed image, especially the interference fringe contrast, is influenced by mode interaction due to multimode transmission with its own polarization state for each mode. Thus, a coherent single mode fiber bundle should be used for image transmission. In order to gain a sufficient pixel density in the transmitted image a new type of fiber with a small cladding in the order of the core diameter (typically a few micrometers) has to be developed. Commercially available at the moment are coherent single mode fiber bundles which have claddings melted together for easier coherent order and fracture resistance. Besides improvement of maintaining mode structure, these fiber bundles show an increased lateral resolution due to higher pixel density compared to conventional multimode fiber bundles. For holographic endoscopic purposes we used a coherent single mode imaging bundle with 20,000 pixels and melted claddings (cladding diameter = center to center distance of adjacent pixels of 4.5 micrometers) within an imaging bundle diameter of 770 micrometers.

While there are quite a number of publications even in the non-classified literature on technical developments and industrial applications of holographic endoscopy[15], medical investigations using this technology are rather rare.

Endoscopic investigations for early recognition of cancerous indurations in the wall of the urinary bladder using holographic interferometry are proposed in Ref. 16. A special holographic endoscopic arrangement is outlined and experiments on technical objects are performed by means of a commercially available, rigid endoscope. In-vitro investigations on excised urinary bladders of rabbits without endoscopic imaging demonstrate that holographic interferometry is capable of differentiating between regular and tumorous wall tissue when applying intracavity liquid pressure changes.

A standard laryngoscope was applied to investigate deformations inside samples of macerated human calvaria using holographic interferometry and a photothermoplastic film as recording material.[17] These experiments were performed to study the mechanical properties of the human skull bone.

In-vitro investigations on the elasticity of the fallopian tube wall were performed by means of holographic interferometry.[18] So far, excised fallopion tubes of rabbits have been studied using a laparoscope. Interference fringes due to intratubal fluid pressure change could be recorded with good contrast without coating of the tube samples (Fig. 2). The aim of this study is to investigate the diagnostic capability of holographic endoscopy for function analysis of the fallopian tube.

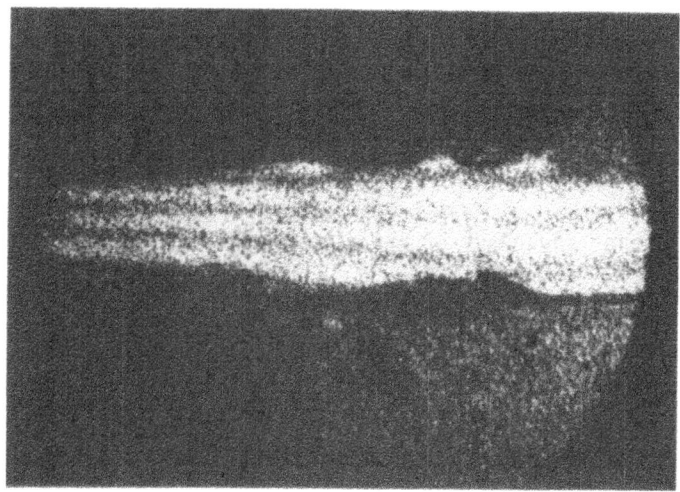

Fig. 2. Double-exposure holographic interferogram of a fallopian tube of a rabbit recorded through a laparoscope; interference fringes characterize the expansion of the tube due to pressure changes of a liquid inside the tube cavity.

In-vivo single exposure holographic recording of vocal cords using a Q-switched ruby laser were taken from an anesthetized dog.[19] The illuminating object beam was guided via a multimode fiber while the reference beam was transmitted through an articulating arm.

The first in-vivo holographic endoscopic recordings of human tympanic membranes in patients are described in Ref. 20. While this group used an otoscope in a rigid holographic set-up, our investigations were performed using a holographic endoscopic camera attached to an otoscope as well as object and reference beam delivery via optical fibers (Fig. 1). In this way vibration analysis of human tympanic membranes for non-invasive differential diagnosis of middle ear diseases can be performed.[21]

HOLOGRAPHIC MOTION ANALYSIS

Holographic analysis of motion can be realized according to the desired (amplitude) resolution either by sequential recording of single holograms of the process under study or by the methods of holographic interferometry.

The former technique, known as cineholography, has been used for microscopic investigations of living marine plankton organisms. Holograms have been taken by stroboscopic illumination using a pulsed Argon laser synchronized to the recording sequence of a camera. Because of the large field of depth it was possible to investigate the microscopic subjects as they moved freely in the object space. A similar technique has been applied to studies of Bend's decompression sickness of deep sea combat divers, which is caused by bubbles forming in the blood vessels.

If two holograms of different states of object motion are superposed on the same holographic plate before development, patterns of dark and bright fringes (interference fringes) are visible in the reconstructed image.[6] These interference fringes connect loci of equal motion amplitude. From these lines the motion of the illuminated object surface can be deduced quantitatively with a resolution of a fraction of the used laser wavelength. The motion may be a periodic vibration or a transient phenomenon such as a deformation. Thus, these methods, known as holographic interferometry, provide the possibility of a high resolving, non-contactive, three-dimensional, non-destructive analysis of alterations in position (motion) but also in shape or structure of the object under study.

The most commonly used holographic interferometric techniques are time-averaged, real-time, and double-exposure holography. They differ in the way of superposing the two holograms and, accordingly, can be used for different applications.

Since for time-averaged holographic recordings the object has to move periodically, this technique is used for vibration analysis e.g. of the tympanic membrane. Due to the somewhat cumbersome repositioning of the reference hologram, real-time holography is restricted to investigations on stable objects like macerated bones or medical technical appliances. These restrictions can be relaxed to some extent if a quasi-real-time technique with video frame repetition rate (electronic speckle pattern interferometry) is used in combination with stroboscopic laser pulsing.

Deformation analysis by double-exposure holography is used especially for biomechanical investigations of the human locomotor system. Since there are possibly adverse influences on osteosythesis by unphysiological load, e.g. after implantation of protheses or fixation by plates after fractures, the mechanical properties of bones have to be known. Comparative holographic investigations have been carried out in real-time on the human femur in-vitro before and after implantation to optimize hip joint protheses, as well as on the human tibia after fracture fixation by compression plates. Latest studies include the technique of external dynamic fixation.

Similar experiments have been carried out on human macerated skulls in order to find zones of stress under load. These investigations can help to increase the knowledge on typical fractures in accident research. It is also expected that the generation of hearing impairment after head traumata will find further detailed explanations in this way.

Using a Q-switched ruby laser for recording of double-exposure holograms - besides periodic vibrations - fast, non-periodic processes can be studied. This possibility was discovered first during military oriented investigations of bullets in flight using a ruby laser which accidentally emitted two short pulses. The reconstructed images showed fringe patterns according to the propagation of the shock wave. An interesting application of this method, which has been used in biomedical research, is the study of transient processes. Thus, movements of tympanic membranes subjected to acoustic impulses have been investigated in in-vitro experiments on guinea pigs by superposition of a hologram recorded at rest and a second hologram taken on the same holographic plate at a certain time after the acoustic event. The aim of these experiments was the study of generation of lesions on the tympanic membrane caused by acoustic impulses such as bursts emanating from weapons. The same technique can be used for clinical diagnostics in patients. Since pathological changes of the mechanical properties of the middle ear have an influence on the vibratory pattern of the tympanic membrane, a vibration analysis may provide the possibility of a differential diagnosis of dysfunctions without opening the tympanic cavity.

Successful in-vivo experiments to investigate motions of teeth and bridge-work to optimize the design of prosthodontic appliances have been carried out on patients, releasing the laser pulses at certain masticatory force levels. Human chest motions have been investigated in-vivo during inhalation with the aim of lung diagnosis, as well as by triggering the laser pulses in relation to heart action to test the possibility for detection of heart diseases by double-pulsed holography. A similar method has been applied to record the vibrations of the frontal part of the human neck for functional analysis of the vocal organ as well as non-invasive early detection of vocal cord cancer.

"STAR WARS MEDICINE"

Although only some of the numerous examples could be mentioned, this review demonstrates that holography has its place in medical research, today. Yet, clinical applications are still rare and none of the holographic techniques have been used really routinely in clinical diagnostics, up to now. Thus, intensive interdisciplinary research activities between physicists, engineers, and biological and medical experts are still necessary in order to make available the advantages and potentials of holography for everyday clinical practice.

On the other hand, publications on holographic applications in medicine demonstrate also, how far science and the military, even medical research and arms development, are already interconnected. Concentration of financial means and personel within giant military programs like SDI intensify such developments and lead to dependencies and constraints of science and scientists. As an alarming sign for the threat of freedom of our science may be regarded the popular term for laser applications - including holography - in medicine: "Star Wars Medicine".[5,7,22]

REFERENCES

1. P. Greguss (ed.),"Holography in Medicine", IPC Science and Technology Press (1975).

2. G. von Bally (ed.), "Holography in Medicine and Biology", Springer-Series in Optical Sciences Vol. 18 (1979).

3. G. von Bally and P. Greguss (eds.), "Optics in Biomedical Sciences", Springer-Series in Optical Sciences Vol. 31 (1982).

4. G. von Bally, Holography in biomedical sciences, SPIE 673: 327 (1987).

5. W.S. Andrus, Fibers and lasers: The "Star Wars" medical team, Photonics Spectra 71: (1985).

6. N. Abramson, "The making and evaluation of holograms", Academic Press (1981).

7. G. von Bally, Holography in medicine and biology - state of the art and the problem of increasing militarization, in: Soares, O. (ed.), "Optical Metrology", NATO ASI Series E: Applied Sciences 131: 441 (1987).

8. J. Tsujiuchi, Multiplex holograms and their applications in medicine, SPIE 673: 312 (1987).

9. G. von Bally, D. Dirksen, Y. Zou, and E. Kraetzig, Micro-Optics in holographic endoscopy, SPIE 1136: 158 (1989).

10. H. Bjelkhagen, Holographic endoscopy through optical fibers, L.I.A. Vol. 50 ICALEO (1987), 134 (1988).

11. C. Sieger, Suppression of disturbing light reflexes in holography applied to practical recording problems in medicine and technology, in: G. von Bally and P. Greguss (eds.): Optics in Biomedical Sciences, Springer-Series in Optical Sciences 31: 125 (1982).

12. G. von Bally, W. Schmidthaus, H. Sakowski, and W. Mette, Gradient-index optical systems in holographic endoscopy, Appl. Opt. 23: 1725 (1984).

13. G. von Bally, E. Brune, and W. Mette, Holographic endoscopy with gradient-index optical imaging systems and optical fibers, Appl. Opt. 25: 3425 (1986).

14. G. von Bally, Otoscopic investigations by holographic interferometry: a fiber endoscopic approach using a pulsed ruby laser system, in: G. von Bally and P. Greguss (eds.): "Optics in Biomedical Sciences", Springer-Series in Optical Sciences 31: 110 (1982).

15. H. Bjelkhagen, E.J. Wesly, J.C. Lin, M.E. Marhic, M. Epstein, Holographic interferometry through imaging fibers, using CW and pulsed lasers <u>SPIE</u> 746: 201 (1987).

16. U. Gruenewald, H. Wachutka, A. Hofstetter, and R. Boewering, Interferometric investigations of the rabbit urinary bladder, <u>in:</u> "Holography in Medicine and Biology", G. von Bally (ed.), Springer-Series in Optical Sciences 18: 147 (1979).

17. H. Podbielska, G. von Bally, and H. Kasprzak, Mechanical reaction of human skull bones to external load examined by holographic interferometry, <u>SPIE</u> 573: 321 (1988).

18. G. von Bally, Holographic endoscopy, L.I.A. Vol. 60 ICALEO 1987, 127 (1988).

19. G. Raviv, M. Marhic, E. Scanlon, S. Sener, and M. Epstein, In-vivo holography of vocal cords, <u>J. Surgical Oncology</u> 20: 213 (1982).

20. W. Fritze, H. Kreitlow, and D. Winter, On holographic-interferometric investigations of the Membrana Tympani (living man), <u>in:</u> "Holography in Medicine and Biology", G. von Bally (ed.), Springer-Series in Optical Sciences 18: 205 (1979).

21. G. von Bally, Otological investigations in living man using holographic interferometry, <u>in:</u> "Holography in Medicine and Biology", G. von Bally (ed.), Springer-Series in Optical Sciences 18: 198 (1979).

22. G. von Bally, Scientists, scientific societies, and military research, <u>SPIE</u> 370: 26 (1984).

LASER REFLECTANCE SPECTROSCOPY OF TISSUE

Brian C. Wilson,[1] and R. Rox Anderson[2]

[1]Ontario Cancer Treatment and Research Foundation
Hamilton, Ontario L8V 1C3 Canada

[2]Department of Dermatology, Massachusettes General Hospital
Boston, USA

INTRODUCTION

The optics of living tissue are complicated by gross and microscopic inhomogeneous structure, dynamic changes in blood content, shape, multiple absorbing compounds, and extreme local and individual variations. This degree of complexity is necessary for life itself, and presents both the major challenge for diagnostic optical techniques and a major justification for developing noninvasive techniques.

The specular component of tissue reflectance follows Fresnel's equations (Nishioka et al, 1988), and therefore depends on incidence angle, polarization , and refractive index. Because specular reflectance preserves the polarization of incident light, whereas multiple scattering does not, it is possible to readily separate specular from diffuse reflectance, using polarized incident light. At near-normal incidence from external medium of air, the specular reflectance of skin is 4-7% (Anderson and Parrish, 1981), being slightly higher in the UV than the IR due to variation of index with wavelength. This reflectance decreases slightly the light entering tissue.

A major and opposite effect of specular reflectance upon the dosimetry within tissue, however, is to greatly increase irradiance within the tissue near the surface by internal reflection of back-scattered radiation (Anderson et al, 1990). For semi-infinite irradiation on a planar tissue interface, the increase in subsurface irradiance is given by:

$$I/I_0 = 1 + \frac{2(1+r_d)R}{(1-r_d)} \, ,$$

where I_o is the incident irradiance, R is the diffuse reflectance, and r_i is the internal reflectance of diffuse radiation at the tissue-external medium interface. Thus, it is possible to deduce the internal irradiance near the surface simply from knowledge of the diffuse reflectance and refractive index. Since r_i is typically 0.5 for air-tissue interfaces, and R can approach 0.8 in the red and near infrared spectrum, internal reflectance can increase irradiance within tissues by five-fold or more.

Numerous studies have used diffuse reflectance spectroscopy for analysis of tissue pigments, most notably bilirubin and hemoglobins. For example, neonatal serum bilirubin levels can be predicted from diffuse reflectance changes around 460 nm (Haunemann et al, 1978), within approximately 2 mg% under controlled conditions. However, variations in skin pigmentation and infant maturity have not been fully compensated for, and the fractions of albumin-bound and free bilirubin have yet to be well correlated with reflectance measurements. It is the free bilirubin concentration which is best related to brain damage (kernicterus) in premature infants. "Pulse oximetry", in which the small variations in tissue reflectance due to arteriolar pulsations with each heart cycle are analyzed, has become a routine patient monitor for both heart rate and oxygenation (Mendelson and Ochs, 1988). Reflectance has been combined with fluorescence of NAD to monitor oxidation states in brain (Haselgrove et al, 1990).

The usefulness of conventional reflectance measurements has been limited, however, by the inability to determine true tissue absorption coefficients from diffuse reflectance alone. Laser reflectance spectroscopy can potentially determine absolute absorption coefficients in tissue by at least two methods.

Quantitative Absorption Spectroscopy in Tissue

The true absorption spectrum of tissue, $\mu_a(\lambda)$, may be obtained non-invasively from measurements of either (1) total, spatially-resolved or (2) temporally-resolved diffuse reflectance spectrum. This is of interest in applications where the concentration of specific chromophores, changes in the chromophore absorption spectrum, or, possibly, the distribution of chromophores in the tissue are required. Qualitatively, it is clear that the diffuse reflectance, which is comprised of photons remitted back through the irradiated tissue surface after single or multiple scattering, depends on both the absorption and scattering properties of the tissue at the measured wavelength. Increasing the absorption reduces the diffuse reflectance, since the probability is less that photons will escape the surface before being absorbed. Conversely, increasing the probability of scatter, as specified by the scattering coefficient, increases the diffuse reflectance for a given absorption. The angular distribution of scattering also affects the probability that photons will propagate back through the tissue surface. For most purposes, the angular dependence can be described adequately by a single parameter, the scattering anisotropy, $g = <\cos \theta>$, which represents the average cosine of the single scatter angle, θ. Except in the case of time-resolved reflectance at very short times ($< 10^2$ ps), the scattering coefficient and anisotropy may be combined into a single transport scattering coefficient, $\mu'_s(\lambda) = \mu_s(\lambda) [1-g(\lambda)]$. The diffuse reflectance depends also on the tissue refractive index, n. This determines the total internal (specular) reflection coefficient for scattered light at the issue surface in the case of a refractive index mismatch with the external medium (e.g., air-tissue). It also determines the speed of light in tissue, $c' = c_o/n$, where c_o is the speed in vacuo (3×10^8 ms^{-1}), which is relevant in the case of time-resolved measurements.

In the discussion below, several simplifying assumptions will be made, as follows. Only linear absorption is considered, that is, the fluence rate at all points in the tissue, $\Phi(\underline{x})$, must be low enough that the local rate of energy absorption is equal to the product of μ_a and Φ,

$$P(\underline{x},\lambda) = \Sigma_i \epsilon_i(\lambda).C_i(\underline{x}).\Phi(\underline{x},\lambda) = \mu_a(\lambda). \Phi(\underline{x},\lambda) \qquad (1)$$

where $\epsilon i(\lambda)$ is the (molar) extinction spectrum of the i-th chromophore present in (molar) concentration, $C_i(\underline{x})$. Except where otherwise stated, it is also assumed that the tissue is macroscopically homogeneous in its optical properties and effectively semi-infinite in extent so that no photons exit the tissue other than through the irradiated surface where they will contribute to the measured diffuse reflectance signal. Specular reflectance of the incident light is also ignored. This is typically a few percent and can simply be subtracted from the diffuse reflectance measurement if necessary. Scattering in tissue is assumed to be elastic, the reflectance being measured at the incident light wavelength, and the individual scattering and absorption interactions are independent so that the radiation transport model of light propagation, as described by the Boltzmann equation, is valid.

Most applications of diffuse reflectance spectroscopy in tissue have been carried out in the red and near-infrared regions of the optical spectrum, where the total absorption of soft tissues is lowest and the diffuse reflectance is highest [Svaasand and Gomer, 1989]. Thus, we will generally assume that the scattering-to-absorption ratio in the tissue is high, so that, in many cases, diffusion theory may be used to investigate the relationships between diffuse reflectance and the tissue optical properties, specifically μ_a, μ'_s, and n. In this optical window, so-called because the diffuse transmittance of light tissue is also at its maximum, the refractive index varies only by a few percent between different soft tissues [Bolin, et al, 1989]. We will generally assume n = 1.4, independent of the tissue and wavelength.

As illustrated in Figure 1, three classes of diffuse reflectance measurements can be made:

i.   total diffuse reflectance, $R_t(\lambda)$,

ii.   local, i.e., spatially-resolved, reflectance, $R(\rho,\lambda)$ at radial distance along ρ the tissue surface from an incident (collimated) beam of small diameter,

iii.   time-resolved reflectance R (t,λ) or $R(\rho,t,\lambda)$ at time t following a short input light pulse.

A further possibility is to measure the diffuse reflectance as a function also of the exit angle, α, of the backscattered photons. This has been studied only to a very limited extent, and so will not be discussed further here.

Classes (i) and (ii) represent measurements made under effectively steady-state conditions, with either continuous-wave illumination or where the light pulse and/or integrating time of the reflectance photodetector are long compared with the propagation time of light in the tissue. Class

75

(iii) refers to photon time-of-flight techniques, where there is measurable temporal spreading of a short incident pulse due to differing propagation times as photons follow different optical paths through the tissue before exiting the surface. Picosecond or nanosecond time scales are involved.

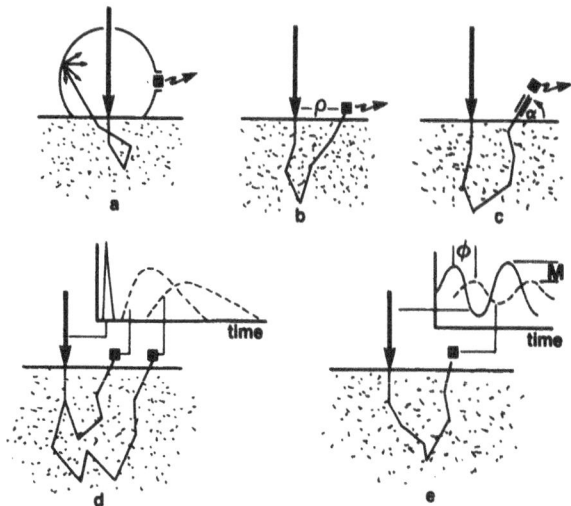

Figure 1. a) total reflectance; b), c) spatially-resolved, d) time domain, e) frequency domain.

1)   Total, Steady-State Reflectance, R (λ)

According to diffusion theory [Flock, et al, 1989], R_t is given by

$$R_t = \frac{a'}{1+2k(1-a')+(1+2k/3)\sqrt{1-a'}} \qquad (2)$$

where a' is the transport albedo, $a' = \mu'_s/(\mu_a + \mu'_s)$. The factor $k = (1 - r_d)/(1 + r_d)$, where r_d is the specular reflection coefficient for internal reflection of diffuse light at the tissue surface, accounts for the total internal reflection. For an index-matched boundary, $r_d = 0$ and $k = 1$, while for an air-tissue boundary with n = 1.4, $r_d = 0.53$ and $k = 0.3$. The dependence of R_t on the transport albedo given by equation (2) has been checked both experimentally and by Monte Carlo modelling of radiation transport in tissue [Flock, et al, 1989a,b] for a' > 0.5 and for anisotropy, g, between 0 (isotropic scattering) and 0.99 (highly forward-peaked scattering). There was no significant dependence on g. In the red/near-IR region, a' is typically greater than 0.9 with g > 0.7 for soft tissues, so that equation (2) should be an accurate description for the total diffuse reflectance.

Equation (2) implies that R_t can only be used to determine a' at any wavelength, and cannot separate the absorption and scattering coefficients. This is illustrated in Figure 2, where $R_t(\lambda)$ has been calculated from equation (2) for a hypothetical absorption spectrum and different, fixed values of the transport scattering coefficient. It is seen that the scattering distorts the true absorption spectrum, both in magnitude and in shape. Conventionally, the size of the dip in the reflectance spectrum associated with a particular chromphore peak is defined (Figure 2b) as

$$A = - \log_{10} (R_p/R_b) \qquad (3)$$

The basis for this is that, if the average optical path length in the tissue of the diffusely reflected photons is l, then

$$R_t = e^{-\mu a l} \qquad (4)$$

so that

$$A = 0.434 (\mu_{ap} - \mu_{ab})l \qquad (5)$$

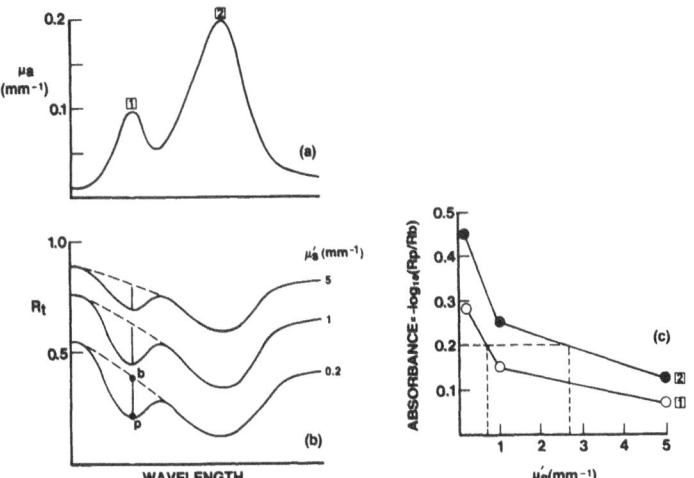

Figure 2. a) Hypothetical absorption spectrum, b) R_t spectra for different scattering, c) Absorption for the two chromophone Peaks.

The optical path length distribution depends on the scattering, so that, for fixed μ'_s, i.e., fixed l, A is proportional to the true chromophere absorption, $\mu_a = \mu_{ap} - \mu_{ab}$, correcting the specific chromophore peak for the tissue background absorption. Such absorbance measurements are used in applications where the relative absorption (i.e., relative concentrations) of known chromophores, or changes in a specific chromophore, such as in the case of oxygen-dependent hemoglobin absorption, are of interest. It should be noted that l depends on both μ'_s and μ_a so that l is only independent of changes in absorption if these changes are small. In addition, the absolute absorption (spectrum) cannot usually be obtained, as illustrated in Figure 2c, since l is unknown in most cases. An exception is where the concentration (and extinction) of one chromophore is known and can be used to calibrate the absorbance value of a second chromophore of unknown

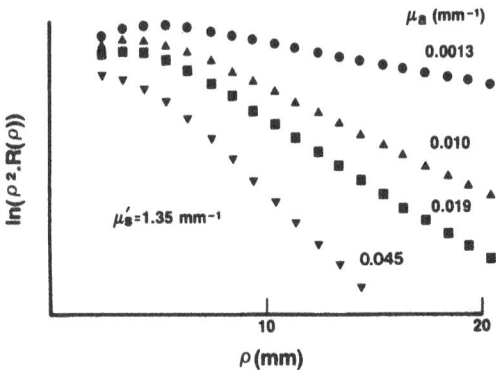

Figure 3. Local reflectance versus radial distance. Experimental data measured in a tissue-simulating medium of fixed scattering and different absorption.

concentration. The 970 nm absorption peak of water has been suggested for this purpose [Delpy, et al, 1988], since the water content (and hence, μ_a) of most soft tissues is known to a good accuracy. Thus, for chromphore i, at peak wavelength λ_i

$$\mu_{ai}(\lambda_i) = A_i(\lambda_i)/A_w(970).\mu_{aw}(970) \tag{6}$$

and measuring A_i and A_w then gives the chromophore absorption μ_{ai}. There are, however, two limitations to this calibration method. First, the 970 nm water peak is rather weak and broad in most tissues, so that it is difficult to determine A_w accurately. A different, say, exogenous chromophore could be used, but some independent measure of its concentration in the tissues is required. The second problem is that equation (6) assumes $l(\lambda i) = l (970)$, whereas, in practice, μ'_s varies approximately as $1/\lambda$ in the optical window and the slope may be different for different tissues [Karagiannes, et al, 1989; van Gemert, et al, 1989; Parsa, et al, 1989; Peters, et al, in press]. Note that this limitation applies also to measurements of relative absorbance: if μ'_s is not constant, then the curves shown in Figure 2c are not parallel and A_1/A_2 is not independent of the scattering coefficients.

These limitations of total steady-state reflectance spectroscopy motivate the examination of spatially or temporally-resolved reflectance measurements for in vivo optical absorptiometry .

2)   Spatially-Resolved, Steady-State Reflectance, R(ρ,λ)

For radial distances large compared with the scattering transport mean-free path ($\rho \gg 1/\mu'_s$), both diffusion theory [Roenhuls, et al, 1983; Patterson, et al, 1989a] have shown that

$$R(\rho) \propto \frac{1}{\rho m} e^{-\rho/\delta} \tag{7}$$

where δ is the effective penetration depth of light in the tissue, given by

$$\delta^2 = 1/3\mu_a(\mu_a+\mu'_s) \tag{8}$$

and m is a constant. The predicted value of m has varied in different studies between 0.5, 1 and 2 depending on the details of the model used, and on whether $R(\rho)$ is taken as being integrated over all exit angles, α, or only normal to the surface ($\alpha = \pi/2$). Here, we will use m = 2, as predicted by diffusion theory and as confirmed by Monte Carlo calculations and by experiments in tissue-simulating media [Patterson, et al, 1989a; and in vivo [Wilson, et al, 1990a] for the angle-integrated signal.

It is of considerable practical interest for in vivo optical dosimetry, as applied, for example, in photodynamic therapy [Wilson and Patterson, 1990] or laser hyperthermia [Svaasand, et al, 1989],that δ also describes the depth (z) dependence of the fluence rate, ρ, within the tissue irradiated by a broad beam source,

$$\Phi(z) = f_c(z,\mu_a,\mu_s,g,n) \cdot \Phi_o \cdot e^{-z/\delta} \tag{9}$$

The "coupling function", f_c, tends to a constant at large z (z> several transport mean-free paths), but depends on the particular tissue optical properties close to the surface [Flock, et al, 1988]. Thus, to within an overall constant, the fluence-depth dependence in the tissue can be determined non-invasively from measurements of the local, steady-state reflectance on the surface.

There are two strategies to determine $\mu_a(\lambda)$ (and $\mu'_s(\lambda)$) from the steady-state reflectance measurements. For the first, both R_t and $R(\rho)$ are used [Patterson, et al, 1989a]. R_t yields the transport albedo, a', and $R(\rho)$ versus ρ at large ρ yields δ. Then, μ_a and μ'_s may be calculated from

$$\mu_a = \delta(a'/3)^{1/2} \tag{10a}$$

and

$$\mu'_s = \delta(a'/1-a')(a'/3)^{1/2} \tag{10b}$$

A practical limitation of this approach, particularly for high albedo, is that the absolute value of R_t is required, so that all the back-scattered light must be collected.

The alternative strategy, for which the initial tests have been encouraging [Wilson, et al, 1990a] is to utilize the shape of the $R(\rho)$ curve over the whole range of ρ values, as illustrated in Figure 3. To a first approximation,

$$R(\rho) = \frac{z_o}{2\pi} \frac{e^{-\rho'/\delta}}{(\rho')^2} [1/(\rho')^2 + 1/\delta^2] \tag{11}$$

where $(\rho')^2 = \rho^2 + z^2_o$ and $z_o = 1/\mu_s'$. This was derived from diffusion theory, assuming that incident pencil beam is equivalent to a first-scatter isotropic point source at depth z_o, and with the approximate boundary condition of zero fluence at the tissue surface. We have found recently [T. Farrell, private communication] that a more rigorous condition, setting the fluence to zero at an extrapolated boundary (at $z = -z'_o$) is required in order to fit the detailed shape of $R(\rho)$ for small ρ values in tissue. With the more exact form, μ_a (and μ'_s) can be obtained from the shape of $R(\rho)$ versus ρ, with an arbitrary overall scaling factor, so that absolute reflectance meaurements may not be required. For typical soft tissues in the optical window, the range of $\rho = 1$ to 10 mm appears adequate for accurate determination of μ_a and μ'_s. A prototype fiber-optic based instrument based on this method had been described [Wilson, et al, 1990a] and initial tests have been successful, both in determining the optical absorption and scattering spectra of tissues and for measuring the time-course of uptake of photosensitizer in tissue in vivo.

Other aspects of the local reflectance signal have also been examined by random-walk [Bonner, et al, 1987], Monte Carlo [Jacques, 1989] and diffusion theory [Patterson, et al, 1990]. Bonner et al [1987] showed that the average number, $<n>_\rho$, of scattering interactions undergone by photons exiting at radial distance ρ, and hence the average optical path length in the tissue, is linearly proportional to ρ. This is in agreement, in the limit of large ρ, with the prediction of diffusion theory

$$<n>_\rho = \frac{(\rho'/\delta)^2}{2(1+\rho'/\delta)} \cdot \frac{\mu_s}{\mu_a} \qquad (12)$$

and for $\rho'>>\delta$, $<n>_\rho = (\rho/2\delta) \cdot (\mu_s/\mu_a)$. For example, for $\mu'_s = 1$ mm^{-1}, $g = 0.95$, $\mu_a = 0.5$ mm^{-1} and at $\rho = 40$ mm, $<n> \sim 3000$. Likewise, the average maximum depth of penetration in the tissue was found by Bonner et al to vary as $\rho^{1/2}$. These results are in some disagreement with Monte Carlo results, [R. L. Barbour, private communication], which gave quadratic and linear dependences, respectively. Further studies are required to resolve this discrepancy. However, qualitatively, both studies demonstrate that the photons remitted at large radial distance have, on average, "probed" the tissue to greater depth. This suggests that some degree of depth profiling may be possible, and this has been examined in the case of a simple 2-layer model by Nossal et al (1988). It was shown that the surface profiles, $R(\rho)$ were sensitive to the presence of a deeper-lying layer of different optical absorption. When the top layer was the more absorbing, it appeared possible to determine the thickness of this layer and the absorption coefficients of both layers, since there was a distinct break in the slope of the $R(\rho)$ versus ρ curves. The reverse situation was more ambiguous. Gush et al [1984] have observed distinct features in experimental measurements of $R(\rho)$ on the finger in vivo, and attributed this to differences in the radiance distributions in the epidermis and dermis. This is clearly an area which requires considerable further investigation, particularly because of the potential applications in dermatology, and time-resolved or time-gated reflectance techniques may contribute significantly to such applications.

3). Time-Resolved Reflectance, R(t,λ) and R(ρ,t,λ)

Photons propagate in tissue at finite speed, $c' = c/n = 0.21$ mm/ps = 21 cm/ns. If, therefore, light is incident as a short pulse, the diffusely reflected photons are spread in time due to their different propagation path lengths, $l = c' t$. As seen above, l increases with the distance between the incident source and detector positions on the tissue surface, and depends on μ_a and μ'_s, so that the time delay also depends on these parameters. The probability that an individual photon will survive to be detected at time t is, on average,

$$p = e^{-\mu_a c' t}$$

Using time-dependent diffusion theory, it can be shown [Patterson, et al, 1989b] that $R(\rho,t)$, for index-matching at the surface, is given by

$$R(\rho,t) = (4\pi Dc')^{-3/2} z_0 t^{-5/2} e^{-\mu_a c' t} e^{-\rho^2(4Dc't)} \tag{13}$$

where ρ and z_0 are as previously defined, and the diffusion coefficient, $D = 1/3 (\mu_a + \mu'_s)$. The total time-dependent diffuse reflectance is then

$$Rt(t) = \int_0^\infty 2\pi\rho R(\rho,t)d\rho \tag{14}$$

$$= (4\pi Dc')^{-1/2} z_0 t^{-3/2} e^{-\mu_a c' t} e^{-z_0^2/(4Dc't)}$$

Figure 4 shows the time-dependence of the local and total reflectances for typical optical parameters. Note that the time-scale in this diffusion regime ($\rho \gg z_0$) is in the order of a few nanoseconds: thus, for example, photons detected at $\rho = 40$ mm and $t = 3$ ns have an optical path length in the tissue of about 60 cm.

From the point of view of absorption spectroscopy in tissue, a critical observation is that

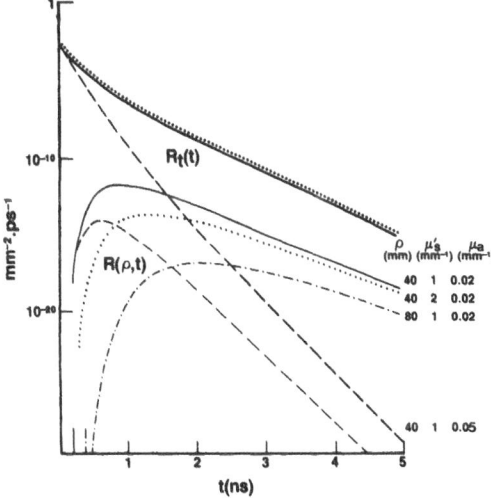

Figure 4. Local and total reflectance time spectra in the diffusion regime, as given by equations 13 and 14, for an offocially homogeneous, semi infinite medium (refractive index matching at the surface).

$$\lim_{t \to \infty} \{\delta/\delta t \ \ln(R(\rho,t))\} = -\mu_a c'$$ (15)

Thus, the final slope of the logarithmic local reflectance-time curve gives directly the true tissue absorption coefficient, <u>independent of the scattering</u> and, at least within the range of validity of the diffusion model, independent of the source-detector distance. This finding was first made experimentally by Chance et al [1988] in tissue-simulating media and in human brain in vivo, and used to determine the hemoglobin oxygenation of the tissue by measuring μ_a at isobestic and non-isobestic wavelengths. Further modelling and experimental studies have confirmed this simple and important aspect of time-resolved measurements. The issues which remain to be addressed with regard to the general applicability of the method include: the time required to reach the terminal slope for a given accuracy in determining μ_a; the tissue volume effectively probed and the minimum volume required for the diffusion model (equation 13) to be valid; the effects of refractive index mismatch at the tissue surface; the effects of tissue layering or other specific or random macro-heterogeneities in the tissue optical properties. Some preliminary investigations of these questions have been performed.

Using the diffusion model and Monte Carlo simulation, Jacques [1989] showed that the error in deriving μ_a from the slope of $R_t(t)$ is proportional to $1/t$, and for typical tissue properties ($\mu'_s = 1 \ mm^{-1}$, $\mu_a = 0.1 - 0.3 \ mm^{-1}$), time delays greater than about 0.1 - 0.2 ns were required for reliable results. At shorter times, diffusion theory was significantly in error compared to the Monte Carlo results for $R_t(t)$. As would be expected, the diffusion theory converged more rapidly to the Monte Carlo (true) values with higher scattering. The local reflectance also takes longer to reach the final slope than does the total reflectance (see figure 4), especially at large radial distances. As μ_a increases, the final slope is obtained earlier in all cases.

Jacques [1989] also showed that the tissue volume required for the diffusion conditions to be satisfied has a radius, $r_{min} = (6Dc't)^{1/2}$. Thus, for example, if measurements are made at t = 1ns, then the required tissue volume to ensure that the slope does give the value $-\mu_a.c'$ is a hemisphere of radius 4-5 cm for $\mu'_s = 2 \ mm^{-1}$. Even at times as low as 200 ps, where diffusion theory probably starts to break down, sampling dimensions are in the range 1-3 cm radius. In practice, there is a trade-off between the sampling volume and the time resolution needed in the source-detector system. Typical equipment used to date [Chance et al, 1988 Wilson, et al, 1989] has consisted of a frequency-doubled, mode-locked Nd:YAG laser pumping a cavity-dumped dye laser, to produce pulses of a few ps at 10-100 MHz repetition rate, together with either a streak camera (typically a few ps resolution) or a time-correlated single photon counter with a multichannel-plate photomultiplier tube (typically 30-50 ps). Thus, measurements in the region of a few hundred ps or greater can be made with adequate time resolution and tissue depths of, say, 5 cm diameter or greater. Indeed, with a streak camera, measurements below the diffusion theory limit can be made, but more studies are required to determine how to extract the tissue optical properties from such reflectance spectra [Wilson and Jacques, in press; Wilson, et al, 1990].

For applications involving nanosecond range propagation times, such as in the brain, large muscles or other large, optically homogeneous organs, a simpler technological approach may be to perform time-dependent diffuse reflectance measurements in the equivalent frequency domain, using a light source modulated at high frequency ($>10^2$ MHz), and phase-sensitive detection. This is analogous to techniques such as frequency-domain, time-resolved fluorescence [Larowicz, et al, 1988] where the time delay, and hence phase shift between the input and output signals is caused by the fluorescence decay behavior. Here, the shift is the result of the spread in propagation time of the photons through the tissue. The feasibility of this has been demonstrated recently, both in light scattering media and in vivo [Lacowicz, et al, 1990], and Patterson, et al, [1990] have derived the equivalent diffusion theory formulation relating the signal amplitude modulation and phase to the parameters μ_a, μ'_s and ρ. Currently, this is being tested against the available experimental data: for example, the change in phase shift due to oxygenation in a yeast/hemoglobin model (754 nm, ρ = 50 mm, μ_s = 0.5 mm^{-1}, μ_a (HbO$_2$, 12 μM) = 0.0024 mm^{-1} μ_a (Hb, 12 μm = 0.0031 mm^{-1}) as measured by Chance (private communication) was about 20° for f = 220 MHz, and the diffusion theory prediction was in reasonable agreement at 15°.

For the case where $2\pi f \ll \mu_a c'$, the phase shift, Φ, is proportional to f.

$$\Phi(\rho,t) = \frac{\pi(\rho'/\delta)^2 \, f}{(1+\rho'/\delta) \, \mu_a c'} \tag{16}$$

Note that the phase shift is characteristic of the mean photon propagation time, (Compare equations 12 and 16: $\Phi = 2\pi f$, $\langle n \rangle = \bar{t} c' \mu_s$).

The dependence of the phase Φ on the tissue absorption is shown in figure 5, for a range of radial distances and scattering coefficients, at a "standard" frequency of 100 MHz. The phase shift decreases with increasing absorption, but increases with the scattering coefficient and distance. A major advantage of working in the frequency domain is that measurements may be made continuously, to follow dynamic changes in the tissue properties. In addition, as can be seen in figure 5, the typical phase shifts are substantial even at frequencies as low as the 100 MHz region. This may permit the use of either multiple modulated diode lasers at discrete wavelengths, or possibly even conventional white light sources, which could give detailed absorption spectra. In each case, both the light sources and detection system are relatively simple and inexpensive, so that time-dependent diffuse reflectance spectroscopy in vivo may well become clinically feasible and affordable.

It should be noted, as illustrated quantitatively in the insert of figure 5, that the amplitude modulation is not very sensitive to changes in the tissue optical properties at these low frequencies, and that frequency scanning is needed in order to normalize the relative signal amplitude (M -> 1 as f -> 0). On the other hand, as seen by equation (16), at low frequency there is no advantage to measuring at more than one frequency as far as determining the phase is concerned (apart from having a larger, and hence, more accurately measurable value at higher frequency): since $\Phi \propto$ f, the slope, and hence the dependence on optical properties, is determined by measurement at a single frequency. Further, it is rather simple to measure the phase very accurately (to within a fraction of a degree). Thus, it is worth examining what optical property information can be obtained by a phase-only measurement technique.

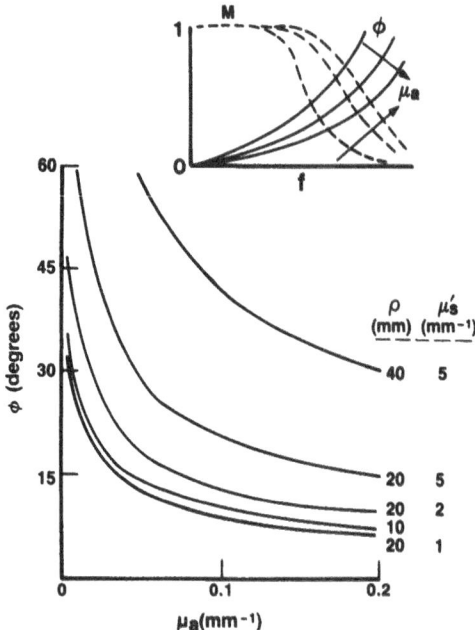

Figure 5. Dependence of phase-shift on tissue absorption, for different
tissue scattering and radial distances (f = 100 MHz). The insert
shows the general behavior of the phase shift and the amplitude
modulation as a function of frequency.

If Φ is measured at two different ρ values, then

$$\frac{\Phi_1}{\Phi_2} = \frac{(\rho_1)^2(1+\rho_2/\delta)}{(\rho_2)^2(1+\rho_1/\delta)} \qquad (17)$$

yielding the value of δ. If each measurement is made at two wavelengths, then the relative absorption coefficient can again be determined,

$$\mu_{a1}/\mu_{a2} = \delta_2{}^2/\delta_1{}^2 \qquad (18)$$

assuming $\mu'_{s1} = \mu'_{s2}$. This has been used by Chance [et al, in press] for in vivo hemoglobin oxygenation studies.

Patterson et al [1990] have shown in principle that a single phase measurement at some distance ρ can be combined with a separate, steady-state, i.e., $R(\rho)$, determination of δ, as in (2) above, to obtain the absorption coefficient from equation (15). Further, for typical tissue properties, the two measurements are orthogonal at low modulation frequency, so that the μ_a value may be determined accurately and unambiguously.

Summary

In conclusion, each of the types of diffuse reflectance measurements, total, spatially-resolved and time-resolved (time or frequency domain) gives, non-invasively, quantitative information on the tissue absorption and scattering properties. For absorption spectroscopy, the total, steady-state reflectance yields only relative data, except in special circumstances. Adding the spatial-dependence of the reflectance permits absolute absorption measurements and is technically straightforward. Further, it may have some limited depth profiling capability. In the regime of large tissue volumes, time-resolved reflectance in the time domain provides direct determination of μ_a independent of the scattering, so that scattering heterogeneity does not affect the result. However, the technology is relatively expensive and cumbersome. Studies in shorter time ranges (10-100 ps) allow the probing of smaller tissue volumes, and may be amenable to endoscopic or even catheter-based clinical applications. Matching the time-resolution and optical power and sensitivity of laser sources and photodetection systems, respectively, remains a problem, and a rigorous understanding of the reflectance characteristics in this region is lacking. Little has been done yet to examine the depth profiling capabilities, or the potential for regional tissue absorptiometry in either the long or short time regimes. Finally, frequency-domain techniques promise technically simpler and cheaper instruments for in vivo absorption spectroscopy, but are likely to be restricted to the large tissue-volume situation. The optimum combinaton of these various techniques is largely unexplored and is likely to be a fruitful area for future research and development.

References

Anderson, R. R., Parrish, J. A., " The optics of human skin", J Invest
        Dermatol 77:13-19, 1981.

Anderson, R. R., Beck, H., Bruggemann, U., Farinelli, W., Jacques, S. L., Parrish, J. A., " Pulsed photothermal radiometry in turbid media: internal reflection of backscattered radiation strongly influences optical dosimetry", Appl Optics 28:2256-2262, 1989.

Bolin, F. P., L. E. Preuss, Taylor R. C., and Ference, R.J., "Refractive index of some mammalian tissues using a fiber optic cladding method", App. Opt., 28, 2297-2303, 1989.

Bonner, R. F., Nossal, R., Havlin, S., and Weiss, A. H., "Model for photon migration in turbid biological media", J. Opt. Soc. Am., 27, 423-432, 1987.

Chance, B., M. Maris, Sorge J., Zhang, M. Z., "A phase modulation system for dual wavelength difference spectroscopy in hemoglobin deoxygenation in tissues", Proc. S.P.I.E., 1204, 1990, in press.

Chance, B., Nioka, S., Kent, J., McCully, K., Fountain, M., Greenfeld, R., and Holtom, G., "Time resolved spectroscopy of hemoglobin and myoglobin in resting and ischemic muscle", Anal. Biochem. 174, 690-707, 1988.

Delpy, D. T., Cope, M., van der Zee, P., Arridge, S., Wray S., and Wyatt, J., "Estimation of optical path length through tissue from direct time-of-flight measurements", Phys. Med. Biol., 33, 1433-1442, 1988.

Flock, S. T., Patterson, M.S., Wilson, B. C., and Wyman, D. R., "Monte Carlo modelling of light propagation in highly scattering tissue - I: Model predictions and comparison with diffusion theory", IEEE Trans Biomed. Eng. 36, 1162-1168, 1989a.

Flock, S. T., Wilson, B. C., Patterson, M. S., "Monte Carlo modelling of light propagation in highly scattering tissues - II: Comparison with measurements in phantoms", IEEE Trans. Biomed. Eng. 36, 1169-1173, 1989b.

Flock, S. T., Wilson, B. C., Patterson, M.S., "Hybrid Monte Carlo - diffusion theory modelling of light distribution in tissue", Proc. S.P.I.E. 908, 20-28, 1988.

Groenhuis, R. A. J., Ferwerda, H. A., Bosch, J. J. Ten, "Scattering and absorption of turbid materials determined from reflection measurements", 1: Theory, App. Opt. 22, 2456-2462, 1983.

Gush, R.J., King, T. A.,, and Jayson, M.I.V., "Aspects of laser light scattering from skin tissue with application to laser Doppler blood flow measurements", Phys. Med. Biol. 29, 1463-1476, 1984.

Haunemann, R. E., DeWitt D. P., Weichel J. F., "Neonatal serum bilirubin from skin reflectance", Pediat Res 12:207-210, 1978.

Haselgrove, J. C., Bashford C.L., Barlow C. H., Quistorff B., Chance B., Mayevsky A., "Time resolved 3-dimensional recording of redox ratio during spreading depression in gerbil brain", Brain Res 506:109-114, 1990.

Jacques, S. L., "Time-resolved reflectance spectroscopy in turbid tissues", IEEE Trans. Biomed. Eng. 36, 1155-1161, 1989.

Karagiannes, J. L., Zhang, Z., Grossweiner, B., Grossweiner, W. I., "Applications of the 1-D diffusion approximation to the optics of tissues and tissues phantoms", Appl. Opt. 28, 2311-2317, 1989.

Lacowicz, J. R., Berndt, K. W., Johnson, M. L., "Photon migration in scattering media and in tissue", Proc. S.P.I.E., 1204, 1990, in press.

Lacowicz, K. W., Laczko, G., Gryczynski, I., Szmacinski, H., and Wiczk, W., "Gigahertz frequency-domain fluorometry: resolution of complex decays, picosecond processes and future developments", J. Photochem. Photobiol. B2, 295-311, 1988.

Mendelson, Y., Ochs, B. D., "Noninvasive pulse oximetry utilizing skin relectance photoplethysmography", IEEE-Trans Biomed Eng 35:798-805, 1988.

Nishioka, N. S., Jacques, S. L., Richter J.M., Anderson R.R., "Reflection and transmission of laser light from the esophagus: the influence of incident angle", Gastroenterol 94:1180-5, 1988.

Nossal, J. Kiefer, Weiss, G. H., Bonner, R., Taitelbaum, H., and Havlin, S., "Photon migration in layered media", <u>Appl. Opt.</u> <u>27</u>, 3382-3391, 1988.

Parsa, P., Jacques, S. L., and Nishioka, N. S., "Optical properties of rat liver between 350 and 2200 nm", <u>Appl. Opt.</u> <u>28</u>, 2325-2330, 1989.

Patterson, M. S., Chance, B., and Wilson, B. C., "Time resolved reflectance and transmittance for the non-invasive measurement of tissue optical properties", <u>Appl. Opt.</u> <u>28</u>, 2331-2336, 1989b.

Patterson, M. S., Moulton, J. D., Wilson, B. C., and Chance, B., "Applications of time-resolved light scattering measurements to photodynamic therapy dosimetry", <u>Proc. S.P.I.E.</u>, <u>1203</u>, 62-75, 1990.

Patterson, M. S., Schwartz, E., and Wilson, B. C., "Quantitative reflectance spectrophotometry for the non-invasive measurement of photosensitizer concentration in tissue during photodynamic therapy", <u>Proc. S.P.I.E.</u>, <u>1065</u>, 115-122, 1989a.

Peters, V. G., Wyman, D. R., Patterson, M. S., Frank, G. L., "Optical properties of normal and diseased human breast tissues in the visible and near infrared", <u>Phys. Med. Biol.</u>, in press.

Svaasand, L. O. and C. J. Gomer, "Optics of tissue, <u>Proc. S.P.I.E.</u>, <u>IS5</u>, 114-132, 1989.

Svaasand, L. O., Gomer, C. J., and Welch, A. J., "Thermiotics of tissue", <u>Proc. S.P.I.E.</u>, <u>IS5</u>, 133-145, 1989.

van Gemert, M. J. C., Jacques, S. L., Sterenborg, H.J.C.M., and Star, W. M., "Skin optics", <u>IEEE Trans Biomed. Eng.</u> <u>36</u>, 1146-1154, 1989.

Wilson, B. C., Farrell, T. J., Patterson, M. S., "An Optical fiber-based reflectance spectrometer for non-invasive investigation of photodynamic sensitizers in vivo,", <u>Proc. S.P.I.E.</u> <u>IS5</u>, 219-232, 1990.

Wilson, B. C. and Jacques, S. L., "Optical reflectance and transmittance of tissues: Principles and applications", <u>IEEE J. Quant. Electr.</u>, in

Wilson, B. C., Park, Y., Hefetz, Y., Patterson, M., Madsen, S., Jacques, S. J., "The potential of time-resolved reflectance measurments for the noninvasive determination of tissue optical properties", PRoc. S.P.I.E., <u>1064</u>, 97-106, 1989.

Wilson, B. C. and Patterson, M. S., "The determination of light fluence distributions in photodynamic therapy", In: d.Kessel (ed), <u>Photodynamic Therapy of Neoplastic Disease</u>, Vol, I, CRC Press, Boca Raton, 1990, pp. 29-145.

Wilson, B. C., Patterson, M. S., Flock, S. T., Wyman, D. R., "Tissue optical properties in relation to light propagation models and in vivo dosimetry", In: Chance (ed), <u>Photon Migration in Tissue</u>, Plenum Press, N.Y., 1990, pp. 25-42.

MONITORING AND IMAGING OF TISSUE BLOOD FLOW BY COHERENT

LIGHT SCATTERING

Gert E. Nilsson, Anneli Jakobsson and Karin Wårdell

Division of Biomedical Instrumentation
Department of Biomedical Engineering
Linköping University, Linköping, Sweden

INTRODUCTION

Monitoring and visualization of tissue perfusion have for a long time been an intriguing task both to the medical scientist wanting to explore the basic nature of microcirculation and the clinical practitioner diagnosing diseases known to affect peripheral circulation. Because of the extreme sensitivity of the microvascular network to almost any kind of stimuli, methods designed for these purposes must be based on principles that do not influence tissue perfusion. Measurement principles utilizing the interaction of light and perfused tissue have been successfully used for the recording of changes in microvascular blood volume (Challoner, 1979) and tissue oxygenization (Tremper and Baker, 1989). Laser Doppler flowmetry (LDF) , a technique intended for the direct recording of tissue perfusion, is based on the fact that photons undergo a frequency shift according to the Doppler principle when scattered by a moving object such as a blood cell (Holloway and Watkins, 1977, Nilsson et. al. 1980a, b).

The first report on recording the Doppler shift of diffusely scattered monochromatic light in tissue in order to study perfusion, was published by Stern in 1975 (Stern, 1975). He noticed that the characteristic static speckle pattern, generally observed when light from a He-Ne laser is scattered in a non-moving object, could not be seen by the naked eye when the beam was scattered in living tissue. The reason for this is the rapid fluctuation of the speckle pattern caused by scattering processes involving Doppler shifts. When perfusion is reduced, the frequency of fluctuations in the speckle pattern decreases. A static speckle pattern can again be observed when perfusion is completely arrested. This fundamental discovery illustrates the operating principle of laser Doppler flowmetry, and has been used during the last decade for the construction of versatile perfusion monitors and more recently also for the development of perfusion imagers.

The purpose of this survey is to give a brief introduction to the topic of dynamic light scattering in tissue, to present the operating principle of the laser Doppler flowmeter together with some of its medical applications, and, finally, to give an outlook on how coherent light scattering in tissue can be utilized to create images of the tissue perfusion.

DYNAMIC LIGHT SCATTERING IN TISSUE

A monochromatic and collimated light beam from a laser becomes diffusely scattered, partially absorbed and spectrally broadened when scattered in the tissue matrix. The migration of the individual photons through the tissue is determined by the scattering and absorption coefficients as well as by the phase function at each scattering site (Van Gemert et. al., 1989). These parameters are dependent on tissue characteristics and determine the penetration depth of light in the tissue at the wavelength in question. This penetration depth can be estimated by the use of Monte Carlo simulation (Jakobsson and Nilsson, 1990). In Caucasian skin tissue, the average measuring depth may be calculated to be around 200-450 μm, the exact value being dependent on the geometry of the probe.

Spectral broadening of the scattered light is caused by single or multiple scattering in moving blood cells of a portion of the incident photons. These scattering events generate a single or successive Doppler shifts. The Doppler shift generated can be described by the equation

$$\omega = q \, v = |g| \; |v| \; \cos(\alpha) \qquad (1)$$

where ω is the Doppler frequency, v is the velocity vector of the moving blood cell and q is the scattering vector (Fig 1).

The scattering vector q can be expressed as:

$$q = k_i - k_s \qquad (2)$$

where k_i and k_s are the wave vectors of the incident and the scattered photon, respectively.

From Fig. 1 and equation 1, it can be concluded that the Doppler shift occurring at a single scattering event is dependent on 1) the absolute velocity $|v|$ of the scatterer, 2) the magnitude of the scattering vector $|q|$ and 3) the cosine of the angle α between q and v. The magnitude of the scattering vector is determined by the relationship between the dimension of the scatterer and the wavelength. During scattering of light from a He-Ne laser in red blood cells, the dimension of the scatterer is an order of magnitude larger than the wavelength. Most photons are therefore scattered within 5° in a forward direction (Steinke and Shepherd, 1988). Since light migrates through tissue in a diffuse way, it can be assumed that the directions of the different k_i-vectors are independent and randomized. The average Doppler shift will therefore be proportional to the average red cell velocity but virtually independent of the direction of the moving cells.

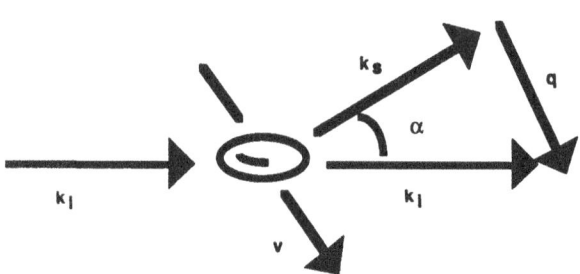

Fig. 1. Scattering of light in a blood cell.

The higher the concentration of moving red cells within the scattering volume, the more photons will be Doppler-shifted. Up to a certain concentration, this magnitude scales linearly with red cell concentration (Nilsson, 1984). At very high concentrations of red cells, however, multiple Doppler shifts occur more frequently, giving rise to non-linear effects. By analysing both the average Doppler frequency and the magnitude of the Doppler signal, information on both the average velocity and concentration of the moving red cells within the scattering volume may be obtained.

By appropriate signal processing, an output signal may be derived that scales linearly with tissue perfusion, defined as the average blood cell velocity (vel_{av}) multiplied by blood cell concentration (conc).

$$Tissue\ perfusion = vel_{av} * conc \tag{3}$$

In addition, signals proportional to the average blood cell velocity and blood cell concentration may be derived and shown separately.

Since the penetration depth of light in tissue is dependent on tissue absorption and scattering properties, the scattering volume differs from one tissue to another. In highly transparent tissue, the increased penetration depth will cause a higher fraction of the eventually backscattered photons to be Doppler-shifted. This results in a higher flowmeter output signal than that recorded in less transparent tissue, even though the two tissues may have identical degrees of perfusion (Ahn, 1986). Laser Doppler flowmeters can therefore generally not be calibrated in absolute units (e.g. ml min^{-1} $100g^{-1}$), and comparisons of results obtained in different kinds of tissue are difficult to make.

OPERATING PRINCIPLE OF LDF

In laser Doppler flowmeters, light is generally guided from a low-power (2 - 5 mW) He-Ne laser operating at 632.8 nm or a solid state laser (most often radiating in the infrared region) to the tissue and back to the instrument by optical fibres (Fig. 2).

In modern LDF equipment, these fibres have a core diameter ranging from 50 to 120 µm. Depending on the application, the probe may be designed in a variety of ways. Needle probes with an outer diameter of less than 0.5 mm facilitate the recording of deep muscular perfusion with a minimal influence on the microvasculature, while integrative probes record the average blood flow in a 1 cm^2 area (Salerud et. al., 1986). In order to record deep-tissue blood flow during an extended period of time, a single-fibre laser Doppler flowmeter has been developed (Salerud and Öberg, 1987) and used successfully (Kvernebo and Salerud, 1987). The signal-to-noise ratio of the instrument may be substantially improved by means of a differential detector (Nilsson et. al.,1980a). This improvement is most important in applications where a low perfusion value is to be separated from the zero baseline level. Movement of the fibre lines during measurement generally results in a false output signal that mimics the recording of blood flow (Nilsson, 1989). This phenomenon is caused by the dynamic interference of different optical fibre modes on the detector surface and cannot be separated from the blood flow-related interference pattern generated by Doppler-shifted and frequency-unshifted photons. A way to overcome this sensitivity to fibre line movement is to place a solid state laser and detectors inside the probe facing the tissue (de Mul et. al., 1984), thereby rendering the optical fibres superfluous.

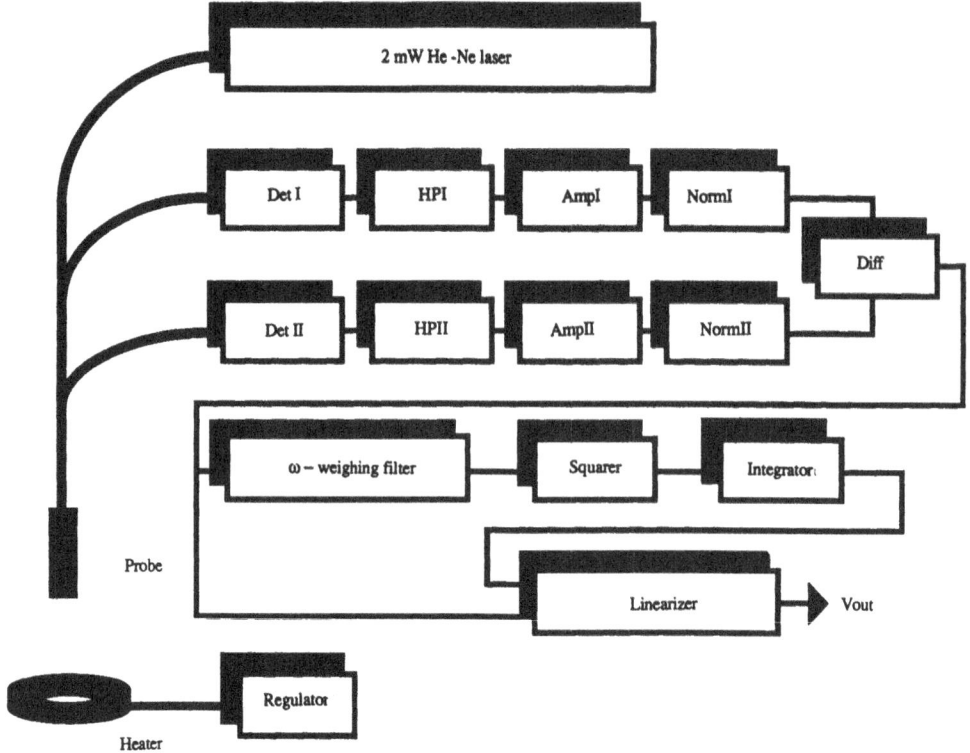

Fig 2. Block diagram of the laser Doppler flowmeter.

This solution also makes possible the recording of integrated blood flow over an area corresponding to the active surface of the detector, which generally exceeds the area of the optical fibres by several orders of magnitude. On the other hand, positioning the laser and photodetectors inside the probe restricts the usefulness of the flowmeter to applications where measurements are made on easily accessible surfaces, such as skin. Furthermore, movement artefacts that have their origin at the skin-tissue interface or in the tissue matrix itself, may still be a significant problem in some applications.

In the signal-processing unit of the instrument, the fluctuating portion is first extracted from the total photocurrent by means of a high-pass filter, then it is amplified (Fig. 2). To make the blood flow-related Doppler signal independent of light intensity, a normalizing circuit divides the Doppler signal by a voltage proportional to the total photocurrent. In the differential channel laser Doppler flowmeter, the difference of the output signal from the two divider circuits is calculated in the differential amplifier. This normalized Doppler signal is then fed through a ω- weighing filter network that gives a proportionally higher weight factor to higher frequencies, corresponding to Doppler shifts generated by blood cells moving at high velocities. The output signal of this network is squared and integrated, giving a total output signal that theoretically scales linearly with the perfusion for low and moderate tissue hematocrits, as defined in equation (3).

To compensate for the non-linear response of laser Doppler flowmeters occurring at a high tissue hematocrit, which implies multiple scattering in more than one moving scatterer, a linearizer has been added to the basic signal processor (Nilsson, 1984). The overall flowmeter output signal can then be expressed as:

$$V_{out} = f(c) \int_{\omega_1}^{\omega_2} \omega P(\omega) d\omega \qquad (4)$$

where P(ω) is the power spectral density of the Doppler signal, ω is the frequency, ω_1 and ω_2 are the lower and upper frequency limits of the processor bandwidth and f(c) is a factor derived from the original Doppler signal, which compensates for nonlinearities due to a high tissue hematocrit.

Since perfusion especially in the skin is dependent on temperature, a heater 30 mm in diameter that brings the skin surface temperature to a preset and regulated value is incorporated in some laser Doppler flowmeters. During measurements this heater is attached to the skin by double adhesive tape. The probe is positioned in the center of the heater.

The sensitivity of the instrument is calibrated by placing the probe in a solution of microspheres undergoing well-defined Brownian motion prior to measurement. The zero blood-flow value is attained by placing the probe in a probeholder attached to a piece of semitransparent plastic material, from which the backscattered light has undergone no Doppler shifts.

During measurement of blood flow the probe should be placed in a probeholder attached to the skin by a double adhesive tape, to avoid relative movements between the probe and the tissue.

EVALUATION

The choice of an ω-weighting filter as the core of the signal processor is based on mathematical models of dynamic light scattering in tissue (Bonner and Nossal, 1981). To keep the complexity of these models within reasonable limits, a number of assumptions must be made. In fact, far from all the phenomena involved in the scattering process can be accounted for. To verify the accuracy of these models, flow simulators have been designed to experimentally ascertain the relationship between flow and laser Doppler flowmeter output signal. Aspects of dynamic light scattering in tissue that are not included in these models can be elucidated by the use of *in vivo* experiments.

In the flow simulator (Fig. 3) red blood cells passed through a number of 300 μm fine channels in a semitransparent polyacetal disc, which served as a static scatterer and thus randomized the propagation vectors k_i of the photons (Nilsson et. al., 1980). The combination of moving and static scatterers yielded diffusive scattering similar to that in real tissue. The center of the disc was illuminated by the laser light, and the surrounding receiving fibers received a fraction of the scattered and Doppler-broadened light. The average velocity of the blood cells was controlled by a syringe pump. Different hematocrits were produced by diluting the blood cells in saline. Within the velocity range (0 - 15 mm/sec) and hematocrit range (0 - 0.6 %), a linear flowmeter response was verified. This linearity persisted regardless of whether the blood cell velocity or the hematocrit was changed. In a successive study (Nilsson, 1984) in which a linearizer was included in the signal processor, a linear relationship was established up to hematocrits of 1.0%, thus covering the entire hematocrit range that can be expected in real tissue.

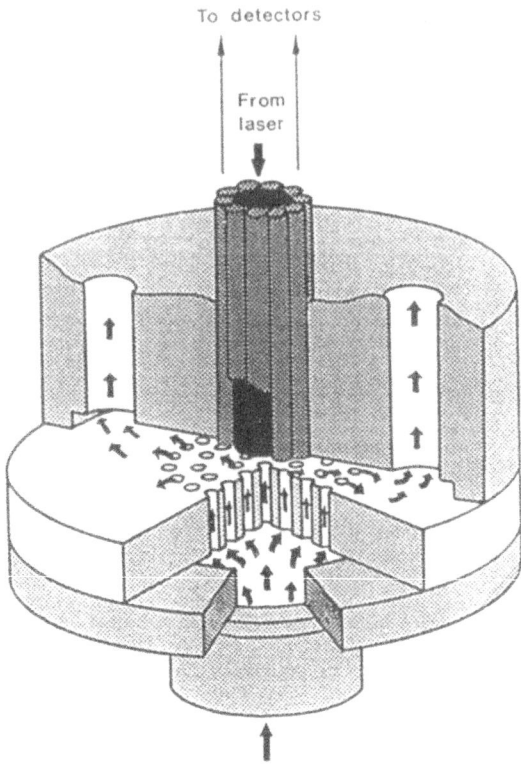

To detectors

From laser

From syringe pump

Fig. 3.   The flow simulator (Nilsson, G.E., Tenland, T. and Öberg, P.Å. (1980)
Evaluation of a laser Doppler flowmeter for measurement of tissue blood flow.
IEEE Trans, BME-27, 597-604, copyright 1980 IEEE).

The results of these experiments agree in general with the predictions of the
mathematical models. A flow simulator, however, does not take into account the fact
that in real tissue the blood cells are confined within vessel walls, thereby violating the
assumption of a homogeneous distribution of moving scatterers. Therefore, various
investigations with animal models have been performed, most often comparing the laser
Doppler output signal with another method for simultaneously measuring perfusion in
the same tissue.

The feline intestine has proven to be a suitable animal model for the study of
performance of laser Doppler flowmeters. It has a uniform and easily controlled blood
perfusion. This model is particularly useful because a specific segment of the intestine
is drained through a single vein, the blood flow through which can be accurately
determined by a drop counting technique (Ahn et. al., 1985). A linear relationship
between flowmeter output signal and intestinal blood flow has been established for
perfusion values ranging from zero to over 300 ml min^{-1} 100g^{-1} (Ahn et. al., 1987).
However, a calibration in absolute flow units is valid only for the conditions under
which it has been obtained, because optical properties may vary from one tissue to
another.

The results with flow simulators and *in vivo* models generally confirm the
mathematical model for dynamic light scattering in tissue proposed by Bonner and

Nossal. Multiple Doppler shifts produced by several sequential scattering events inside a vessel do not seem to violate the linear relationship predicted by theory. Neither do internal tissue movements seem to contribute substantially to the output signal, except for situations where measurements are made on motile tissues.

CLINICAL APPLICATIONS OF LDF

Since many diseases, drugs and medical proceduresinfluence tissue perfusion, it is not surprising that LDF has found applications in a wide variety of experimental and clinical disciplines. In this summary only a few of the more primary fields of application will be reviewed. A more detailed guide to the titles of full papers published on this topic will be found in Perimed Literature Reference List, No. 9 (Perimed, 1989).

In dermatology, LDF was early recognized as a useful tool in the objective assessment of the degree of skin irritability in patch testing procedures (Nilsson et. al., 1982). The irritant is normally applied to the volar forearm in different concentrations under occlusion for 24h. The blood flow is then recorded at three different times (26, 48 and 72h) following application. The degree of skin irritation is interpreted in terms of elevation in blood flow above the resting flow condition recorded on normal skin.

In pharmacology, LDF has been successfully used in recording the microvascular effects of vasoactive substances and drugs. The maximal response generally appears after a specific time delay following the application of a substance or intake of a drug. The blood flow should therefore be recorded at regular intervals or, even better, continuously during the estimated time of the vascular response. It is generally expressed in percentage changes from a resting value obtained prior to intake of the drug. It must be emphasized, however, that such percentage changes are very much dependent on the resting values obtained and may give ambiguous results if the microvascular network of the skin is constricted or dilated when the control values are recorded.

In plastic surgery, postoperative changes in the blood flow of a graft have proven to be a parameter that reacts before clinical signs of failure appear (Svensson and Svedman, 1982, Jones and Mayou, 1982, Jenkins et. al.,1988). Postoperative LDF monitoring may therefore constitute an early warning technique that automatically warns the nursing staff when the blood flow of the graft falls below a set level. A special version of a laser Doppler flowmeter (Periflux PF3[R], Perimed AB, Stockholm), which incorporates a level-triggered alarm, has been exclusively designed for this and similar applications.

LDF in combination with endoscopy allows the investigation of intestinal and gastric blood flow in conscious patients (Kvernebo et. al., 1986). In this application the fibre lines are introduced through the biopsy channel of the endoscope. The exact position of the probe tip facing the mucosa may be controlled through the viewing channel of the endoscope. The results of studies of gastric mucosal blood flow indicate that its reduction may be an important factor in the development of gastric ulcers (Lunde and Kvernebo, 1988). The extent of this blood flow impairment in patients with chronic radiation injury may be determined by LDF prior to surgery (Johansson, 1988).

The use of LDF in assessing myocardial perfusion in conjunction with cardiac surgery is a relatively recent field of application (Ahn et. al., 1988a and 1988b). A series of studies has shown that myocardial blood flow may be recorded with LDF in the arrested heart, while recordings obtained from the beating heart are difficult to interpret due to the influence of Doppler signals generated by the moving tissue.

Myocardial perfusion in the anteroapical part of the left ventricle in the arrested heart has been monitored intraoperatively with LDF after grafting the internal mammary artery to the left anterior descending artery (Ahn et. al., 1988c). The blood flow in the heart muscle after declamping only the mammary artery averaged about 60% of the perfusion found after declamping the aorta also, and thereby the coronary vessels. In one out of the ten patients investigated, no flow could be recorded in the heart muscle after declamping the graft alone, indicating a poor supplying function of the vessel.

Moreover, LDF has proven to be of value in other applications, including intraoperative measurement of cerebral and tumor blood flow (Arbit et. al., 1989, Fasano et. al., 1988) as well as bone blood flow (Schlehr et. al., 1987). Further applications include assessment of burn wound depth (O'Reilly et. al., 1989), Raynaud's phenomenon (Wollerheim et. al.,1988, Bunker et. al., 1988, Ekenvall et. al., 1988) and changes in microvascular blood flow in association with diabetes (Tooke et. al., 1985, Rayman et. al., 1986 and Newrick et. al., 1988).

Now, after more than 10 years of practical experience in recording tissue perfusion with laser Doppler flowmetry, the main advantages and drawbacks of the technique may be listed as follows:

Advantages

- Tissue perfusion can be recorded continuously and *non-invasively* without any discomfort to the patient.

- LDF directly senses both the velocity and concentration of the blood cells, quantities which make up the perfusion value.

- Very small probes based on thin and flexible optical fibres facilitate the recording of blood flow in organs difficult to access by other methods.

- The laser Doppler flowmeter is an easy instrument to use, although the results may sometimes be difficult to interpret.

Drawbacks

- Perfusion measurements are made within a limited tissue volume (about 1 mm^2).

- Fibre line movement artefacts may obscure the results.

- Perfusion recorded in motile tissues is difficult to interpret due to tissue movement artefacts.

- LDF output signal is presented in relative rather than absolute units.

FUTURE OUTLOOK

LDF in its present stage constitutes a versatile technique for continuous and *non-invasive* monitoring of blood flow in different kinds of tissue. Due to the short migration distance of photons in tissue, the method essentially records blood flow at a single point. The pronounced spatial variations of skin blood flow, for example, mean that moving the probe just a few millimeters laterally may give a totally different value of the recorded perfusion. Therefore, in many applications tissue perfusion should preferably be assessed by an imaging technique rather than a one-point method.

To attain this goal of creating an image, a tissue perfusion imager based on scattering of coherent light in tissue has recently been developed (Nilsson et. al., 1989). Light from a 2 mW He-Ne laser sequentially scans 4,096 measurement points over the tissue under study, covering a surface area of 12 * 12 cm (Fig. 4).

At each measurement site, the incident monochromatic light beam penetrates the tissue to a depth of a few hundred microns. In the presence of moving blood cells, the light is Doppler-broadened and partially backscattered. A fraction of the backscattered and Doppler-broadened light is received by a photodetector positioned about 20 cm over the tissue surface. The signal from the photodetector is processed similarly to ordinary LDF equipment and successively fed into a computer also controlling the position of the laser beam. After further processing, an image of tissue perfusion composed of 4,096 pixels, corresponding to a resolution of approximately 4 mm^2, may be generated.

In a preliminary evaluation of the perfusion imager, the arrested flow distal to a finger occlusion was easily revealed, as was the extension of hyperaemia surrounding an experimental burn. In a clinical evaluation study, the influence of argon laser treatment on blood flow in hemangioma was successfully investigated. Images were recorded prior to and immediately after treatment as well as after 24 and 48h.

This new approach to perfusion measurement seems promising, not only because it produces an image of tissue perfusion, but also because it eliminates the need for fibre lines and allows the recording of tissue blood flow without touching the object. The latter characteristic implies elimination of sterilization problems, and should facilitate clinical investigation of perfusion in diagnosis and treatment of peripheral vascular disease.

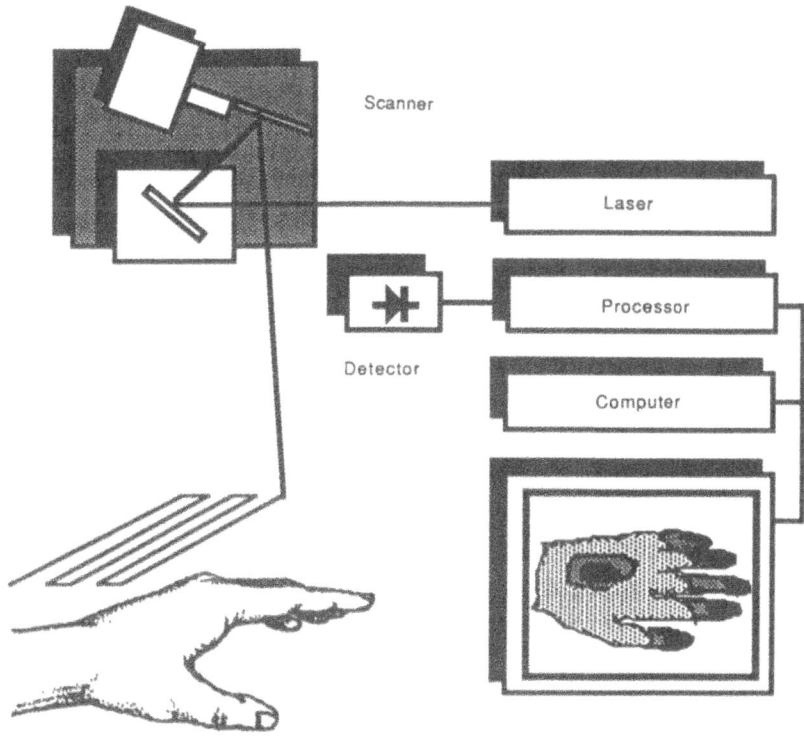

Fig. 4. A block diagram of the Perfusion Imager.

REFERENCES

Challoner, A.V.J., 1979, Photoelectric plethysmography for estimating cutaneous blood flow, in: "Non-Invasive Physiological Measurements", Vol 1., P. Rolfe, ed., Academic Press.

Tremper, K.K. and Barker, S.J., Pulse Oximetry, 1989, Anesthesiology, 70:98.

Holloway, G.A. and Watkins, D.W., 1977, Laser Doppler measurement of cutaneous blood flow, J.Invest Dermatol, 69:306.

Nilsson, G.E., Tenland, T. and Öberg, P.Å., A new instrument for continuous measurement of tissue blood flow by light beating spectroscopy, 1980a, IEEE Trans., BME-27:12.

Nilsson, G.E., Tenland, T. and Öberg, P.Å., 1980b, Evaluation of a laser Doppler flowmeter for measurement of tissue blood flow, IEEE Trans., BME-27:597.

Stern, M.D., In vivo evaluation of microcirculation by coherent light scattering, 1975, Nature, 254:56.

Van Gemert, M.J.C., Jacques, S.L., Sterenborg, H.J.C.M. and Star, W.M., 1989, Skin optics, IEEE Trans., BME-36:1146.

Jakobsson, A. and Nilsson, G.E., 1990, A Monte Carlo model for simulation of light scattering in laser Doppler flowmetry, XVI European Conference on Microcirculation, Zurich, August 26-31.

Steinke, J.M. and Shepherd, A.P., 1988, Comparison of Mie theory and the light scattering of red blood cells, Appl. Optics, 27:4027.

Nilsson, G.E., 1984, Signal processor for laser Doppler tissue flowmeters, Med. & Biol. Engng. & Comput., 22:343.

Ahn, H.C., 1986, Measurement of gastrointestinal blood flow with laser Doppler flowmetry, Linköping University Medical Dissertation No. 226, Linköping, Sweden.

Salerud, E.G and Nilsson, G.E., 1986, Integrating probe for tissue laser Doppler flowmeters, Med. & Biol. Engng. & Comput., 24:415.

Salerud, E.G. and Öberg, P.Å., 1987, Single-fibre laser Doppler flowmetry: A method for deep tissue perfusion studies, Med. & Biol. Engng. & Comput., 25:329.

Kvernebo, K. and Salerud, E.G., 1987, Single fibre laser Doppler flowmetry in the evaluation of human muscle blood flow, Fourth World Congress for Microcirculation, Tokyo, Japan, July 26-30.

Nilsson, G.E., 1989, Perimed's laser Doppler blood flowmeter, In : "Laser Doppler Blood Flowmetry", P. Shepherd and P.Å. Öberg, eds., Kluwer Academic Publishers (in press).

de Mul, F.F.M., van Spijker, J., van der Plas, D., Greve, J., Aarnoudse, J.G. and Smits, T.M., 1984, Mini laser Doppler (blood) flow monitor with diode laser source and detection integrated in the probe, Applied Optics, 23:2970.

Bonner, R. and Nossal, R., 1981, Model for laser Doppler measurements of blood flow in tissue, Appl. Optics, 20:2097.

Ahn, H., Lindhagen, J., Nilsson, G.E., Salerud, E.G., Jodal, M. and Lundgren, O., 1985, Evaluation of laser Doppler flowmetry in the assessment of intestinal blood flow in cat, Gastroenterology, 88:951.

Ahn. H., Johansson, K., Lundgren, O. and Nilsson, G.E., 1987, In vivo evaluation of signal processors for laser Doppler tissue flowmeters, Med. & Biol. Engng. & Comput., 25:207.

Perimed Literature Reference List No 9. Perimed AB, Stockholm, Sweden, 1989.

Nilsson, G.E., Otto, U. and Wahlberg, J.E., 1982, Assessment of skin irritancy in man by laser Doppler flowmetry, Contact Dermatitis, 8:401.

Svensson, H. and Svedman, P., 1982, On postoperative monitoring of the circulation in flaps, Int. J. Microcirc.: Clin.& Exp., 1:326.

Jones, B. and Mayou, B.J., 1982, The laser Doppler flowmeter for microvascular monitoring: a preliminary report, Brit. J. Plast. Surg., 35:147.

Jenkins, S., Sepka, R. and Barwick, W.J., 1988, Routine use of laser Doppler flowmetry for monitoring autologous tissue transplants, Ann. Plast. Surg., 21(5):423.

Kvernebo, K., Lunde O.C., Stranden E. and Larsen, S., 1986, Human gastric blood circulation evaluated by endoscopic laser Doppler flowmetry, Scand. J. Gastroenterol., 21:6851.

Lunde, O.C. and Kvernebo, K., 1988, Gastric blood flow in patients with gastric ulcer measured by endoscopic laser Doppler flowmetry, Scand. J. Gastroenterol., 23:546.

Johansson, K., 1988, Gastrointestinal application of laser Doppler flowmetry, Thesis No.269, Linköping University.

Ahn, H.C., Ekroth, R., Nilsson, G.E. and Svedjeholm, R., 1988a, Assessment of myocardial perfusion with laser Doppler flowmetry, Scand. J. Thor. Cardiovascular. Surg., 22:45.

Ahn, H., Ekroth, R., Hedemark, J., Nilsson, G.E. and Svedjeholm, R., 1988b, Assessment of myocardial perfusion in the empty beating porcine heart with laser Doppler flowmetry, Cardiovascular Research, 22:719.

Ahn, H.C., Ekroth, R., Nilsson, G.E., Svedjeholm, R. and Thelin, S., 1988c, Laser Doppler flowmetry estimating myocardial perfusion after internal mammary grafting, Scand. J. Thor. Cardiovasc. Surg., 22:281.

Arbit, E., DiResta, G.R., Bedford, R.F., Shah, N.K. and Galicich, J.H., 1989. Intraoperative measurement of cerebral and tumor blood flow with laser-Doppler flowmetry, Neurosurgery, 24:166.

Fasano, V.A., Urciuoli, R., Bolognese, P. and Mostert, M., 1988, Intraoperative use of laser Doppler in the study of cerebral microvascular circulation, Acta Neurochir. (Wien), 95:40.

Schlehr, F.J., Limbird, T.A., Swiontkowski, M.F. and Keller, T.S., 1987, The use of laser Doppler flowmetry to evaluate anterior cruciate blood flow, J. Orthop. Res., 5:150.

O'Reilly, Spence R.J., Taylor, R.M. and Scheulen, J.J., 1989, Laser Doppler flowmetry evaluation of burn wound depth, J. Burn. Care. Rehabil., 10:1.

Wollerheim, H., Reyenga, J. and Thien, T., 1988, Laser Doppler velocimetry of fingertips during heat provocation in normals and in patients with Raynaud's phenomenon, Scand. J. Clin. Lab. Invest., 48:91.

Bunker, C.B., Lanigan, S., Rustin, M.H. and Dowd, P.M., 1988, The effects of topically applied hexil nicotinate lotion on the cutaneous blood flow in patients with Raynaud's phenomenon, Br. J. Dermatol., 119:771.

Ekenwall, L., Lindblad, L.E., Carlsson, A. and Etzell, B.M., 1988, Afferent and efferent nerve injury in vibration white fingers, J. Auton. Nerve. Syst., 24:261.

Tooke, J.E., Lins, P.E., Östergren, J., Adamson, U. and Fagrell, B., 1985, The effects of intravenous insulin infusion on skin microcirculatory flow in type 1 diabetes, Int. J. Microcirc.: Clin. Exp. 4:69.

Rayman, G., Hassan, A. and Tooke, J.E., 1986, Blood flow in the skin of the foot related to posture in diabetes mellitus, Brit. Med. J., 292:87.

Newrick, P.G., Cochrane, T., Betts, R.P., Ward, J.D. and Boulton, A.J., 1988, Reduced hyperaemic response under the diabetic neuropathic foot, Diabetic Med., 5:570.

Nilsson, G.E., Jakobsson, A. and Wårdell, K., 1989, Imaging of tissue blood flow by coherent light scattering, IEEE 11th Ann EMBS Conference Proc., Seattle, Nov 9-12.

TRANSILLUMINATION IMAGING

P.C. Jackson, H. Key, and P.N.T. Wells

Department of Medical Physics
Bristol General Hospital
Guinea Street
Bristol BS1 6SY
United Kingdom

Introduction

The use of transillumination imaging as a non-invasive tumour detection technique of the breast was first described by Cutler (1929). In this early work an electric torch was held against the underside of the breast and the observer viewed the regionally attenuated light from the opposing surface of the breast. The transmitted intensities of light were low since the limiting factor for the light output used was the necessity to avoid burning the patient's skin or causing discomfort. This resulted in the examination being performed in a darkened room and the observer having to undergo dark adaptation. The technique, however, demonstrated that blood vessels and tumours would preferentially absorb more light than surrounding tissue and project 'dark areas'. Conversely, cystic structures appeared to transmit light more readily than normal tissue and resulted in 'light areas' being visualised. These observations were the basis for all subsequent work and mainly applied to investigation of the breast. Although Cutler demonstrated the potential for breast transillumination as a diagnostic technique it was not until Gros et al (1972) developed a water cooled lamp with a filter to remove wavelengths less than 0.6 µm, that high intensity light transmissions were possible which enabled photographs to be satisfactorily recorded.

Transillumination imaging has also been used to examine inflammatory diseases of the sinuses and differentiation between solid tumours and hydrocele of the testis, but most subsequent developments and evaluations during the 1980s have been directed towards the breast (Ohlsson et al 1980; Carlsen 1982; Watmough 1982, 1983; McIntosh 1983; Bartrum & Crow 1984; Marshall et al 1984; Merritt et al 1984, Drexler et al 1985; Geslien et al 1985), as an aid to diagnosis or screening. However, following from the early work of Bright (1831) the technique of transillumination imaging has been used also to investigate more quantitatively pathological

Optronic Techniques in Diagnostic and Therapeutic Medicine
Edited by R. Pratesi, Plenum Press, New York, 1991

101

conditions associated with the infant skull (Martin et al 1977; Donn et al 1975; Johns 1979; Arridge et al 1985) and organs of the abdomen (Martin et al 1977).

Much of the equipment used for breast examination has been based on a technique called' telediaphanography' and this remains the most commonly employed technique. Images are viewed in real time using a infra-red sensitive silicon vidicon or newvicon camera. The sequence of images can be recorded on video tape for further reference or the video signal digitised for appropriate digital image processing. The method of transillumination imaging has relied upon an intense light source transilluminating an organ of interest and forming 'shadows' visualised by either direct viewing, film or low light level video camera. The image formed represents the absorption of light by different tissues of the body and the mechanisms involved, in relation to wavelength dependence, have received consideration (Cartwright, 1930; Hardy & Muschenheim, 1936; Derksen & Monahan, 1952; Clark et al, 1953; Hardy et al 1956; Watmough, 1982). It has been suggested that the increased vascularity in a breast tumour together with the relatively high absorption due to the presence of oxyhaemoglobin is a primary factor in the detection of tumours (Watmough, 1982).

Physical basis of transillumination imaging

Light transmission imaging is possible because there exists a 'window' in the absorption spectrum of biological tissues from about 0.5 to 1.4 μm. This transillumination window includes green, yellow and red light and extends into the near infra-red region of the electromagnetic spectrum with the boundary between visible and infra-red regions normally taken as 0.75 μm. The absorption is ascribed either to low energy electronic or high energy vibrational excitations. The short wavelength limit to optical transmission occurs when electronic absorption in body pigments such as haemoglobin, myoglobin and melanin becomes very intense. The upper limit to light transmission occurs at wavelengths longer than 1.4 μm because of the strong vibrational absorption in water molecules.

Especially important for transillumination imaging is the absorption of the haeme proteins, which owe their red colouration to the absorption bands in the region of 500-600 nm. The visible and near infra-red absorption spectra of the blood pigment haemoglobin are illustrated (Fig 1) both in oxygenated and deoxygenated forms, showing maxima at 542 nm and 577 nm when oxygenated and 555 nm when deoxygenated. In relation to differentiation of vascular and cystic structures infra-red absorption bands are associated with transitions between molecular vibrational modes. The fundamental vibrational resonant frequencies are centred in the intermediate infra-red spectrum at 2.92 μm and 6.11 μm, but progressively weaker overtones and combinations fall into the near infra-red spectrum at 1.46, 1.20, 0.97 and 0.76 μm (Fig 2).

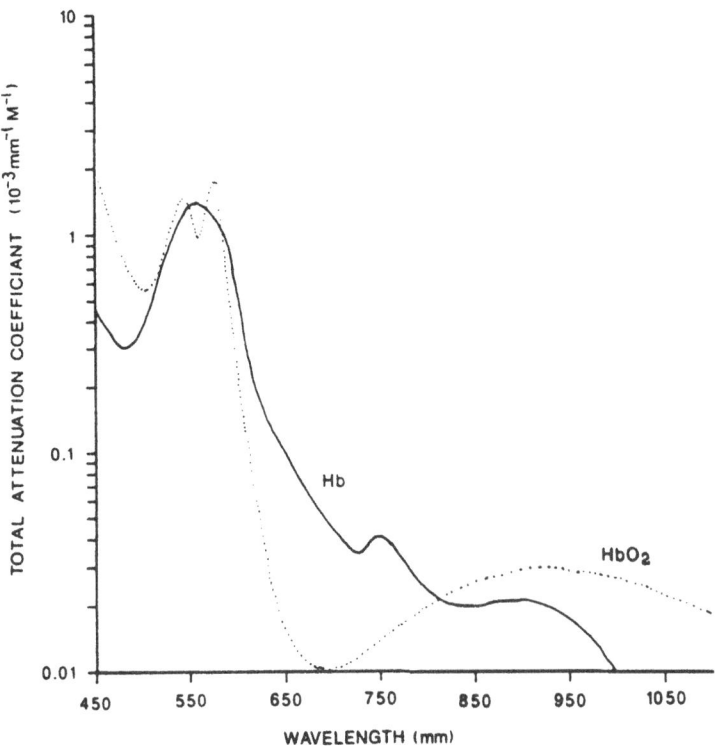

Fig. 1.   Absorption spectra of haemoglobin (Hb) and oxyhaemoglobin (HbO_2).

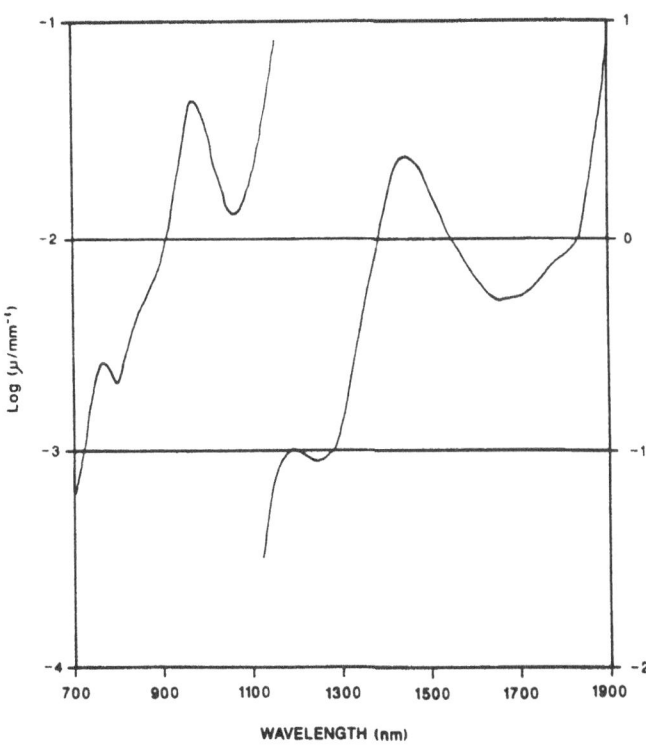

Fig. 2.   Near infra-red absorption spectrum for pure water where (μ) is the total attenuation coefficient.

Within the transillumination window, however, absorption processes are greatly exceeded by scattering. The scattering arises at inhomogeneities in the refractive index of the medium and dominates light propagation in tissue because there are many inhomogeneities of cellular structures and particles that are of the order of an optical wavelength. Hence, there appear to be fundamental attenuation problems in the choice of suitable wavelengths for transmission through tissue, which because of the structure of tissue give rise to scattering of the transmitted light. For a successful transmission imaging technique the requirements are for significant differences in the absorption characteristics of different biological tissues with minimal spatial scattering of the constituent tissues.

There has been a sparsity of good published work in relation to the quantitative interaction characteristics of light with various tissues. Investigators have measured the effective attenuation coefficient (μ_{eff}) in tissues experimentally by interstitial measurement either in vivo or post mortem, or from thick section photometry (Table 1). The vital consideration as to whether optical characteristics measured in vitro or in vivo are different has been studied (Wilson et al 1985). There appear to be significant post mortem changes in μ_{eff} at wavelengths less than 600 nm which may be ascribed to blood loss and deoxygenation during sample preparation. However, differences of μ_{eff} at higher wavelengths appears insignificant. Few investigators have measured or deduced the fundamental scattering and absorption coefficients which would aid the development of transillumination imaging devices (Table 2). However, recent work has been undertaken (Key, 1989) which has provided further measurements related to breast tissue in order to assess the potential for developing new imaging techniques by transillumination.

Current clinical status of telediaphanography

Investigators have assessed the effectiveness of telediaphanography in detecting breast cancer by comparison with mammography. The studies have been useful but no trial has been adequately designed to quantify accurately the diagnostic capability of the method. The patients selected for telediaphanography have often been taken from small, biased and unrepresentive cohorts. The trials have often consisted of examining women in whom the likelihood of disease is great and few 'blind' evaluations have been conducted.

Many of the studies have been performed in 1980s and the findings of some of the important evaluations undertaken are summarised in Table 3. The early results of Carlsen (1982) have not been reproduced in subsequent studies and mammography achieves substantially better performance levels, especially in the detection of small, non-palpable tumours. With current and commercially available equipment it is unlikely that telediaphanography would even provide a useful adjunct to mammography

Table 1. Measurements of the effective attenuation coefficient (μ_{eff}) of light at different wavelengths in tissues from animal and human models.

Reference	Tissue type	Wavelength (nm)	μ_{eff} (mm^{-1})	Technique
Doiron, et al (1983)	Brain (cat)	630	0.44 - 0.98	Interstitial, in-vivo
	Muscle (cat)	630	0.16 - 0.23	
Eichler, et al (1977)	Liver (human)	600	0.42	Thick section photometry
		1000	0.55	
	Kidney (human)	600	1.22	
Strax (1980)	Brain (adult human)	660	0.6 - 0.8	Interstitial, post mortem
		1060	0.15 - 0.34	
	Tumours (human brain)	660	0.33 - 0.40	
		1060	0.15 - 0.33	
Wilson et al (1985)	Muscle (rabbit)	630	0.27 - 0.34	interstitial — post mortem
			0.26 - 0.48	in vivo
	Brain (pig)		0.43 - 0.56	post mortem
			0.37 - 0.45	in vivo
Wilson & Muller (1986)	Brain (human)	630	0.59 - 1.0	Measured during PDT, in vivo
Wilson & Patterson (1986)	Tumours (human brain)	630	0.22 - 0.67	Measured during PDT, in vivo

Table 2. Measurements of the fundamental attenuation characteristics of light in tissues where the following notation is used: Wavelength (λ), linear absorption coefficient (μ_a) reduced scattering coefficient (μ_s*) and mean cosine of scatter (g)

Reference	Tissue Type	λ(nm)	μ (mm^{-1})	μ_a (mm^{-1})	μ_s* (mm^{-1})	g	Notes
Navarro & Profio (1988)	Breast (human)	695	–	0.075	0.29	–	Deduced using diffusion theory
		950	–	0.067	0.26	–	
	Carcinoma (human breast)	695	–	0.12	0.33	–	
		950	–	0.12	0.30	–	
Key (1989)	Fibroglandular (human breast)		15	–	1.2	0.92	
	Fat (human breast)	700	25	–	1.3	0.95	
	Carcinoma (human breast)	similar to fibro-glandular tissue			1.8	0.88	

106

Table 3. Summary of results from prospective blind evaluation of telediaphanography and X-ray Mammography

Reference	Year performed	Size of cohort	Prevalence of cancer (%)	SENSITIVITY (%)		SPECIFICITY (%)	
				light scan	mammography	light scan	mammography
Carlsen	1982	3293	2.4	92	82	99	-
Bartrum & Crow	1984	1200	2.8	74 (40)	94 (90)	83	93
Marshall et al	1984	1000	3.4	77	85	96	98
Sickles	1984	1239	7.5	53 (19)	96 (94)	-	-
Drexler, et al	1985	1476	1.6	58 (40)	96 (100)	-	-
Geslien, et al	1985	1265	2.6	58 (30)	97 (100)	-	-
Dowle, et al	1987	285	14	87	83	-	78
Monsees, et al	1987	1110	2.2	58	88	93	96
Monsees. et al	1988	88	26	(30)	-	(88)	-

(Figures in brackets refer to detection of non-palpable breast carcinoma)

107

in a screening programme since it has been shown to be ineffective in reducing the number of negative biopsies subsequent to suspicious images obtained from mammography (Monsees et al, 1987). The detection of early breast cancer in women under 50 years of age might prove to be the most valuable role for an improved technique of transillumination imaging since it has been noted that the sensitivity of telediaphanography appears independent of the radiological density of the breast whilst mammography is found to be less effective in the dense breast of younger women (Bundred et al, 1987).

The ability of current telediaphanography techniques to assist in a screening programme or as an adjunct to mammography has been hindered by practical considerations. These factors include the need to perform the examination in a darkened room, to make multiple images of each breast, to tolerate the relatively lengthy investigation time and the inseparable role of operator and image interpreter. However, transillumination imaging is safe, non-invasive, acceptable to patients and can detect some early stage carcinomas. Various research programmes have been undertaken to explore further the potential of transillumination imaging.

Developments in transillumination imaging

In an attempt to improve the diagnostic performance of transillumination imaging a system for achieving, transillumination computed tomography was constructed (Jackson et al 1987). The system consisted of a tungsten halogen lamp (150 W) delivering light via glass fibre optic bundles and a 3 mm aperture for illuminating small volumes of the object. The transmitted light was collected by an opposing collimator and fibre optic system, and detected by a photomultiplier tube with a S20 photocathode. The light transmission and receiving modules moved in a translate-rotate sequence similar to first generation x-ray CT scanners (Fig 3,4). The system was evaluated using paraffin wax test objects with embedded "lesions" in an attempt to simulate the reflective, absorptive and transmission properties of tissue (Fig 5). The prototype scanner was successful in producing images through the simple test objects although opaque objects embedded in the wax were poorly resolved because of the extensive light scattering. Also the images contained artefacts ascribed to the reflection of light at the test object surface (Fig 6). The results indicated the large dynamic range of data to be digitised, the excessive scattering of light and the need to couple the light source to the object in order to minimise specular reflection at the surface.

Another development of current interest is the combination of telediaphanography with Doppler ultrasound detection of blood flow. Doppler ultrasound is successful in detecting early breast cancer but its practical application in screening for pre-clinical disease is severely impaired by the need to examine every volume element of the breast.

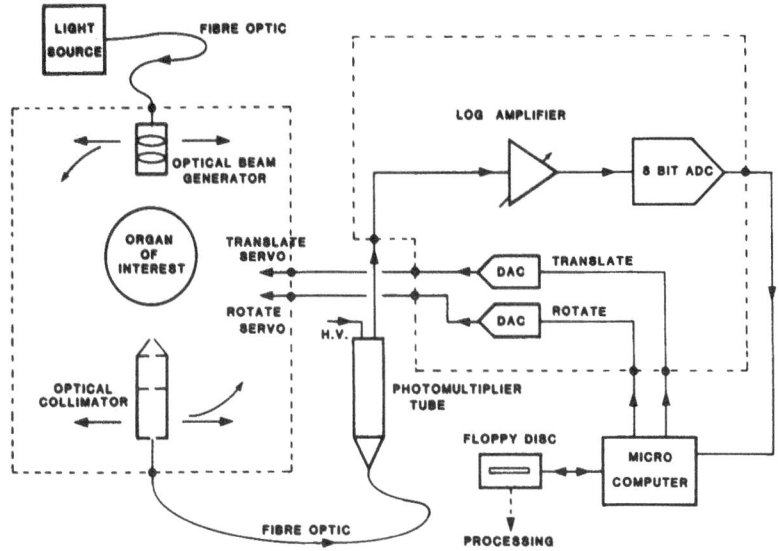

Fig. 3.   A schematic diagram of equipment capable of performing
transillumination computed tomography

Fig. 4.   A system designed to perform transillumination computed tomography
with test objects in situ.

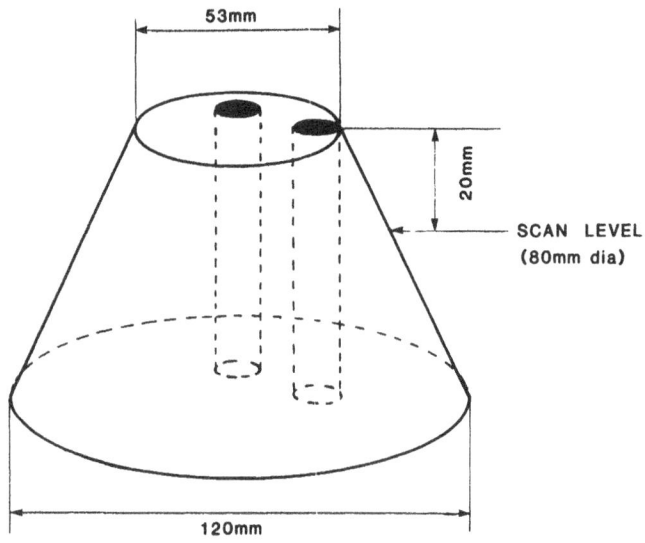

Fig. 5. A schematic diagram of a test object used to assess
transillumination computed tomography containing two 'lesions'.

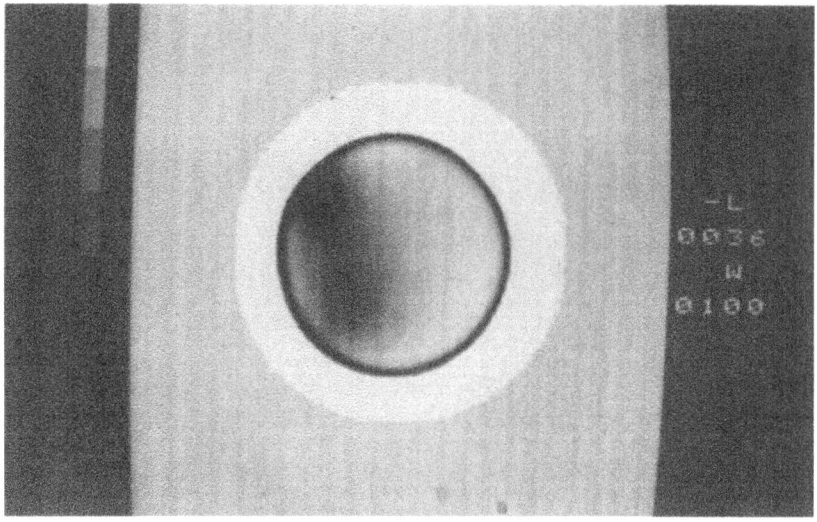

Fig. 6. A computed tomographic image of a test object containing two
lesions.

Watmough et al.,(1988) have assessed the effectiveness of combined instrumentation in which telediaphanography is used to identify suspicious regions of tissue which are subsequently examined by a Doppler ultrasound probe.

Another simple development for improving the accuracy and effectiveness of telediaphanography would be to compress the breast between parallel transparent plates (Key et al., 1988; Watmough et al., 1988). With compression the depth of lesions beneath the skin would be reduced together with the scattering path length of photons, which should increase object detectability. In addition, a more uniform tissue thickness would be obtained than when the breast is compressed by hand, hence reducing one source of artefact. However, imaging of breast tissue close to the chest wall would remain a problem together with possible discomfort of the patient.

Recent work by Key (1989) has examined the optical properties of the fundamental attenuation characteristics of breast specimens at visible and near infra-red wavelengths. Within the transmission window for tissue of 0.6 to 1.5 μm the measurements demonstrated a scattering coefficient greatly exceeding the absorption coefficient (Table 2) with a scattering phase function that was highly forward peaked. In addition Key (1989) has noted differences in the absorption spectra of carcinoma and uninvolved breast tissues at green wavelengths (Table 4) and in first differential values of absorption spectra at certain near infra-red wavelengths (Table 5).

Future developments for transillumination imaging

From the work of Key (1989) and the use of Monte-Carlo modelling techniques it appears that because of the wide spatial point spread funtions there is a clear advantage in acquiring an image on a 'point by point' basis with synchronous scanning of a small area source and light collector. This should enable the pixel value to relate more closely to the absorption on the projection axis than if a wide beam light source and video camera are used.

Time of flight gating could be a useful technique for improving spatial resolution of the imaging system. The Monte-Carlo modelling of Key (1989) has shown that in a 40 mm of breast tissue a 400 ps time gate triggered on the leading edge of an incident pulse might reduce the mean value of the point spread function from 15 mm to about 11 mm. Such time of flight analysis might usefully be incorporated into a scanned projection system where the breast is compressed to minimise the scattering path length. This type of system would require sufficiently powerful sources and sensitive detectors at appropriate wavelengths. Laser diodes are a low cost light source that can be operated to produce pulses at approximately 10 W peak power. The lasers are, however,

TABLE 4

Values of the linear absorption coefficient
calculated at selected wavelengths

The values stated here were calculated from Monte Carlo relationships that assumed μ_s = 0.75 mm $^{-1}$. They are mean values (± σ) derived from six specimens: the units are mm $^{-1}$.

	Wavelength (λ)		
	0.58 μm	0.85 μm	1.30 μm
Carcinoma	0.45±0.08	0.04±0.05	0.05±0.08
Adjacent tissue	0.26±0.11	0.03±0.02	0.08±0.06
Paired t - test t	3.2	0.60	1.38
p	0.023	0.58	0.23

TABLE 5

The first differential of the transmitted power
spectrum calculated at selected wavelengths

The values stated here are of $(\Delta S/\Delta\lambda)/S_o$, where S_o was the signal at the central wavelength and S was measured over an increment, $\Delta\lambda$, of 20 nm. They are mean values (± σ) derived from six specimens: the units are nm^{-1}.

	Wavelength (λ)			
	1μm	1.18 μm	1.23 μm	1.30 μm
Carcinoma	7.0±1.3	0.4±1.4	5.0±0.7	-5.6±0.3
Adjacent tissue	0.7±1.7	-12.4±2.2	25.5±5.4	-0.1±1.1
Paired t - test t	12.7	12.1	8.5	13.5
p	<0.001	<0.001	<0.001	<0.001

the highest powers obtained between 0.82 and 0.91 μm, and at longer wavelengths centred about 1.3 μm and 1.55 μm. These may not be optimal wavelengths for diagnostic imaging.

Conclusions

Transillumination imaging in the form of telediaphanography has not yet provided an acceptable technique for detecting and evaluating breast disease. There is still a requirement to investigate more fully the optical properties of tissue in vivo in both diseased and normal state in the spectrum from 0.5 to 1.4 μm. However, as further progress is made in the development of electro-optical technology, it should be possible to design systems with improved geometry, transmitting pulsed light at optimal wavelengths and with sufficient power to improve on current transillumination imaging results.

References

Arridge, S.R., Cope, M., Van der Zee, P., Hillson, P.J., and Delpy, D.T., 1985, Visualisation of the oxygenation state of brain and muscle in newborn infants by near infra-red transillumination, in: "Information Processing in Medical Imaging," S.L. Bacherach, ed., Martinus Nijhoff, Dordrecht), pp 155 - 176.

Bartrum, R.J., and Crow, H.C., 1984, Transillumination light scanning to diagnose breast cancer: a feasibility study, American Journal of Roentgenology, 142:409.

Bright, R., 1831, "Reports of medical cases selected with a view of illustrating the symptoms and cure of diseases by a reference to morbid anatomy", vol 2, part 1, Longman, Rees, Orme, Brown, Green, Highley London, pp 431 - 435.

Bundred, N., Levack, P., Watmough, D.J., and Watmough, J.A., 1987, Preliminary results using computerised telediaphanography for investigating breast disease, British Journal of Hospital Medicine, 38:70.

Carlsen, E., 1982, Transillumination light scanning, Diagnostic Imaging, 3:28 and 60.

Cartwright, C.H., 1930, Infra-red transmission of the flesh, Journal Optical Society of America, 20:81.

Clark, C., Vinegar, R., and Hardy, J.D., 1953, Goniometric spectrometer for the measurement of diffuse reflectance and transmittance of skin in the infra-red spectral region, Journal of the Optical Society of America, 43:993.

Cutler, M., 1929, Transillumination as an aid in the diagnosis of breast lesions, Surgery, Gynecology and Obstetrics, 48:721.

Derksen, W.L., and Monahan, T.I, 1952, A reflectometer for measuring diffuse reflectance in the visible and infra-red regions, Journal of the Optical Society of America, 42:263.

Doiron, D.R., Svaasand, L.O., and Profio, A.E., 1983., Light dosimetry in tissue, in: "Porphyrin Photosensitization," Kessell D., Dougherty, T.J., eds., Plenum, New York, pp 63 - 76.

Donn, S.M., Sharp, M.J., Kuhns, L.R., Uy, J.O., Knake, J.E., and Duchinsky, B.J., 1979, Rapid detection of neonatal intracranial hemorrhage by transilluimination, Pediatrics, 64:843.

Dowle, C.S., Caseldine, J., Tew, J., Manhire, A.R., Roebuck, E.J., and Blamey, R.W., 1987, An evaluation of transmission spectroscopy (light scanning) in the diagnosis of symptomatic breast lesions, Clinical Radiology, 38:375.

Drexler, B., Davis, J.L., and Schofield, G., 1985, Diaphanography in the diagnosis of breast cancer, Radiology, 157:41.

Eichler, J., Knof, J., and Lenz, H., 1977, Measurements on the depth of penetration of light in tissue, Radiation and Environmental Biophysics, 14:235.

Geslien, G.E., Fisher, J.R., and Delaney, C., 1985, Transillumination in breast cancer detection: screening failures and potential, American Journal of Roentgenology, 144:619.

Gros, C.M., Queeneville Y., and Hummel Y., 1972, Diaphanologie mammaire. J. Radiol. Electrol., 53:297.

Hardy, J.D., Hammel, H.T., and Murgatroyd, D., 1956, Spectral transmittance and reflectance of excised human skin, Journal of Applied Physiology, 9:257.

Hardy, J.D., and Muschenheim, C., 1936, Radiation of heat from the human body, V. The transmission of infra-red radiation through skin, Journal of Clinical Investigation, 15:1.

Jackson, P.C., Stevens, P.H., Smith, J.H., Kear, D., Key, H., and Wells, P.N.T., 1987. The development of a system for transillumination computed tomography, British Journal of Radiology, 60:375.

Johns, M., 1979, Transillumination of the infant skull, Journal of Audiovisual Media in Medicine, 2:140.

Key, H., 1989, The modelling of light attenuation and transmission in biological tissues, PhD Thesis, University of Bristol UK.

Key, H., Jackson, P.C., and Wells, P.N.T., 1988, New approaches to transillumination imaging, Journal of Biomedical Engineering, 10:113.

Marshall, V., Williams, D.C., and Smith, K.D., 1984, Diaphanography as a means of detecting breast cancer, Radiology, 150:339.

Martin, A.J., Kuhns, L.R., Gutowski, D., and Poznanski, A.K., 1977, Production of permanent radiographic record of transillumination of the neonate, Radiology, 122:540.

McIntosh, M.F., 1983, Breast light scanning: a real time breast imaging modality. Journal de l' Association Canadienne des Radiologistes, 34:288.

Merritt, C.R.B., Sullivan, M.A., Segaloff, A., and McKinnon, W.P., 1984, Real time transillumination light scanning of the breast. Radiographics, 4:989.

Monsees, B., Destouet, J.M., and Gersell, D., 1987, Light scanning evaluation of non-palpable breast lesions, Radiology, 163:467.

Monsees, B., Destouet, J.M., and Gersell, D., 1988, Light scanning of non palpable breast lesions: re-evaluation, <u>Radiology</u>, 167:352.

Monsees, B., Destouet, J.M., and Totly, W.G., 1987, Light scanning versus mammography in breast cancer detection, <u>Radiology</u>, 163:463.

Ohlsson, B., Gunderson, J., and Nilsson, D.G., 1980, Diaphanography: a method for evaluation of the female breast, <u>World Journal of Surgery</u>, 4:701.

Sickles, E.A., 1983, Breast cancer detection with transillumination and mammography. <u>American Journal of Roentgenology</u>, 142:841.

Strax, P., 1980, An overview of breast cancer: defining the problem, <u>Medical Progress Technology</u>, 7:131.

Watmough, D.J., 1982, Diaphanography mechanism responsible for the images, <u>Acta Radiologica Oncology</u>, 21:11.

Watmough, D.J., 1983, Transillumination of breast tissues: factors governing optimal imaging of lesions, <u>Radiology</u>, 147:89.

Watmough, D.J., Moran, C., and Watmough, J.A., 1988, 'Son et lumiere': a new combined optical and Doppler ultrasound approach to the detection of breast cancer, <u>Journal of Biomedical Engineering</u>, 10:119.

Wilson, B.C., Jeeves, W.P., and Lowe, D.M., 1985, *In-vivo* and *post mortem* measurements of the attenuation spectra of light in mammalian tissue, <u>Photochemistry and Photobiology</u>, 42:153.

Wilson, B.C., and Muller, P.J., 1986, An update on the penetration depth of 630 nm light in normal and malignant human brain tissue *in-vivo*, <u>Physics Medicine and Biology</u>, 31:1294.

Wilson, B.C., and Patterson, M.S., 1986, The physics of photodynamic therapy, <u>Physics Medicine and Biology</u>, 31:327.

FLUORESCENCE TECHNIQUES IN DIAGNOSTIC MEDICINE WITH SPECIAL

CONSIDERATION OF OPTOELECTRONIC METHODS

Eberhard Unsöld

GSF - Zentrales Laserlaboratorium
D - 8042 Neuherberg / Munich, FRG

INTRODUCTION

Several substances have the following distinctive property: when irradiated with light (photons) or - in a broader sense - fed with energy, they emit radiation which usually has a longer wavelength than the excitation light or a lower photon energy than the excitation energy. This emission phenomenon is called luminescence. Depending on the method of feeding energy to the sample under investigation, the following types of luminescence can be distinguished:

- bio-	luminescence
- cathodo-	luminescence
- chemo-	luminescence
- electro-	luminescence
- photo-	luminescence
- radio-	luminescence
- thermo-	luminescence
- tribo-	luminescence

Although all these are of some importance in biomedical research or medical diagnosis, this article will only deal with photoluminescence, i.e. light-induced phenomena, and its use in biomedical analysis and diagnosis.

The light absorption and emission properties of a substance deliver analytical data primarily on the substance itself but also partly on its environment, e.g. the influences of the solvent, the pH-value, temperature, etc. . The absorption and emission characteristics can be summarized in a Jablonski diagram (Fig. 1).[1] This shows the simplified energy level diagram of a molecule and its several radiational and nonradiational transitions. Two types of luminescence can be distinguished, fluorescence deactivating singlet states and phosphorescence deactivating triplet states. Usually the temperature dependence of both phenomena allows for their discrimination (Fig. 2); phosphorescence is observed mainly at low temperatures, while fluorescence dominates at room or body temperature, the biomedically relevant temperature domain.

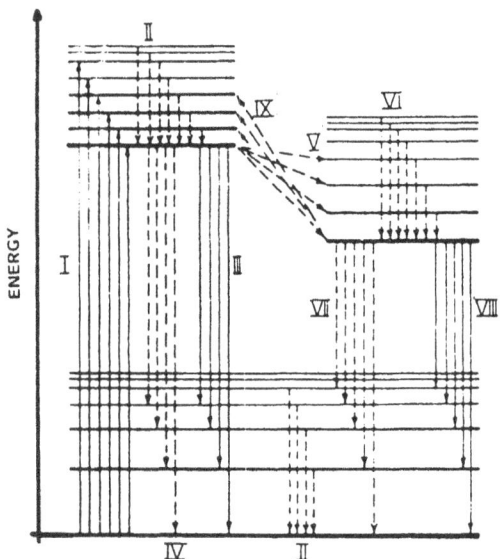

Figure 1. Energy level diagram of a typical organic molecule, including only ground singlet, first excited singlet (left terms), and its corresponding triplet state (right terms). Solid lines indicate radiational transitions, and dashed lines indicate nonradiational transitions[1].

Process I: <u>Absorption</u>, II: Vibrational deactivation, III: <u>Fluorescence</u>, IV: Quenching of excited singlet state, V: Intersystem crossing, VI: Vibrational deactivation, VII: Quenching of first triplet state, VIII: <u>Phosphorescence</u>, IX: Intersystem crossing to excited singlet state.

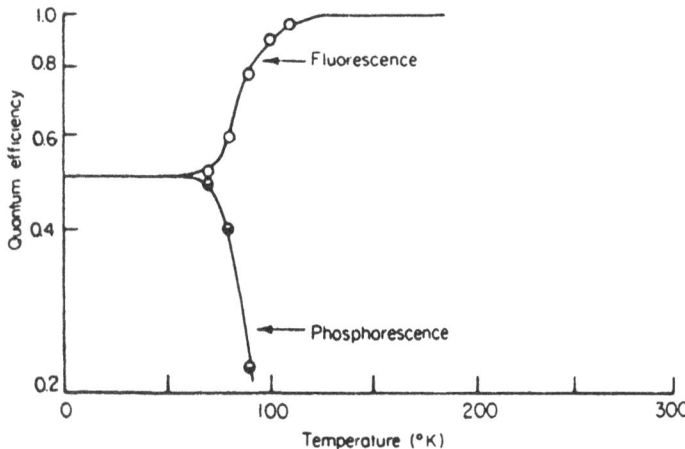

Figure 2. Comparison of variation in quantum efficiencies (= number of photons absorbed/number of photons emitted): Fluorescence and phosphorescence as a function of temperature with no quenching.[1]

In certain cases, however, investigations at low temperatures might be of great value, e.g. for examination of tissue sections deep frozen in liquid nitrogen. Thermal broadening of spectral lines should be reduced tremendously at low temperatures, but frequently cooling with liquid helium is necessary to attain observable effects.

An additional parameter for distinguishing between phosphorescence and fluorescence and, of course, for obtaining further molecular information is luminescence decay time, which, due to the statistical nature of luminescence decay, is closely related to the mean lifetime in the excited state[2, 3, 4]. Rough orders of lifetime value magnitude are presented in Tab. 1, and the technical effort necessary for their measurement is indicated ; these values can vary tremendously with temperature, due mainly to the thermal interaction between the molecules.[1]

Table 1. Mean lifetime of molecules in the excited state
(only orders of magnitude are given; see text)

Electronic transition (singlet - singlet)	10^{-8} s
Electronic transition (triplet - singlet)	10^{-6} s
Vibrational transition	10^{-3} s
Rotational transition	10^{-2} s

When molecules are excited by polarized light, the luminescence radiation will also be polarized to a certain degree. The change in polarization is due to mainly the disorientation of the molecules caused by Brownian motion and is correlated to several parameters of the molecule and its environment, such as molecule dimension, solvent viscosity, and temperature.

The optical parameters mentioned so far are obtained from non-destroyed samples. If "partial destruction" of the molecules or samples are also taken into account, e.g. photodecomposition or bleaching and its recovery, then luminescence techniques deliver further molecular parameters through optical investigation.

LUMINESCENCE TECHNIQUES IN BIOMEDICINE

In order to make fluorescence and phosphorescence, as briefly summarized above, useful tools in biomedical research and analysis and in medical diagnosis,[5] the following possibilities exist:

- Many biological samples intrinsically have luminescent properties; they show primary or autofluorescence.[6, 7, 8]

- A great number of samples, however, show only little or no autofluorescence. In many cases staining with specific chromophores allows for selective photomarking of certain parts of the sample for further luminescence investigation with artificially induced secondary fluorescence. These chromophores, familiar in biomedical routine, are dyes such as Na-fluorescein, acridine orange, and rhodamine, but there are also antibiotics such as tetracycline.[5]

Recently tumor-selective dyes with additional photodynamic or therapeutic properties have become highly interesting[9-19]. The application of these drugs promises new possibilities in oncology (see below).

- Coupling chromophores to "vehicles", e.g. antibodies, liposomes, or biochemical substrates (enzymes) can enhance the selectivity and specificity of staining. This enables detection and identification of pathogenic organisms (germs) via an antigen-antibody reaction,[5] photomarking of tumors for diagnostic and therapeutic purposes,[20] or registration of enzymatic defects in single embryonic cells[21,22].

- Further methods for artificially increasing selective tissue fluorescence signals are known in histo- and cytochemistry.[6,23]

- There are fluorescent dyes, which significantly change their optical properties due to electrochemical changes in their environment; such substances can be used as "probes" for local pH-values or indicators of electric potential.[24]

Biomedical examination through qualitative examination of primary and secondary luminescence phenomena has been well known for a long time and will not be treated in this article. Microscopic and macroscopic techniques are now highly developed and, with the aid of optoelectronic registration, also allow for quantitative fluorometry in daily biomedical routine.[1,5,6,23] Progress in the development of microelectronics, lasers,[25] optoelectronics and electrooptics have added new dimensions to luminescence (fluorescence and phosphorescence) techniques for basic research, analysis, and diagnosis in biology and medicine.

This will be demonstrated in the following presentation, using examples of applications which could only be realized by aid of the modern techniques mentioned above. The first examples are based on microscopic methods concerning phenomena in single cells. Then the extension of these methods to in situ and in vivo applications will be shown, and finally a macroscopic fluorescence imaging technique will be presented.

FLUORESCENCE METHODS ON A MICROSCOPIC SCALE

Time-Resolved Micro-Spectrofluorometry

Lasers have proved to be versatile tools for biomedical research. Highly focalized through a microscope lens, they allow for fluorescence probing on a cellular and subcellular scale. A broad spectrum of monochromatic emission is also a good prerequisite for fluorescence analysis. Continuous wave (cw) emission or pulsed emission at pulse durations in the nano- to picosecond range and at repetition frequencies in the Hz - MHz range enables a wide variety of time resolved kinetic measurements.

A) Registration of Pharmacokinetics on a Cellular Level (Time Scale in the Range of Hours)

Stabilized laser-assisted and optoelectronically controlled and registered fluorometry delivers information on the uptake and release of the photosensitizing drug mentioned above in the context of photodynamic therapy (PDT). Due to the high reproducibility of this technique rather longterm measurements on single cells have become possible. Time constants have been derived from the kinetics obtained, which - although not yet significantly - indicate that the rate of uptake is about equal for all cells investigated so far, but release is faster from normal cells than from tumorous cells (Tab. 2). This indicates that the transient tumor selectivity of the photosensitizer, which is not yet fully understood, might be based on cellular mechanisms.

Table 2. Time constants of cellular uptake and release for the hematoporphyrin-containing photosensitizer Photofrin II

Cells: n=100 per kinetic

Photosensitization: Photofrin II, 10 μg/ml medium (uptake) Sensitizer-free medium (release)

		Uptake	Release
Cell line	Species	Time constant [h]	Time constant [h]
Osteosarcome	mouse	9 (-2, +5)	50 (-25, +50)
Muscle	mouse	8 (-2, +6)	25 (-10, +10)
Osteosarcome	rat	8 (-2, +5)	50 (-20, +50)
Fibroblast	rat	13 (-4, +7)	30 (-10, +70)

B) Fluorometric Registration of Neuron Activity
(Time Scale in the Millisecond Range)

The time scale of switch processes in neurons is mainly in the millisecond range. Usually such processes are registered via electric leads through iontophoretic microprobes inserted into the cell. This technique is time-consuming and allows for investigations on single cells. Signal artifacts are frequently observed due to probe-induced membrane destruction. Membrane fluorescence, dependent on its potential, induced by Cohen's dye, cw laser excitation, and optoelectronic registration, makes the switching state optically accessible (Fig. 3).[24] Since only changes of <1% in fluorescence intensity should be registered, the laser has to be actively stabilized to a higher degree by electrooptic means; the variation of laser output so far could be reduced to $\leq 10^{-5}$.

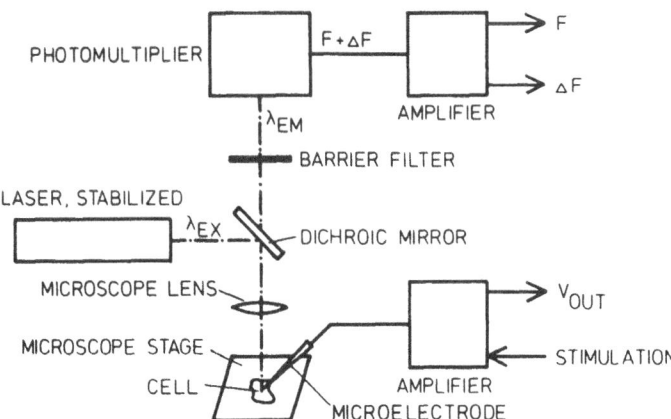

Figure 3. Epi-fluorometric set-up for optical and electrical registration of neuronic activity[24]

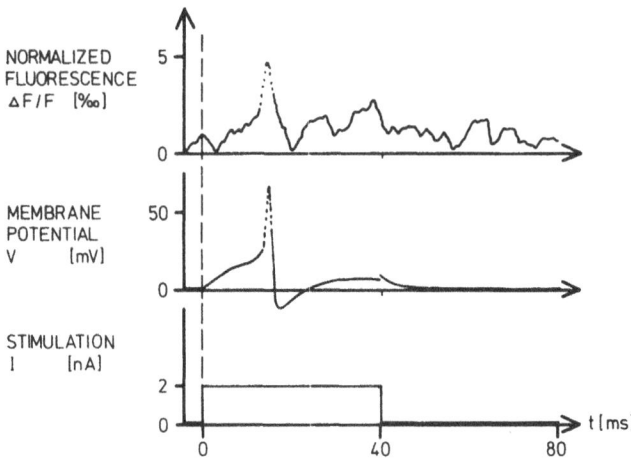

Fig.4. Neuronic activity of a sensoric neuron from a segmental ganglion of *Hirudo medicinalis*: Optical registration (top), electrical lead (middle), triggering current (bottom). The curves are averaged from 64 registrations.

Comparison of signals deduced optically and electrically from sensoric neurons (Fig. 4) shows a good temporal coincidence and indicates that the laser-assisted optical technique can successfully replace electrical probes.

The advantages are obvious: in a short time great numbers of cells are accessible to this measuring technique. Neuronic cells may now be used as sensitive test objects on the influence of environmental pollution, trace element poisoning etc.. Rapid scanning of the probing laser focus over a two-dimensional neuronic network in vitro is in progress and will give data on the two-dimensional exchange of neuronic information, an important step on towards the understanding of three-dimensional structures and the brain.

Whereas the time scale in these examples was chosen with respect to investigation of the entire object "cell", the time resolution in the following examples will be reduced to the molecular level, i.e. below the values given in Table 1.

C) Radiophotoluminescence/(RPL) dosimetry of ionizing radiation
    (Time scale in the nano-/picosecond range)

Ionizing radiation (γ-radiation) can induce color centers in silver-phosphate glass. Their number is proportional to the dose over several orders of magnitude and can be determined rather easily by fluorometric methods; fluorescence intensity becomes a function of dose. Small glass samples can be used as simple dosimeters for ionizing radiation and have several advantages over conventional film or ionization chamber-do-simeters used so far in personal dosimetry. For such application, however, an overlaying "predose" signal kept the lower detection limit of the RPL-glass dosimeters too high.[26]

Excitation of the glass through laser pulses in the nano- to picosecond range[25] and time-resolved fluorescence detection[4, 26, 27] showed that "predose"- and RPL signals occur successively; both signals can be separated in time easily. By computer-assisted evaluation of only the RPL-signal the former detection limit of RPL glass dosimetry,[26] obtained through cw-evaluation, was reduced by several orders of magnitude. Averaging over a great number of events due to the high repetition rate of the pulsed laser adds to the reliability of the measurement. Personal dosimetry is now possible. The dose limit corresponds approximately to the amount received through environmental radioactivity for 2-3 months.

Further increase in time resolution to picoseconds opens up new insights to molecular processes in biology and medicine.[28,29]

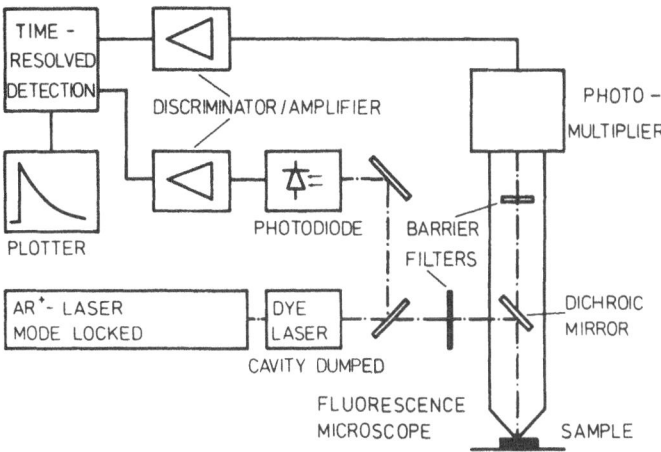

Figure 5.   Schematic diagram of a micro-fluorometer with picosecond laser excitation and fluorescence detection by time-resolved single/multiple photon counting[4]

D) Prenatal Diagnosis of Genetic Disorders
   (Time scale in the picosecond range)

Quantitative assays of enzyme activity allow for early prenatal diagnosis of genetic disorders. Today about 65 diseases, e.g. Niemann-Pick, Gaucher's, Farber's, or Pompe's disease can be detected through amniotic samples.[21]   The number of fetal cells or volume of amniotic fluid necessary for such a test is so high that general screening of gravidae is not tolerable. This problem is overcome by the time-resolving technique described above. Nanoliter-assays and enzyme quantities produced by single cells become accessible to fluorometry (Fig. 5). The relevant fluorochrome methyl umbelliferone (MU) can be detected to an absolute quantity of only 10^{-15} moles.[22] The corresponding fluorescence is discriminated in time from parasitic fluorescence and (excitation) stray light. The rate of increase of MU fluorescence is proportional to the cellular enzyme activity, which is a measure for the genetic disorder of the embryonic cell. The rate of increase at fluorescence has to be determined over hours. Automatization of the fluorometric set-up for registration of such kinetics and for large sample numbers of single cells should extend the method described to routine screening in prenatal diagnosis for genetic disorders.

E) Underline Cellular Distribution of Substances
   Underline Time scale in the picosecond range, spatial resolution in the cellular and subcellular region

The high spatial resolution of laser-assisted microfluorometry and the increase in sensitivity due to time-resolving techniques nowadays allows for the determination of intracellular distribution and kinetics of substances. The significance of such measurements for biomedical and especially pharmacological research is obvious.[12,27,30,31]

FLUORESCENCE METHODS ON A MILLIMETER SCALE

Remote fluorometry

The microscopic methods described above usually require specific preparation of microsamples. Application of optical fibers according to Fig. 6 enables in situ measurements via fiber probes, which otherwise could only be realized with great effort.[25] Sample preparation and correlated artifacts can largely be avoided, and in vivo data readily become available, thus opening up new ways in, for example, intensive care medicine. The optical data can be transferred to one central optical measuring unit which might include all the measuring techniques described.

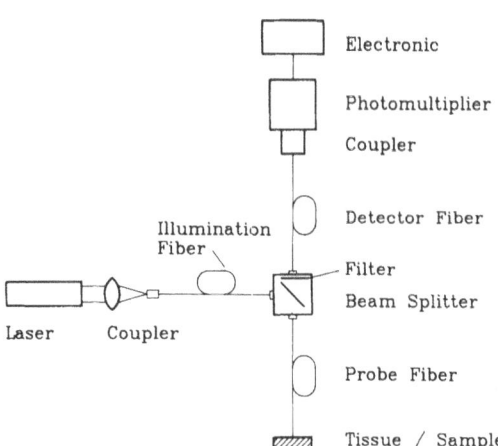

Fig. 6. Schematic diagram of a fluorometric fiber probe for in situ measurements. The set-up is suited for cw or pulsed laser excitation and continuous to time-resolved fluorescence detection.

FLUORESCENCE IMAGING ON A MACROSCOPIC SCALE

With the introduction of tumor-selective (photodynamically active) photosensitizers to oncology, their superficial detection by fluorescence imaging of secondary fluorescence on a macroscopic scale or via endoscopes became of great interest to diagnostic medicine.[10,11,15,16,18,32-34] Several methods have been proposed for excitation of the sensitizer-induced tumor-correlated fluorescence and its detection. All systems so far known have rather poor signal/noise ratios and therefore do not show any great advantage over conventional methods.[3,35-39]

The signal is limited by limitations in the excitation intensity because of bleaching of the sensitizer; the noise results mainly from intrinsic primary or autofluorescence of tissue and to a lesser extent from incomplete tumor selectivity of the substance. Fluorescence excitation at two wavelengths reveals maximal and minimal sensitizer fluorescence emission respectively.[12,19,20] Tuning the excitation intensities for precisely equal background signals allows for their elimination by computer-aided image subtractions.[40,41]

The signal/noise ratio and the image contrast are highly increased (Fig. 7) and make detection of conventionally invisible tumors, i.e. carcinoma in situ, possible. First clinical tests of the system are in progress.

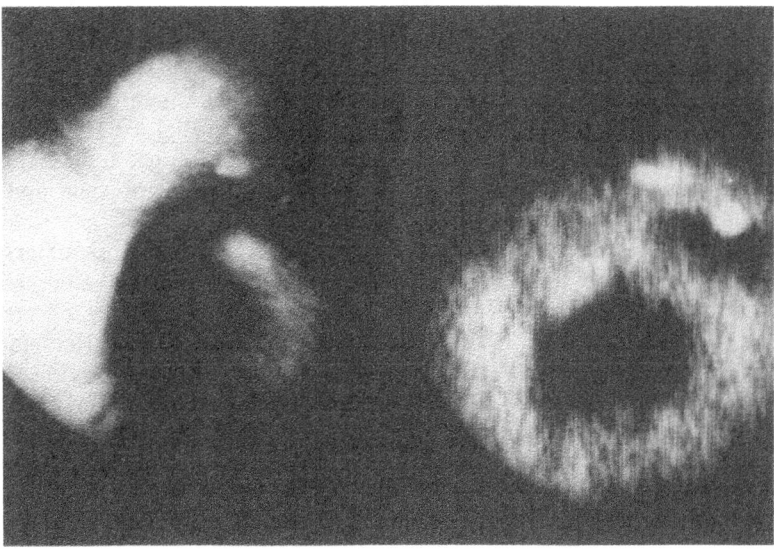

Figure 7. Endoscopic view (left) and fluorescence difference imaging (right) of a bronchial tumor in a dog

In addition sensitivity is so increased that the sensitizer dose can be reduced to about 1/5 - 1/10 of the usual therapeutic dose of 2 mg/kg bodyweight (Di-hemato-porphyrin-containing sensitizer). Correspondingly, the sensitizer-correlated side effects[19,20] are reduced to generally tolerable values, promising the introduction of computer-assisted tumor diagnostics with two-wavelength laser fluorescence excitation to clinical routine.

First successful applications of the two-wavelength fluorescence method on plaque in atherosclerotic vessels in vitro show a further potential for guiding "intelligent" laser in angioplasty, thus directly combining diagnostics and therapy.

CONCLUDING REMARK

The examples presented prove that fluorescence techniques - especially in combination with modern electrooptic, optoelectronic, and quantum electronic methods - are extremly valuable tools in biomedical analysis and research, in (non-invasive) medical diagnosis, and for guiding therapy. The examples should also trigger further efforts to improve the results obtained so far.

REFERENCES

1. S. Udenfriend, "Fluorescence Assay in Biology and Medicine," Academic Press, New 1962/64 and 1969 York-London, Vol. I (1962-64); Vol. II (1969).
2. J.R. Alcala, E. Gratton, and F.G. Prendergast, 1. Resolvability of fluorescence lifetime distributions using phase fluorometry. 2. Fluorescence lifetime distributios in proteins, Biophys. J. 51:587 and 597 (1987).
3. D.R. Doiron, , Fluorescence bronchoscopy for the early localization of lung cancer, PhD-Thesis, University of California, St. Barbara, USA (1982).
4. F. Pauker, H. Schneckenburger, and E. Unsöld, Time-resolved multiphoton counting, J. Phys. E, Sci. Instrum. 19:240 (1986).
5. K.F. Koch, "Fluoreszenzmikroskopie," Ernst Leitz GmbH, Wetzlar (1972).
6. J.F. Danielli, "General Cytochemical Methods," Academic Press, Inc., New York (1958).
7. J. Körbler, Untersuchung von Krebsgewebe im fluoreszenzerregenden Licht, Strahlentherapie 41:510 (1931).
8. A. Policard, Etudes sur les aspects offert par des tumeurs expérimentales examinées à la lumière de Woods, Comp. Rend. Soc. Biol. 91:1423 (1924).
9. R.R. Alfano, D.B. Tata, J. Cordero, P. Tomashefsky, F. Longo, and M.A. Alfano, Laser induced fluorescence spectroscopy from native cancerous and normal tissue, IEEE J. Quant. El. QE-20:1507 (1984).
10. S. Andersson, J. Ankerst, E. Kjellén, S. Montan, E. Sjöholm, K. Svanberg, and S. Svanberg, Tumor localization by means of laser-induced fluorescence in hematoporphyrin derivative (HpD) - bearing tissue, in: "Laser Spectroscopy," T.W. Hänsch and Y.R. Shen, eds., Springer Verlag, Berlin-Heidelberg-New York-Tokyo, (1985) Vol. VII, 401.
11. R. Baumgartner, J. Feyh, A. Götz, D. Jocham, H. Schneckenburger, H. Stepp, and E. Unsöld, Experimental study on laser-induced fluorescence of hematoporphyrin derivative (HpD) in tumor cells and animal tissue, Laser Med. Surg. 2:4 (1986).
12. R. Baumgartner, H. Fisslinger, D. Jocham, H. Lenz, L. Ruprecht, H. Stepp, and E. Unsöld, A fluorescence imaging device for endoscopic detection of early stage cancer - Instrumental and experimental studies, Photochem. Photobiol. 46:759 (1987).
13. F.H.J. Figge, G.S. Weiland, and L.O. Mangianello, Cancer detection and therapy: affinity of neoplastic, embryonic, and traumatized tissues for porphyrins and metalloporphyrins, Proc. Soc. Exp. Biol. Med. 68:640 (1948).
14. J.P. Keene, D. Kessel, E.J. Land, R.W. Redmond, and T.G. Truscott, Direct detection of singlet oxygen sensitized by haematoporphyrin and related compounds, Photochem. Photobiol. 43:117 (1986).
15. R.L. Lipson, E.J. Baldes, and M.J. Gray, Hematoporphyrin derivative for detection and management of cancer, Cancer 20:2255 (1967).
16. G.C. Peck, H.P. Mack, W.A. Holbrook, and F.H.J. Figge, Use of hematoporphyrin fluorescence in biliary and cancer surgery, Am. Surgeon 21:181 (1955).
17. W.R. Potter., and T.S. Mang, Photofrin II levels by in vivo fluorescence photometry, in: "Porphyrin Localization and Treatment of Tumors," D.R. Doiron and C.J. Gomer, eds., Alan R. Liss, Inc., New York, (1984), 177.
18. D.S. Rassmussen-Taxdahl, G.E. Ward, and F.H.J. Figge, Fluorescence of human lymphatic and cancer tissue following high doses of intravenous hematoporphyrin, Cancer 8:78 (1955).
19. E. Unsöld, R. Baumgartner, D. Jocham, and H. Stepp, Application of photosensitizers in diagnosis and therapy, Laser Med. Surg. 3:210 (1987).
20. E. Unsöld, R. Baumgartner, W. Beyer, D. Jocham, and H. Stepp, Fluorescence detection and photodynamic therapy of photosensitized tumours in special consideration of urology, Lasers Med. Sci. 5:207 (1990).

21. P. Hösli, Quantitative assays of enzyme activity in single cells: early prenatal diagnosis of genetic disorders, Clin. Chem. 23:1476 (1977).

22. H. Schneckenburger, and E. Unsöld, Time-resolved ultrasensitive fluorescence detection for enzyme analysis, in: "Trace Element Analytical Chemistry in Medicine and Biology," P. Brätter and P. Schramel, eds., W. de Gruyter, Berlin-New York, vol.II.:1143 (1983).

23. B. de Lerma, "Handbuch für Histochemie I - Die Anwendung von Fluoreszenzlicht in der Histochemie," G. Fischer-Verlag, Stuttgart (1958).

24. P. Saggau, M. Galvan, F. Rucker, and E. Unsöld, Erfassung neuronaler Aktivität mit optischen Methoden, in: "Optoelectronics in Medicine," W. Waidelich, ed., Springer-Verlag, Berlin-Heidelberg-New York-Tokyo, (1984), 233.

25. J. Hermann, and B. Wilhelmi, "Laser für ultrakurze Lichtimpulse - Grundlagen und Anwendungen," Physik Verlag, Weinheim (1984).

26. H. Schneckenburger, D.F. Regulla, and E. Unsöld, Time-resolved investigations of radiophotoluminescence in metaphosphate glass dosimeters, Appl. Phys. A 26, 23 (1981).

27. F. Docchio, R. Ramponi, C.A. Sacchi, G. Bottiroli, and I. Freitas, Time-resolved fluorescence spectroscopy of hematoporphyrin-derivative in human lymphocytes, Chem. Biol. Interact. 50:135 (1984).

28. W. Kaiser, Ultrakurze Lichtimpulse. Ihre Anwendung in Physik und Biologie, Verhandlungen der Gesellschaft Deutscher Naturforscher und Ärzte, 114. Versammlung, München, Wissenschaftliche Verlagsgesellschaft mbH, Stuttgart, 339-347 (1987).

29. H. Schneckenburger, Y. Tsuchiya, M. Miwa, M. Frenz, and L. Schleinkofer, Picosecond fluorescence measurements of hematoporphyrin derivative (HpD) in single cells, Laser Med. Surg. 3:133 (1986).

30. B. Ehrenberg, Z. Malik, and Y. Nitzan, Fluorescence spectral changes of hematoporphyrin derivative upon binding to lipid vesicles, Staphylococcus aureus, and Escherichia coli cells, Photochem. Photobiol. 41:429 (1985).

31. H. Schneckenburger, F. Pauker, E. Unsöld, and D. Jocham, Intracelluar distribution and retention of the fluorescent components of photofrin II, Photobiochem. Photobiophys. 10:61 (1985).

32. J. Ankerst, K. Svanberg, S. Montan, and S. Svanberg, Contrast enhancement in tumor localization using hematoporphyrin derivative laser-induced fluorescence, in: "Technical Digest of CLEO '84," Optical Society of America, Toledo, USA, (1984) 234.

33. G.H.M. Gijsbers, M.J.C. van Gemert, D. Breederverld, J. Langelaar, and T.A. Boon, In vivo fluorescence excitation spectra of hematoporphyrin-derivative (HpD), in: "Porphyrins in Tumor Phototherapy," A. Andreoni and R. Cubeddu, eds., Plenum Press, New York-London, 339 (1984).

34. S. Montan, K. Svanberg, and S. Svanberg, Multicolor imaging and contrast enhancement in cancer-tumor localization using laser-induced fluorescence in hematoporphyrin-derivative-bearing tissue, Opt. Lett. 10:56 (1985).

35. K. Aizawa, S. O'Hata, H. Kato, H. Sakai, K. Nishimiya, M. Saito, K. Kinoshita, T. Hirano, S. Miyaki, M. Yamashita, and Y. Hayata, HpD localization using excimer laser, in: "Photodynamic Therapy of Tumors and Other Diseases," G. Jori and C. Perria, eds., Libreria Progetto, Padova (1985), 199.

36. Y. Hayata, H. Kato, J. Ono, Y. Matsushima, N. Hayashi, T. Saito, and N. Kawate, Fluorescence fiberoptic bronchoscopy in the diagnosis of early stage lung cancer, Rec. Results Cancer Res. 82:121 (1982).

37. T. Hirano, K. Ishida, M. Yasukawa, S. Miyaki, A. Houma, K. Aizawa, H. Kato, Y. Hayata, and M. Yamashita, Cancer diagnosis system using HpD and excimer-dye laser, in: "Photodynamic Therapy of Tumors and Other Diseases," G. Jori and C. Perria, eds., Libreria Progetto, Padova, 325 (1985).

38. S. Ikeda, K. Eguchi, R. Ono, and D. Kato, A new television system for photodynamic images by using argon laser - Clinical application for diagnosis of lung cancer, in: "Laser Tokyo '81," K. Atsumi and N. Nimsakul, eds., The Japan Society for Laser Medicine, Tokyo, paper 16/15 (1981).

39. A.E. Profio, and O.J. Balchum, Fluorescence diagnosis of cancer, in: "Methods in Porphyrin Photosensitization," D. Kessel, ed., Plenum Press, New York-London, (1985), 43.

40. R. Baumgartner, and E. Unsöld, High contrast imaging using two-wavelength laser excitation and image processing, J. Photochem. Photobiol. 1:130 (1987).

41. S. Bommer, Über sichtbare Fluoreszenz beim Menschen, Acta Dermat.Vener. (Stockholm) 10:253 and 391 (1929).

SURVEY OF THE UV AND VISIBLE SPECTROSCOPIC PROPERTIES OF NORMAL AND ATHEROSCLEROTIC HUMAN ARTERY USING FLUORESCENCE EEMS

Rebecca Richards-Kortum[1] Richard Rava[1] Joseph Baraga[1]
MS, Maryann Fitzmaurice[2] John Kramer[3] Michael Feld[1]

[1]G.R. Harrison Spectroscopy Laboratory
Massachusetts Institute of Technology
Cambridge, MA 02139

[2]Department of Pathology
[3]Department of Cardiology
Cleveland Clinic Foundation
Cleveland, OH 44106

INTRODUCTION

Fluorescence spectroscopy has been recently explored as a new technique for the diagnosis of disease in human tissue [1-5]. Several groups have demonstrated that normal and atherosclerotic human artery can be differentiated on the basis of their fluorescence emission spectra [1, 3, 4]. Many have suggested that this represents a new technique for determining the histochemical composition of atherosclerotic plaque *in vivo* and potentially guiding laser angiosurgery catheters [1, 3, 4]. Similarly, it has been shown that neoplastic tissue can be discriminated from non-neoplastic tissue, in a variety of organs, including the colon [2] and lung [5], on the basis of tissue fluorescence emission spectra. Several workers have suggested that this technique may ultimately improve current endoscopic methods for diagnosing neoplasia, as it potentially can be used to identify small areas of neoplastic transformation difficult to detect visually with current endoscopes [2,6].

Despite this recent interest in the application of fluorescence spectroscopy to diagnostic medicine, little attention has been devoted to optimizing the method of collection and analysis of tissue fluorescence spectra for the differentiation of normal and pathologic tissues. We have developed a step-wise, general method which can be used to develop an optimal diagnostic algorithm for pathology based on features of tissue fluorescence emission spectra [7]. This procedure is illustrated in Figure 1. Several of the steps in this procedure have been discussed in detail previously [1, 2, 7-10, 12]. This paper briefly discusses the role that fluorescence excitation-emission matrices (EEMs) play in this procedure. These steps are highlighted in Figure 1.

An EEM is a matrix which contains the fluorescence intensity as a function of both the excitation and emission wavelength [8]. In general, the fluorescence emission lineshape of a single fluorophore in a dilute solution is not a function of the excitation wavelength; only the overall fluorescence intensity changes as the excitation wavelength is varied [11]. Thus, the spectroscopic properties of a solution containing only a single fluorophore can be characterized completely with a single fluorescence excitation spectrum and a single emission spectrum. However, the fluorescence emission lineshape of any sample containing multiple fluorophores is a function of excitation wavelength, as the relative contribution of each fluorophore to the emission spectrum changes as the excitation wavelength is varied [11]. The spectroscopic properties of a multi-fluorophore solution can be characterized by a single excitation spectrum and a single emission spectrum of each component. However, if the identity of the individual fluorophores in the solution is unknown, these can be difficult to obtain. An alternative way to characterize the fluorescence properties of such a solution is via a fluorescence EEM [8].

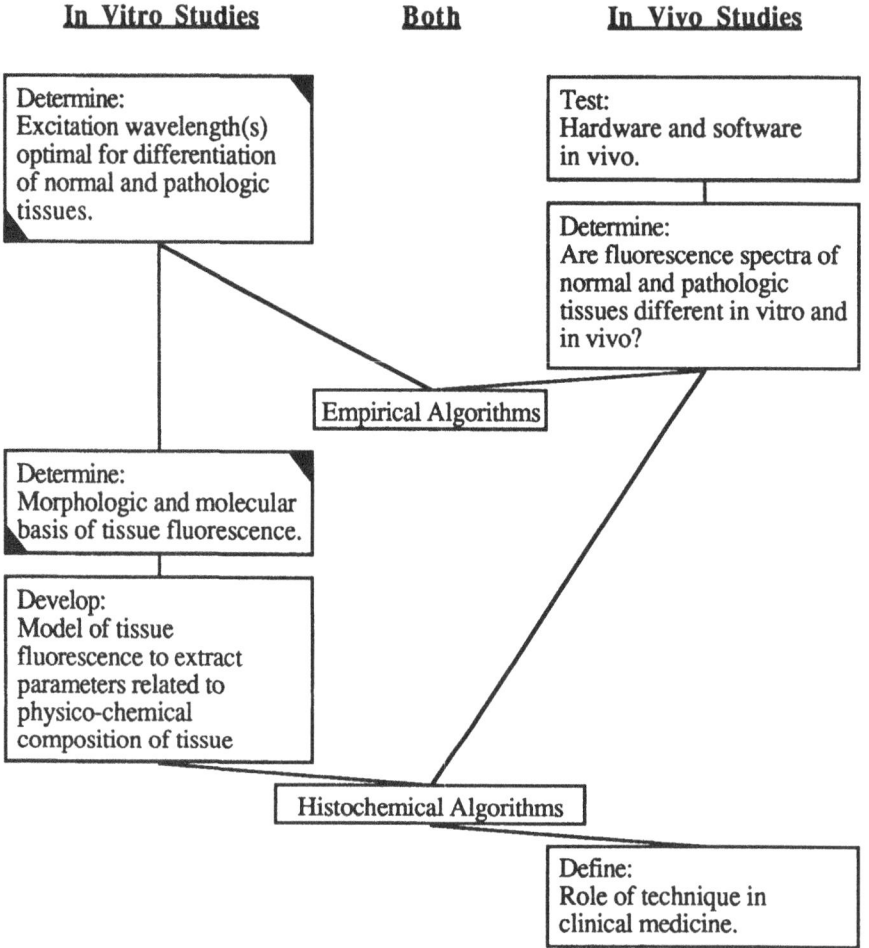

Figure 1. Schematic diagram illustrating the steps involved in our approach to defining spectroscopic algorithms for real time diagnosis of disease *in vivo*.

Wolfbeiss has previously utilized fluorescence EEMs of dilute, multi-fluorophore solutions in clinical diagnostic medicine [12. He showed that fluorescence EEMs could be used to select excitation and emission wavelength pairs to differentiate serum from normal and tumor bearing individuals based on fluorescence intensity [12]. In addition, he demonstrated that the fluorophores in human serum could be identified by comparing the location of the excitation-emission maxima in the fluorescence EEM of serum to those in fluorescence EEMs of individual fluorophores or to those values published in the literature [12].

This method of analysis can also be applied to select an optimal excitation wavelength for the differentiation of normal and pathologic human tissues and to identify fluorophores contributing to tissue fluorescence spectra. However, in contrast to dilute solutions discussed above, human tissue is a highly scattering and absorbing (turbid) medium. The fluorescence spectra of such turbid samples contain contributions from attenuation of the emitted fluorescent light [13]. This paper discusses the application of fluorescence EEMs to turbid, biological samples in order to select the optimal excitation wavelength for the differentiation of normal and pathologic samples. Furthermore, demonstrate that these same EEMs can be an important tool in identifying fluorophores and absorbers which contribute to the fluorescence spectra. These methods are general in nature, but are here illustrated for normal and atherosclerotic coronary artery.

EXPERIMENTAL METHODS

Fluorescence EEMs were recorded from 8 samples of coronary artery *in vitro*. Tissue samples were obtained surgically during heart transplants from 5 recipient hearts. After removal of the heart, the first several centimeters of the main branches of the coronary arteries were blunt dissected. Intact arteries were then frozen in liquid nitrogen and isopentane and stored at -70°C. Before study, samples were thawed to room temperature; during all experiments samples were kept moist with a buffered (TRIS 50 mM), pH 7.4, 140 mM saline solution. The artery was opened, and fluorescence EEMs and total reflectance spectra were obtained from the intimal surface. Spectra were recorded from grossly visible non-calcified atherosclerotic lesions and from surrounding normal appearing areas. Following study, the intimal areas sampled spectroscopically were marked with india ink. Samples were fixed in 4% formalin and submitted for routine histologic analysis. A single pathologist classified all samples according to the classification scheme currently used in the surgical pathology department at the Cleveland Clinic Foundation [14].

Fluorescence EEMs of arterial tissue were constructed from a series of fluorescence emission spectra collected at excitation wavelengths (λ_{exc}) varying from 250 to 550 nm in 10 nm steps. For each emission spectrum, fluorescence intensities were recorded at 10 nm intervals of emission wavelength (λ_{em}) over the range

$$[\lambda_{exc} + 10 \text{ nm}] < \lambda_{em} < [2\lambda_{exc} - 10 \text{ nm}] \text{ or } [700 \text{ nm}].$$

At each emission wavelength, the fluorescence signal was integrated for a 1 s interval.

Emission spectra were collected using a standard spectrofluorimeter. Excitation light was incident perpendicular to the lumenal surface of the tissue; emission light was collected at a 23° angle with respect to the excitation beam. The instrument was modified to yield an excitation beam size of approximately 2 x 3 mm at the tissue. In these experiments, the spectral resolution of the excitation monochromator was 2 nm FWHM, that of the emission monochromator was 4 nm FWHM.

The incident intensity varied with excitation wavelength, but was always less than 40 μW/mm^2. Variations in the incident intensity were corrected for by ratioing the fluorescence emission of the sample to that of a standard 1 cm reference cuvette containing 8 μg/L of Rhodamine B in reagent grade propylene glycol [15]. Collection of an entire fluorescence EEM required approximately one and one half hours. To minimize the exposure of tissue to excitation light during this procedure, a shutter was employed, so that light was incident on tissue only during the time which fluorescence emission was being recorded. Emission spectra were collected in order of increasing excitation wavelength, with the 250 nm excited spectrum collected first. To monitor for photobleaching, a 250 nm excited spectrum was recorded again following collection of the EEM. In all cases, deviations between the first and last 250 nm excited spectrum were less than 5%.

To correct for slight day to day variations in the throughput of the system, absolute fluorescence intensities were calibrated daily using a standard fluorescent filter. Fluorescence intensities are reported here in arbitrary units relative to the intensity of this standard. All spectra have been corrected for the non-uniform spectral response of the spectrofluorimeter. These correction curves were obtained by recording the emission spectrum of a calibrated tungsten filament lamp.

EEMs are presented here as fluorescence contour maps, where emission wavelength is on the abscissa, excitation wavelength is on the ordinate, and contour lines connect points of equal fluorescence intensities.

Total reflectance spectra were collected with a standard absorption spectrophotometer equipped with an integrating sphere. Total reflectance was measured over the region 200 - 800 nm with a resolution of 5 nm FWHM. The incident radiation was normal to the intimal surface of the sample. The intensity of the incident radiation varied, but was always less than 40 μW/mm^2. Collection of a total reflectance spectrum required approximately 3 minutes. The spectral response of the instrument was calibrated using a barium sulfate standard with a total reflectance approaching unity at all wavelengths of interest. Total reflectance is expressed here as the percentage of the incident energy reflected from the tissue surface as a function of incident wavelength.

RESULTS

Of the 8 samples investigated, one was classified as (0) normal, four as (1) intimal fibroplasia, two as (2) atherosclerotic plaque, and one as (4) atheromatous plaque. Average EEMs were calculated for non-atherosclerotic samples of coronary artery belonging to categories (0) and (1), and non-calcified plaques including categories (2) and (3). These average EEMs are shown as fluorescence contour maps in Fig. 2. Average total reflectance spectra were also calculated for these tissue categories and are shown in Figure 3.

The fluorescence EEMs of each tissue type contain similar features: four broad excitation-emission maxima, located near (290, 330 nm), (330, 380 nm), (340, 440 nm), and (450, 520 nm). These peaks are separated by valleys parallel to the excitation and emission axes at 420 nm. More subtle valleys are also present at 540 nm. Although the general features of these fluorescence EEMs are similar, many differences can be appreciated in the fluorescence intensity, location and lineshape of these peaks. To summarize this information and facilitate comparison, these fluorescence EEMs have been characterized by the approximate position and intensity of all local excitation-emission maxima in Table 1. This comparison neglects the small shifts in the positions of these maxima and the fluorescence lineshapes of the associated peaks. In addition, Table 1 includes a potential assignment of the morphologic or molecular basis of each of these peaks. These assignments are based on the results of experiments described later in this paper.

Table 1. Excitation Emission Maxima in the Fluorescence EEMs of Coronary Artery

Approximate Excitation-Emission Maximum	Fluorescence Intensity (Arbitrary Units)		Potential Chromophore/ Chromomorph
	Non-Atherosclerotic Tissue (0,1)	Non-Calcified Plaque (2,3)	
(290, 330 nm)	200	140	Tryptophan
(330, 380 nm)	80	80	Collagen Fibers
(340, 440 nm)	40	20	Elastin Fibers Ceroid
(460, 520 nm)	20	18	Elastin Fibers Collagen Fibers Ceroid

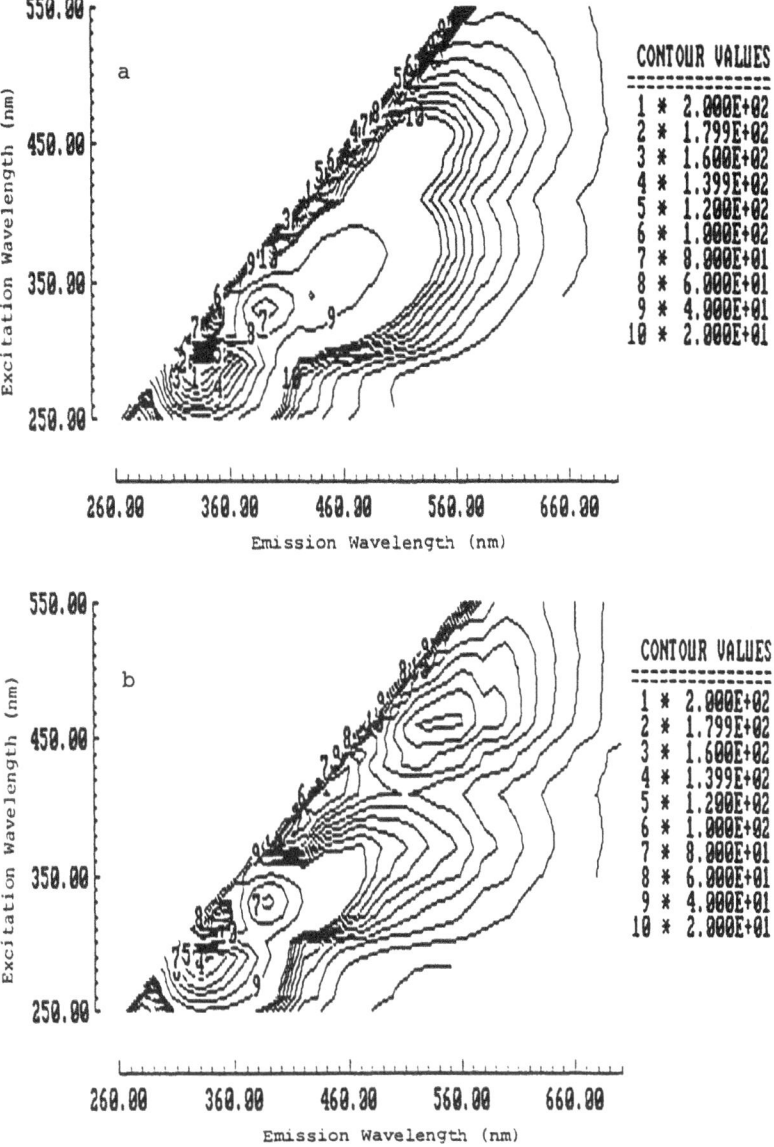

Figure 2. Average fluorescence contour maps of (a) (0,1) non-atherosclerotic coronary artery and (b) (2,3) non-calcified plaque. Two sets of linearly spaced contour lines are shown; ten from 200.0 to 20.0 and nine from 18.0 to 2.0 units, where fluorescence intensity is given in arbitrary units relative to the fluorescence intensity of a standard. The ten contours are labeled as per the figure legend. The same set of arbitrary units is maintained for all EEMs shown in this paper.

Figure 3 shows average total reflectance spectra of (0,1) non-atherosclerotic coronary artery and (2,3) non-calcified plaque. Valleys in the total reflectance spectra indicate peaks in attenuation, and are present in both types of tissue at 420, 540 and 580 nm. These are consistent with the attenuation of oxy-hemoglobin [7, 13, 16]. The EEMs of both types of tissue contain valleys parallel to the excitation and emission axes at 420 and 540 nm. It has been previously demonstrated that attenuation peaks in turbid samples give rise to valleys in fluorescence spectra [1, 7, 13]; thus, the origin of these valleys is the attenuation of oxy-hemoglobin within tissue.

Figure 3. Average total reflectance spectra of (a) (0,1) non-atherosclerotic coronary artery and (b) (2,3) non-calcified plaque.

DISCUSSION OF RESULTS

Selection of Optimal Excitation Wavelengths

From Table 1, it is evident that, for a given peak, the fluorescence intensity varied with tissue type; for example, the (290, 330 nm) peak was more intense in non-atherosclerotic tissues than in non-calcified plaques. The peak at (340, 440 nm) was also slightly more intense in non-atherosclerotic tissues. In addition, although each type of tissue contained four peaks in generally similar locations, there were slight shifts in the exact location of the maxima, as well as slight differences in peak width or fluorescence lineshape. For example, visual inspection of Figure 2 shows that the peaks near (340, 440 nm) and (450, 520 nm) are shifted to longer emission wavelengths in the fluorescence EEMs of non-calcified plaques.

Thus, based on this limited number of samples, it can be concluded that the fluorescence properties of normal coronary artery and non-calcified atherosclerotic plaque differ. Thus, further study, to exploit these differences in a diagnostic algorithm for atherosclerosis was undertaken. The first step in our process of algorithm development is the selection of excitation wavelengths at which this discrimination can best be achieved. We have utilized the fluorescence EEMs to select these wavelengths in coronary artery.

We construct a difference EEM by subtracting the value of each point of an EEM of normal tissue from that of the corresponding point of an EEM of diseased tissue. We use the difference of the average normal and pathologic tissue fluorescence EEMs to choose the optimal excitation wavelengths for diagnosing the presence of disease. The average difference map summarizes all spectroscopic bands where the spectroscopic properties of normal and diseased tissues differ. We wish to obtain the most complete set of diagnostic information with the minimum number of excitation wavelengths. A single excitation wavelength can be represented by a horizontal line on our contour map representation of tissue EEMs. Thus, we can achieve our goal graphically by drawing the minimum number of horizontal lines which intersect (or are in the vicinity of) all local minima and maxima in the average difference map.

Figure 4 shows a contour map representation of the difference of the average non-atherosclerotic tissue EEM and the average EEM of non-calcified plaque. This difference map exhibits maxima at three locations: (290, 330 nm), (360, 460 nm) and (440, 510 nm). The largest absolute difference is at (290, 330 nm). Smaller differences in absolute fluorescence intensity can be observed at (340, 370 nm) and (520, 600 nm). These correspond roughly to the differences observed from a comparison of the fluorescence intensities of the three peaks at (290, 330 nm), (340, 440 nm) and (450, 520 nm), summarized in Table 1. The slight shift in the positions of the excitation emission maxima in the difference maps indicate that there are differences in the exact position and lineshape of the four peaks observed in the average fluorescence EEMs of tissue.

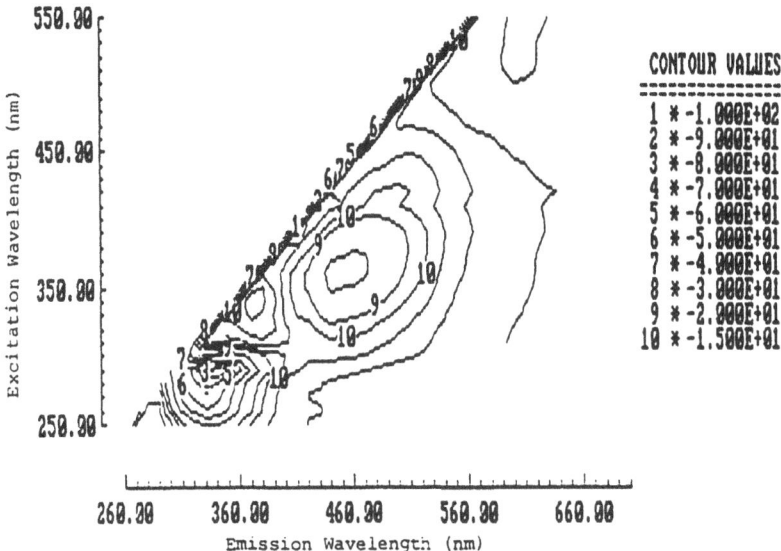

Figure 4. Contour map representation of the difference of the average fluorescence EEMs of non-atherosclerotic tissue and non-calcified plaque. Two sets of linearly spaced contours are shown; nine from -100 to -20 units and six from -15 to 10 units. The same set of arbitrary units is maintained for all EEMs in this paper. Ten contours are labeled as per the figure legend.

Figure 4 shows, that for normal and atherosclerotic coronary artery, all local minima and maxima in the ratio maps can be sampled with only three excitation wavelengths at 310, 350 and 450 ± 10 nm. These wavelengths thus represent the optimal excitation wavelength regions for spectroscopically differentiating these types of tissue.

Interpretation of Arterial EEMs

These excitation wavelengths have been selected on the basis of an empirical comparison of the fluorescence spectra of normal and atherosclerotic coronary artery. However, fluorescence EEMs can also be used to identify the arterial constituents which contribute to the tissue fluorescence spectrum. In this section, we will show that the empirically selected excitation wavelengths sample chromophores which are intimately related to the pathobiology of atherosclerosis.

Wolfbeiss [12] has suggested that a comparison of the location of the excitation-emission maxima in the fluorescence EEM of an unknown, dilute, multi-fluorophore solution to those in EEMs of pure chromophores and those in the literature can result in identification of individual fluorophores. Such a procedure can also be applied to the EEMs of turbid materials such as human tissue. However, the valleys produced by attenuation along the excitation and emission axes can act to significantly shift the excitation emission maxima, making this interpretation more difficult. Here, we illustrate the advantages and disadvantages of this simple method of identification with the EEMs of normal and atherosclerotic coronary artery presented earlier.

Previous studies have shown that at visible excitation wavelengths, the important fluorophores in normal coronary artery are those associated with the structural proteins, collagen and elastin [9]. These fluorophores also contribute to the fluorescence spectra of atherosclerotic coronary artery; but in addition, the fluorescence of oxidized lipoproteins, or ceroid, are also important [9].

In an attempt to further quantify the contribution of the structural proteins, collagen and elastin, to the arterial tissue EEMs, we recorded EEMs of powdered bovine collagen and elastin [Sigma]. These are shown in Figure 5.

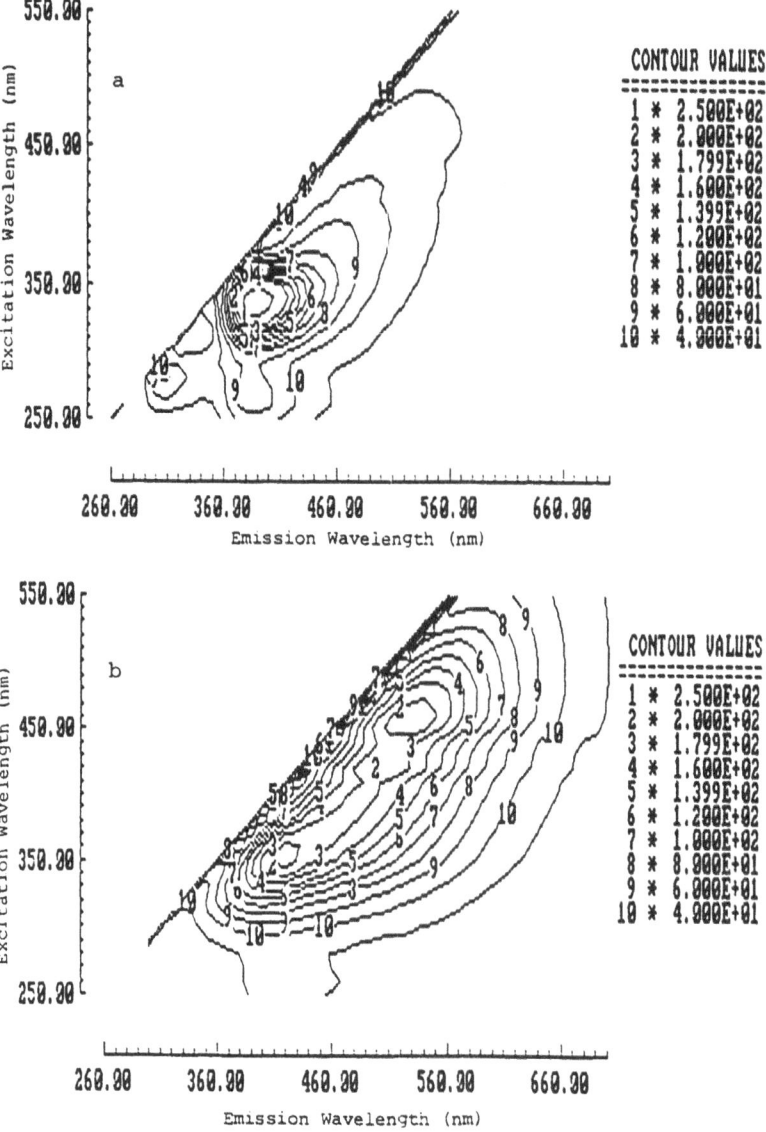

Figure 5. Contour map representation of the fluorescence EEMs of (a) powdered bovine collagen and (b) elastin. Ten sets of contours are shown as per the figure legend. The same set of arbitrary units is maintained for all EEMs shown in this thesis.

The fluorescence EEM of collagen shows peaks at (280, 310 nm), (265, 385 nm), (330, 390 nm) and (450, 530 nm). The largest peaks is at (330, 390 nm). The fluorescence EEM of elastin shows peaks nearly equal in intensity at (350, 420 nm), (410, 500 nm), and (450, 520 nm).

These are similar in location to several of the peaks noted in the fluorescence EEMs of normal and atherosclerotic coronary artery. The peak in the tissue EEMs at (330, 380 nm) is similar in location to that of collagen at (330, 390 nm). That at (340, 440 nm) is similar in location to the peak noted in the EEM of elastin at (350, 420 nm). This peak was most intense in the fluorescence EEM of normal coronary artery and this is consistent with the degeneration of the internal elastic lamina and the increase in intima thickness as atherosclerosis develops. Finally, the peak at (460, 520 nm) in the arterial EEMs is attributable to both the collagen peak at (450, 530 nm) and the elastin peak at (450, 520 nm). Other studies [7, 17] have shown that the other peaks noted in the arterial fluorescence EEMs can be attributed to tryptophan and ceroid fluorescence, as in Table 1.

CONCLUSIONS

In summary, we have shown that fluorescence EEMs can be used to survey the visible and ultra-violet spectroscopic properties of turbid, biologic tissues. In addition, they can be used to select excitation wavelengths optimal for the differentiation of normal and pathologic tissues, and as a tool to help identify the chromophores which contribute to these spectra. Although the techniques discussed here are promising, improved methods, using physical models of tissue fluorescence spectra [1], are required to quantitatively extract information from EEMs of such turbid samples.

REFERENCES

1. Richards-Kortum RR, Rava R, Fitzmaurice M, Tong L, Ratliff NB, Kramer JR, Feld MS: A One-Layer Model of Laser Induced Fluorescence for Diagnosis of Disease in Human Tissue: Applications to Atherosclerosis, IEEE Transactions on Biomedical Engineering, 36:1222-1232, 1989.

2. Cothren RM, Richards-Kortum RR, Sivak MV, Fitzmaurice M, Rava RP, Boyce GA, Hayes GB, Doxtader M, Blackman R, Ivanc T, Feld MS, Petras RE: Gastrointestinal Tissue Diagnosis by Laser Induced Fluorescence Spectroscopy at Endoscopy, in press, Gastrointestinal Endoscopy, 1990.

3. Laifer LI, O'Brien K, Stetz ML, Gindi GR, Garrand TJ, Deckelbaum LI, Biochemical Basis for the Diffference Between Normal and Atherosclerotic Arterial Fluorescence, Circulation, 80:1893-1901, 1989.

4. Leon MB, Lu DY, Prevosti LG, Macy WW, Smith PD, Gransovsky M, Bonner RF, Balaban RS, Human Arterial Surface Fluorescence: Atherosclerotic Plaque Identification and Effects of Laser Atheroma Ablation, JACC, 12:94-102, 1988.

5. Alfano RR, Tang GC, Pradhan A, Lam W, Choy DSC, Opher E, Fluorescence Spectra from CAcnerous and Normal Human Breast and Lung Tissues, IEEE JQE, QE23:1806-11, 1987.

6. Anderson PS, Montan S, Svanberg S, Multispectral System for Medical Fluorescence Imaging, IEEE JQE QE23:1798, 1987.

7. Richards-Kortum RR: Fluorescence Spectroscopy as a Technique for Diagnosis of Pathologic Conditions in Arterial, Urinary Bladder and Gastro-Intestinal Tissues, PhD Thesis, Dept. of Health Sciences and Technology, Massachusetts Institute of Technology, May 1990.

8. Christian GD, Callis JB, Davison ER, in Wehry EL, ed, Modern Fluorescence Spectroscopy, Vol. 4, Plenum Press, New York, Ch. 4, 1981.

9. Fitzmaurice M, Bordagaray JO, Engelmann G, Richards-Kortum R, Kolubayev T, Feld MS, Ratliff NB, Kramer JR: Argon Ion Laser Induced Autofluorescence in Normal and

Atherosclerotic Aorta and Coronary Artery: Morphologic Studies, in press, American Heart Journal, 1989.

10. Harris EK, Albert A, Multivariate Interpretation of Clinical Laboratory Data, Decker, New York, 1987.

11. Schulman SG, Fluorescence and Phosphorescence Spectroscopy: Physico-chemical Principles and Practice, 1st ed., Pergamon Press, Oxford, 1977.

12. Wolfbeiss OS, Leiner M, Mapping of the Total Fluorescence of Human Blood Serum as a New Method for its Characterization, Analytica Chimica Acta, 167:203-15, 1985.

13. Keijzer M, Richards-Kortum RR, Jacques SL, Feld MS: Fluorescence Spectroscopy of Turbid Media: Autofluorescence of Human Aorta, Applied Optics, 28:4286-4292, 1989.

14. Personal communication, M. Fitzmaurice, MD, Dept. of Pathology, Cleveland Clinic Foundation, Cleveland, OH, 1989.

15. Taylor DG, Demas JN, Light Intensity Measurements I: Large ARea Bolometers with μW Sensitivities and Absolute Calibration of the Rhodamine B Quantum Counter, Anal Chem, 51:712-17, 1979.

16. Van Assendelft OW, Spectrophotometry of Haemoglobin Derivatives, Royal Van-Gorcum, Ltd, 1970.

17. Baraga JJ, Ultra-Violet Laser Induced Fluorescence Spectroscopy of Normal and Atherosclerotic Human Artery Wall, MS Thesis, MIT, Dept. of Physics, December, 1989.

DEVELOPMENT OF NEAR-IR ANGIOGRAPHY

Robert W. Flower

The Applied Physics Laboratory and The Wilmer Ophthalmological
Institute of The Johns Hopkins University and Hospital Baltimore
Maryland, USA

INTRODUCTION

Near-infrared ocular angiography was developed originally to address one specific
ophthalmological need, visualization of one of the choroidal circulation of the eye,
however, it appears the technique can be applied generally to examination of the
vasculatures of any organ accessible by use of an endoscope. A brief description of that
original ophthalmoscopic problem will provide insight as to some of the problems
associated generally with visualization of vasculatures lying within tissue.

The sensory retina is unique in that it is sandwiched between two entirely
separate vasculature networks, approximately half of its thickness being supported by
each vasculature. The inner vasculature (the retinal circulation) is the more familiar
of the two, having been routinely observed for more than a 100 years following invention
of the ophthalmoscope by Helmholtz. The outer choroidal vasculature, however, is by
comparison rarely visualized, yet the unit-time volume blood flow of the choroidal
circulation is nearly one hundred times greater than that of the retinal circulation.
Moreover, that part of the retina responsible for most acute vision, the fovea, is entirely
dependent upon choroidal blood flow, there being no retinal vessels through it.

For more than 50 years, ophthalmologist have had limited access to the choroidal
circulation when performing sodium fluorescein dye angiography of the retinal
circulation.[1] This requires intravenous injection of fluorescein dye; then as dye passes
through the retinal blood vessels, it is excited to fluorescence by light from the filtered
xenon flash tube light source of a fundus camera, and the fluorescent light is recorded
on black and white photographic film after it passes through a barrier filter which
excludes the shorter wavelength excitation light. Because of the much greater and more
rapid blood flow through the choroid, fluorescein enters the choroidal vasculature before
entering the retinal circulation. Unfortunately, both the fluorescein excitation and
fluorescence light wavelengths lie in the visible spectrum whose wavelengths are not
readily transmitted by the layer of pigmented epithelial cells lying between the retina
and the choroid nor by pigment distributed throughout the choroid itself; there is also
additional pigment (macular xanthophyll) beneath the all-important macular area of the

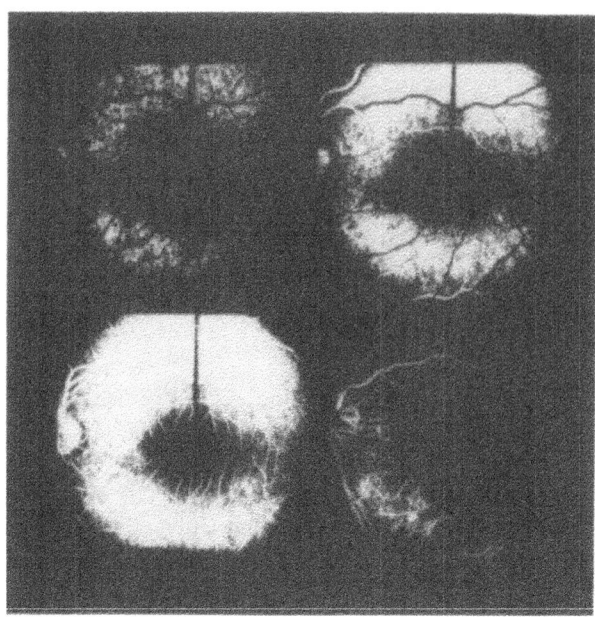

Fig.1.   Four frames of a fluorescein angiogram demonstrating the diffuse dye filling
of the choroidal vasculature. The time sequence runs from left to right and
then top to bottom.

retina virtually blocks all fluorescein fluorescence from that area.   Moreover, the
fenestrated structure of the choroidal capillary vessels permits unbound fluorescein dye
molecules to leak into and stain adjacent choroidal tissue, 40 percent of all fluorescein
injected does not bind to blood protein. As a result, passage of fluorescein through the
choroid usually is seen only as a diffuse fluorescent flush prior to filling the retinal
vessels (see Figure 1).

CHOROIDAL ICG ABSORPTION ANGIOGRAPHY

The solution to the problem of routine visualizing the choroidal circulation began
in 1974 with the use by Koger, et al.[2] of indocyanine green (ICG) dye's light absorption
properties to make so-called false color angiograms of a monkey's choroidal circulation
with high-speed color infrared film. These investigators took advantage of the fact that
wavelengths of near-infrared (IR) light are more readily transmitted than visible light
through the pigmented ocular tissues where they would be absorbed by the ICG dye as
it passed through the choroidal vasculature. Although ICG - perhaps better known by
its trade name, Cardio-Green - had proved through many years of clinical use to be
completely safe, prohibitively large amounts had to be injected intra-arterially, to
achieve successful angiograms; and even then, interpreting the results was not easy
since the angiograms could not be made fast enough to accommodate the rapid dye
passage through the choroid.

We improved that method in 1976 by using high-speed black-and-white infrared
film and illuminating the ocular fundus with a narrow spectrum of light centered at
about 790 nm (where ICG dye maximally absorbs light).[3] In this way, it was possible
to use clinically acceptable amounts of venously injected dye to observe choroidal blood
flow. Unfortunately, because those ICG dye-filled blood vessels having the greatest

cross-sectional diameters absorb the greatest amount of light, the absorption angiographic images showed mostly choroidal vein filling, the much smaller diameter arteries and choriocapillaris absorbing too little light to be seen.

FLUORESCENCE ICG ANGIOGRAPHY

It was only after discovery that the relatively weak fluorescence spectrum of ICG dye in aqueous solution is considerably enhanced when the dye molecules bind to blood protein that a method for visualization of the entire choroidal vasculature became possible. The method was based on the assumption that fluorescence of ICG dye in blood would be more or less independent of the diameter of the blood vessels containing it. Moreover, the affinity of ICG dye molecules for protein results in its nearly complete binding to blood,[4] preventing the extravasation and subsequent choroidal tissue staining which occurs with use of sodium fluorescein dye. Thus, the basic technique for near-IR ocular angiography became analogous to that used in fluorescein angiography, except that it is the near-IR, instead of the visible, region of the spectrum that is used (see Figure 2).

Initial attempts at fluorescence ICG angiography were disappointing since only the retinal vasculature was seen, but eventually the technique was perfected to the point that individual lobules of the choriocapillaris could be seen in angiograms made at the rate of 20 frames per second; the lobules of the choriocapillaris are the functional nutritive unit of the choroidal vasculature and are approximately 400 sq. microns in area (see Figure 3). Getting to that point, however, required considerable instrumentation development which focused mainly on the light source and excitation and barrier filters used in modifying a fundus camera.

Until about 2 years ago, ICG fluorescence angiograms were recorded on high-speed black-and-white infrared-sensitive film. Although its quantum efficiency is not as great as that of many electronic sensors, the spatial resolution and response time of photographic film was the best choice at the time; analog recorded electronic images were of too poor quality to be useful. Since the safe maximum level of retinal irradiance was fairly well established, selection of the optimal combination of light source, filters,

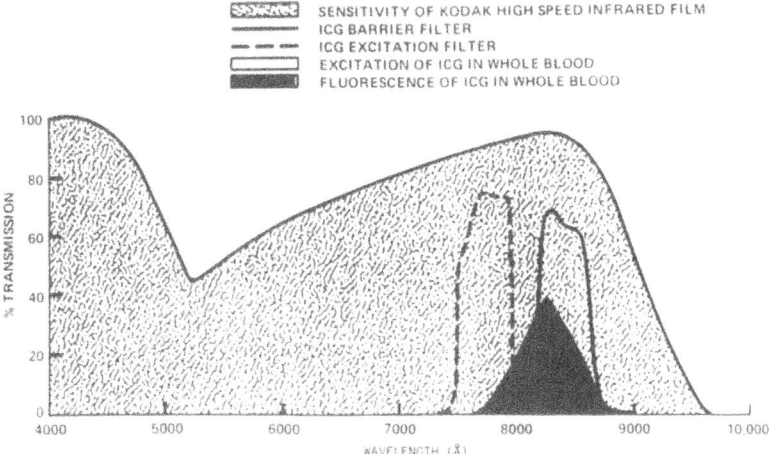

Fig.2.   Spectral characteristics of film, filters and dye used in ICG fluorescence angiogrphy.

and optical system for the ICG choroidal angiography required a considerable effort. The result was use of a hot cathode indium iodide light source and a matched pair of excitation and barrier band-pass interference filters in a modified Zeiss fundus camera. The light source had to burn continually at optimal temperature to achieve maximum brightness, and therefore, it had to be mechanically shuttered in order to safely limit retinal exposure. This configuration restricted retinal exposure to 20 sec., sufficient time to record ICG dye transmit through the choroidal vasculature using a retinal irradiance of about 175 mw/cm.[2] Nevertheless, with this instrumentation, angiograms were routinely produced at a rate of 8 to 10 per second, the speed nominally required to capture complete choroidal dye transit. For light wavelengths at about 820 nm, a spatial resolution of 20 microns on the retina was achieved in the angiograms; this compared favorably with the theoretical diffraction limited resolution of 11 microns imposed by the optics of an eye having a pupillary opening of 1.5 mm (only the central 1.5 mm of the pupillary is used to image light from the retina; the remaining annulus of pupillary opening is used to image illuminating light onto the retina). By comparison, a VHS video tape-recorded image at 820 nm resolved only 89 microns on the retina.

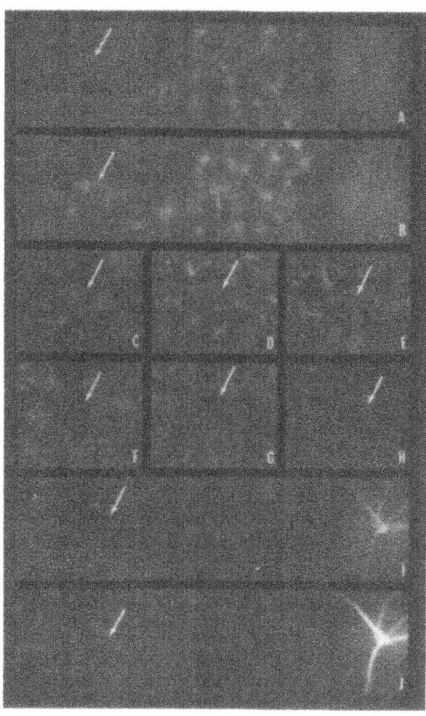

Fig.3. Ten consecutive frames from a 20-frames/sec. ICG angiogram showing the filling and emptying of a choriocapillaris lobule. The arrow indicates the same position in each frame.

CHARACTERISTICS OF FLORESCENCE ICG ANGIOGRAPHY

Compared to visible light sodium fluorescence angiography, near-IR ICG angiography has several advantages, including reduced light scatter by small amounts of extravascular blood, unclear vitreous, or diffuse cataract. Serial angiographic studies can be performed on the patient within very short time periods, and photophobic patients are virtually untroubled by the near-IR light source; in fact, it has been shown possible to perform near-IR angiography without inducing mydriasis.

There is one other characteristic of ICG in blood which perhaps makes it unique amongst presently available bio-compatible injectable dyes. As the concentration of ICG in blood increases up to the level of 0.03 mg/ml, so does fluorescence intensity, but beyond that level fluorescence intensity decreases. This phenomena is known as fluorescence quenching and is thought to be the result of dimer formation.[5] Its occurrence makes it possible to provide a calibration standard which permits quantitative dye concentration measurements. It is possible to reach concentrations as high as 0.08 mg/ml in the human ocular blood vessels by venus injection of a small 50 mg/ml ICG dye bolus. While sodium fluorescein dye also can undergo fluorescence quenching, it occurs at a concentration nearly 1 log unit greater than for ICG. Intravenous injection of sodium fluorescence at a concentration on the order of 1,500 mg/ml would be necessary to cause the phenomena to occur in the ocular vasculatures, but concentrations greater than 250 mg/ml are not available. It is possible, therefore, using ICG to provide a standard for calibrating the dye concentration and fluorescence intensity in the human ocular vasculatures regardless of unknown attenuation caused by intervening ocular tissue.[6]

ANALYSIS OF ICG CHOROIDAL ANGIOGRAMS

Although high quality human choroidal angiograms were made routinely, the problem of interpreting the complex and rapidly changing choroidal dye-filling patterns remained. The same by-eye-only method of pattern recognition applied to the relatively slow 2-dimensional retinal vasculature dye filling is not adequate to cope with the complexities of rapid dye filling of the multi-layered, 3-dimensional choroidal vasculature. Computer-aided image analysis of the large amount of data in a typical choroidal angiogram had been under consideration for some time, but since experience has shown that solutions to problems based on expensive, custom-made instrumentation do not readily find acceptance in the clinical environment, we postponed using that approach until relatively inexpensive, commercially available hardware components became available. The recent advent of such hardware, capable of real-time digital image recording, significantly reduced the loss of spatial resolution inherent with analog recording, making feasible the use of sensors having higher quantum efficiencies than photographic film and, consequently, use of alternative light sources also. As result, we now record ICG angiograms using either a thermoelectrically-cooled CCD video camera or an extended-red sensitive Newvicon camera and a filtered quartz halogen lamp light source. Images are real-time digitally recorded as 420 by 540 pixel arrays at a rate of 15 frames per second over a period of about 2 seconds, resulting in 32 consecutive images.

Since this discussion is intended to focus mainly on technology related to obtaining choroidal angiograms, a detailed description of angiogram analysis is not necessary, but a brief one is appropriate in view of the fact that future changes in that technology will be significantly influenced by the potential clinical relevancy of data currently derived from the angiograms.

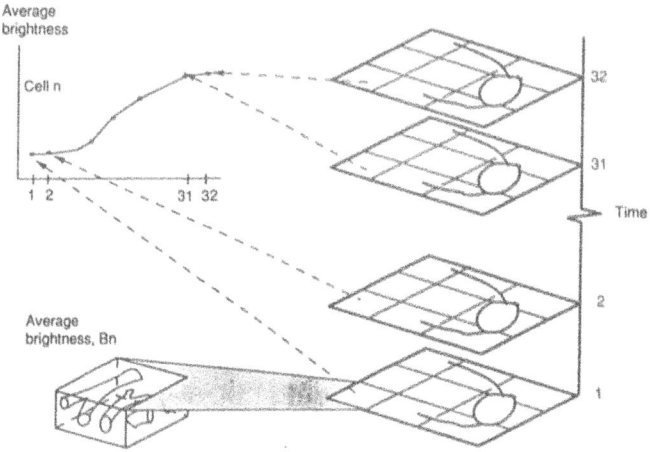

Fig.4. Schematic representation of the algorithm used to perform time sequence analysis of 32 consecutive ICG angiogram frames.

Most of the previous image processing methods - including our own - have concentrated on enhancing static images, but since blood flow is a dynamic process, our software algorithm development focused on time-sequence analysis of angiograms.[7] To do this, each fundus image in the angiographic sequence is divided into a regular grid pattern, each grid corresponding approximately to a 0.25 mm sq. sector of the fundus, approximately equal to the area of a choriocapillaris lobule. Then the average brightness of all such segments in every image is computed, providing a time history of the average brightness for every segment into which the fundus was divided (see Figure 4). The average brightness time history of these segments generally take the form of a sigmoid curve which begins at some initial background level and then increases to a point of maximum brightness as ICG dye fills the vasculature volume within that segment.

The first time derivative of these brightness time-history curve is cyclical, and the frequency of the cyclical variations in these curves corresponds to the frequency of the subject's cardiac cycle. From the time-varying data for all segments of the grid, a three-dimensional surface is generated, the X-Y plane of which corresponds to the fundus area in the angiograms and the topography of which represents the dye-filling dynamics of the area. It has been shown that such surfaces, where the Z-axis represents the instantaneous filling rate during systole (as determined by the cyclical time-varying brightness curve), uniquely characterize the choroidal circulation of the subject eye. That is, such surfaces generated for data from the same eye obtained on different days, separated by as much as a year, are virtually identical (see Figure 5).

Grouping the characteristic three-dimensional surfaces according to various topographical features they have in common can provide, for the first time, a basis for discriminating amongst the various individual choroidal dye-filling patterns. This was recently demonstrated using angiograms made on a colony of 22 diabetic monkeys for which angiograms were made on two different occasions, separated by one year.

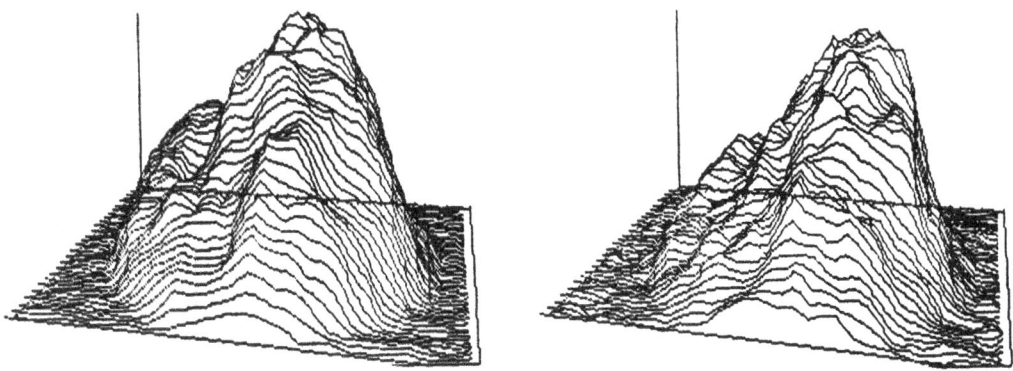

Fig.5.   Two 3-dimensional surfaces characterizing the ICG dye filling of the same
subject's choroidal vasculature on two different occasions separated by
approximately two months. Although the heart rate (94) and blood pressure
(148/84) on the first occasion (left surface) were significantly different from
those on the second occasion (h.r.=56, b.p.=138/74), the two surfaces are nearly
identical.

These three-dimensional surfaces can be used to demonstrate a number of
physiological changes which occur in the choroidal vasculature. For example, significant
changes in the distribution of relative filling rates across the choroid have been
demonstrated following breathing of ten-percent CO_2 in air which dilates vessels;
likewise, opposite changes in the distribution result from breathing 100 percent oxygen
which causes the vessels to constrict (see Figure 6). This approach to analyzing
choroidal angiograms makes it possible to draw conclusions about such physiological
parameters as vasculature compliance and choroidal blood-flow resistance, all of which
are important to understanding the role of the choroidal circulation in etiology of
various retinal diseases.

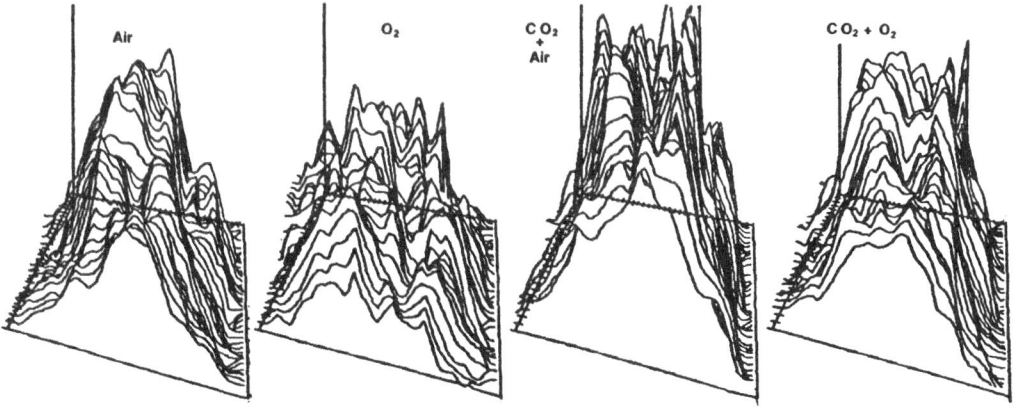

Fig.6.   Characteristic 3-dimensional surfaces demonstrating the relative choroidal
blood flow rates of the same eye as a result of breathing the indicated gas
mixtures.

Potentially, the same characteristics of ICG fluorescence angiography which make possible visualization of the choroidal circulation can be used for visualizing any other vasculature beds which are accessible by endoscopy. An example is visualization of the gut circulation by modifying a standard colonoscope. The object of this effort is to provide a way to identify areas of ischemic bowel and to pinpoint sources of occult bleeding.

Modification of the colonoscope, however, proved to be somewhat more challenging than a straight forward addition of appropriate filters to the optical system of the instrument. The overall efficiency of the colonoscope optics proved to be considerably less than that of a fundus camera because of the use of coherent fiber optics both to provide illumination and to transmit images. In order to compensate for the inefficiencies of the fiber optics, a potentially more efficient source of illumination was investigated, use of a diode laser whose output matches the peak absorption of ICG dye in blood. For this purpose, a continuous operation 0.5 watt diode laser with a 790 nm wavelength output was chosen. Because of the monochromatic output of the laser, no excitation filter is required in the illumination optical path; only a band-pass barrier filter is needed in front of the CCD camera sensor. Preliminary in vitro experiments with this configuration indicate it to be on the order of five times more efficient than the excitation/barrier filter pairs used in the fundus camera modification.

To date this new configuration of the colonoscope has been used only on the intestinal vasculature of a rat, but the results have been extremely promising (see Figure 7). Moreover, the apparently greater efficiency of the laser diode illumination system is being applied to our present fundus camera for viewing the choroidal circulation. In addition, we plan to apply similar modifications to other scopes (e.g., laparascope, gastroscope, and cystoscope) and to make initial evaluations of other internal organ vasculatures.

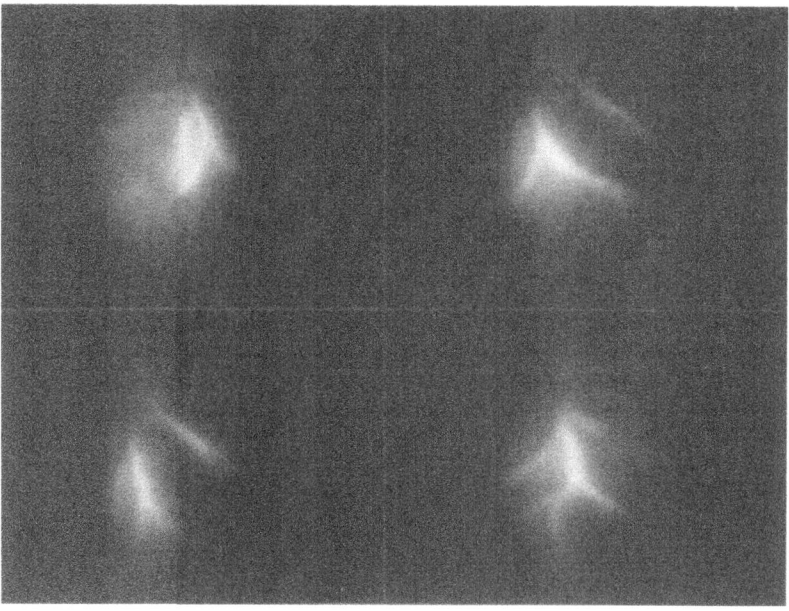

Fig.7.   Four frames of an ICG angiogram demonstrating blood flow through the intestine of a rat.

THE FUTURE OF ICG ANGIOGRAPHY

There are currently several manufacturers who are marketing or planning to market equipment for performing ICG angiography of the choroidal circulation. Two of these use fundus optics modified as described above, but one utilizes a revolutionary approach in a device known as the scanning laser ophthalmoscope (SOP). This device uses a confocal optical configuration which causes a spot of approximately 50 microns to be raster-scanned across the fundus, with reflected or fluorescent light image of the fundus being constructed point-by-point using a single pixel sensor. The theoretical advantage of such a system is that scattered light can be rejected before arriving at the sensor, leaving only directly reflected light to form an image. Such a configuration offers potentially much greater contrast in angiographic images.

There is one additional area of future investigation which should be mentioned: the search for and implementation of angiography using bio-compatible dyes having greater quantum efficiency than ICG. There is at least one such candidate dye currently available (like ICG it is a tri-carbocyanine dye), but before its use can be approved, even on an experimental basis, I believe infrared angiography will have to be first generally accepted as a useful clinical modality. I also believe that given the rapidly increasing interest in choroidal angiography, its clinical usefulness rapidly will become apparent and accepted. Moreover, applying this same technology to examination of other organs beside the eye may well accelerate its acceptance.

REFERENCES

1. P. Bischoff, and R.W. Flower, Ten years experience with choroidal angiography using indocyanine green dye: a new routine examination or an epilogue?, Documenta Ophthalmologica 60:235 (1985).
2. K. Kogure, N.J. David, U. Yamanouchi, et al., Infrared absorption angiography of the fundus circulation, Arch. Ophthalmol. 83:209 (1970).
3. R.W. Flower, and B.F. Hochheimer, A clinical technique and apparatus for simultaneous angiography of the separate retinal and choroidal circulations, Invest. Ophthal., 12:248 (1973).
4. G.R. Cherrick, S.W. Stein, C.M. Leevy, et al., Indocyanine green: observations on its physical properties, plasma decay, and hepatic extraction, J. Clin. Invest. 39:592 (1960).
5. R.C. Benson, and H.A. Kues, Fluorescence properties of indocyanine green as related to angiography, Phys. Med. Biol., 23:159 (1978).
6. R.W. Flower, and B.F. Hochheimer, Quantification of indicator dye concentration in ocular blood vessels, Exp. Eye Res. 25:103 (1977).
7. G.J. Klein, R.H. Baumgartner, and R.W. Flower, An image processing approach to characterizing choroidal blood flow, Investigative & Visual Science 31, No. 4 (1990).

PART 4

THERAPEUTIC TECHNIQUES

ULTRAVIOLET RADIATION LAMPS FOR THE PHOTOTHERAPY OF PSORIASIS

B.L. Diffey

Regional Medical Physics Department
Dryburn Hospital, Durham   DH1 5TW

INTRODUCTION

The first treatment of psoriasis with a source of artificial ultraviolet radiation (UVR) is credited to Sardemann(1) who used a carbon arc lamp of the type developed by Finsen at around the turn of the century. These lamps were unpopular in clinical practice because of their noise, odour and sparks(2), and were superseded by the development of the medium-pressure mercury arc lamp. In the 1960s, a variety of metal halides was added to mercury lamps to improve the emission in certain regions of the ultraviolet and visible spectra(3). Fluorescent lamps were developed in the late 1940s and, since then, a variety of phosphor and envelope materials have been used to produce lamps with different emissions in the ultraviolet region. Today there exists a wide range of different types of lamps which are used for the phototherapy of psoriasis. This chapter will review the physical characteristics of lamps and photoirradiation systems used for so-called UV-B phototherapy of psoriasis. It will not cover sources emitting primarily UV-A radiation which are used in conjunction with photosensitising agents (psoralens) to treat psoriasis.

THE NATURE OF ULTRAVIOLET RADIATION

Ultraviolet radiation covers a small part of the electromagnetic spectrum. Other regions of this spectrum include radiowaves, microwaves, infra-red radiation (heat), visible light, X-rays and gamma radiation. The feature that characterises the properties of any particular region of the spectrum is the wavelength of the radiation.

Ultraviolet radiation spans the wavelength region frcm 400 to 100 nm. Even in the ultraviolet portion of the spectrum the biological effects of the radiation can vary enormously with wavelength and for this reason the ultraviolet spectrum is further subdivided into three regions.

UV-C (290 to 100 nm). The rays in this wavelength interval do not pass through the Earth's atmosphere and so are not present in terrestrial sunlight. Nevertheless, UV-C is produced by many artificial ultraviolet sources and if present can be particularly damaging to the eyes.

UV-B (320 to 290 nm). It is these rays which are primarily responsible for nearly all of the biological effects following exposure to sunlight. UV-B produces sunburn, suntan, and, following many years of exposure, premature ageing of the skin and skin cancer. Exposure of the eyes to UV-B can produce photokeratitis and conjunctivitis. The only well-established benefit of exposure to UV-B in normal skin is the production of vitamin D.

UV-A (400 to 320 nm). These rays are closest to the visible spectrum, can pass through window glass and are least harmful in humans on a dose-for-dose basis. Nevertheless UV-A radiation can produce erythema, tanning and probably skin cancer, but the doses required are about 1000 times greater than those with UV-B.

THE PRODUCTION OF ULTRAVIOLET RADIATION

Ultraviolet radiation is produced artificially by the passage of an electric current through a gas, usually vapourised mercury. The mercury atoms become excited by collisions with the electrons flowing between the lamp's electrodes. The excited electrons return to particular electronic states in the mercury atom and in doing so release some of the energy they have absorbed in the form of optical radiation, that is, ultraviolet, visible and infra-red radiation.

The spectrum of the radiation emitted consists of a limited number of discrete wavelengths (so-called 'spectral lines') corresponding to electron transitions characteristic of the mercury atom, and the relative intensity of the different wavelengths in the spectrum depends upon the pressure of the mercury vapour. For lamps containing mercury vapour at about atmospheric pressure (so-called medium-pressure mercury arc lamps), radiation is emitted with several different wavelengths in the UV-C, UV-B, UV-A, visible and near infra-red(IR-A) regions. By adding traces of metal halides to mercury vapour lamps, both the power and the width of the spectrum emitted, particularly in the UV-A and visible regions, may be enhanced.

Alternatively ultraviolet radiation can be produced by utilising the phenomenon of fluorescence. A fluorescent tube is a low pressure mercury vapour lamp which has a phosphor coating applied to the inside of the envelope. At low pressures in mercury vapour there is a predominant spectral line at a wavelength of 253.7 nm and radiation of this wavelength is efficiently absorbed by the phosphor on the lamp envelope. This results in the re-emission of longer wavelength fluorescence radiation. The wavelength range of the fluorescence radiation will be a property of the chemical nature of the phosphor material and the composition of the lamp envelope. Phosphors are available which produce their fluorescence radiation mainly in the visible (for artificial lighting purposes), the UV-A, or the UV-B regions.

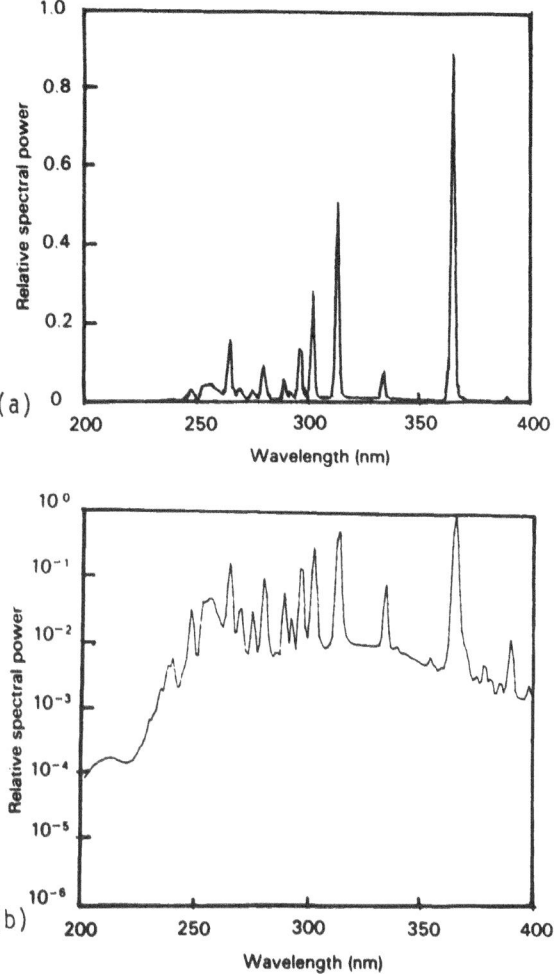

Fig. 1. The spectral power distribution of ultraviolet radiation from a medium pressure mercury arc lamp: (a) linear scale on ordinate; (b) logarithmic scale on ordinate.

Spectral Power Distributions

It is common practice to talk loosely of 'UV-A lamps' or 'UV-B lamps'. However such a label does not characterise adequately ultraviolet lamps since nearly all phototherapy lamps will emit both UV-A and UV-B, and even UV-C, visible light, and infra-red radiation. The only correct way to specify the nature of the emitted radiation is by reference to the spectral power distribution. This is a graph (or table) which indicates the radiated power as a function of wavelength. The data are obtained by a technique known as spectroradiometry(4). Figure 1 shows the spectral power distribution of ultraviolet radiation emitted by a medium pressure mercury arc lamp (commonly called a 'hot-quartz' lamp in the U.S.A.) which has been used for many years for the phototherapy of skin diseases. In Fig 1a, the relative power emission is plotted on a linear scale. Although lamp spectra are commonly plotted using a linear scale, this representation may not be the most appropriate. Since the erythemal sensitivity of normal skin varies by four orders of magnitude(5) over the spectral region 200 to 400 nm, it is helpful to be able to discern components of the spectrum which may be of small amplitude in physical terms but nonetheless important in photobiological terms. Figure 1b shows the spectrum from the same lamp with the relative power plotted on a logarithmic scale. We can see now the characteristic wavelengths present in all mercury lamps are superimposed upon a low level continuous distribution of radiation. The exact shape of the continuum, particularly at wavelengths less than 300 nm, depends on factors including the lamp envelope material and the vapour pressure of the mercury.

Stability of Radiation Output

The radiation output from ultraviolet lamps will fluctuate with factors such as the voltage of the main supply, lamp current stabilization and operating temperature(6,7). Medium and high pressure lamps take several minutes to reach a stable output, since shortly after striking these lamps there is an excess of liquid mercury which vapourizes during the heating process. Also since these lamps cannot be struck again until they have cooled down, it is not feasible to switch them on and off between patients unless there is a gap of about 30 minutes or more. Fluorescent lamps, on the other hand, reach their full output within one minute of switching on, and yield maximum radiation output when the lamp is running in free air at an ambient temperature of about 25°C. As the temperature increases, the output decreases(8). This problem can be particularly severe in irradiation units which incorporate large numbers of fluorescent lamps packed closely together, unless adequate forced-air cooling by fans is incorporated into the unit.

Some medium and high pressure lamps will be located behind optical glass filters in order to selectively remove unwanted components of the emission spectrum. The problem here is that glass filters will change their transmission spectrum as they heat up and this can result in an appreciable change(8) in both the quality and quantity of UVR received during patient irradiation if patients are treated shortly after switching on the lamps. The solution is to run the lamp for 20 minutes or so before exposing patients to allow the glass filters to reach thermal equilibrium.

The output from ultraviolet lamps deteriorates with time. There is a 'running-in' time with all lamps during which the period rate of fall in radiation output is considerably greater than for longer times. For fluorescent lamps the running-in period is about 100 hours, but is only about 20 hours for medium and high pressure lamps. The useful lifetime of most ultraviolet lamps is between 500 to 1000 hours. After this period the output will have fallen to around 80% of the value at the end of the running-in period.

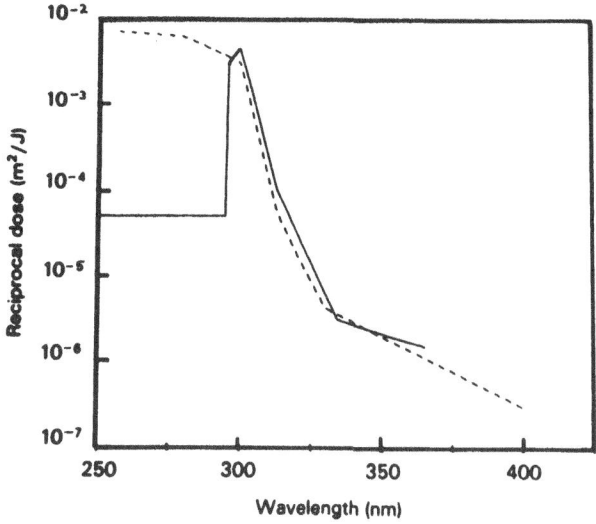

Fig. 2. The action spectrum for the lowest effective daily dose for clearing psoriasis (———) (9), and the action spectrum for minimal erythema at 24 h (- - - -) (10). The horizontal solid line from 250 to 290 nm represents the reciprocal of the highest daily doses tried and found not to be effective in clearing psoriasis (9).

The UV output of medium and high pressure lamps deteriorates more rapidly than the visible light output. With fluorescent lamps, however, the relative decrease in radiation output with usage is more or less independent of wavelength; in other words the spectrum of the radiation remains approximately constant even though the absolute radiation output decreases.

THE BIOPHYSICAL BASIS OF PHOTOTHERAPY

It is the degree of UVR-induced erythema that appears after irradiation which limits the daily exposure to phototherapy lamps and, consequently, it is important to optimize the therapeutic response for an accepted degree of erythema. As both the therapeutic response of psoriasis to UVR and the treatment limiting factor of erythema are wavelength-dependent(9,10), each factor requires consideration in the selection of a radiation source for the treatment of psoriasis.

Spectrum of therapeutic response of psoriasis

Based on the response of psoriasis to monochromatic UVR(11-14), Parrish(9) constructed an action spectrum (Fig. 2) of the reciprocal of the lowest effective daily dose (LEDD) of radiation to clear psoriasis. Even with doses considerably greater than those required for minimal erythema, no therapeutic response was seen with wavelengths less than 290 nm (UV-C). Throughout the UV-B and UV-A spectral regions, the daily

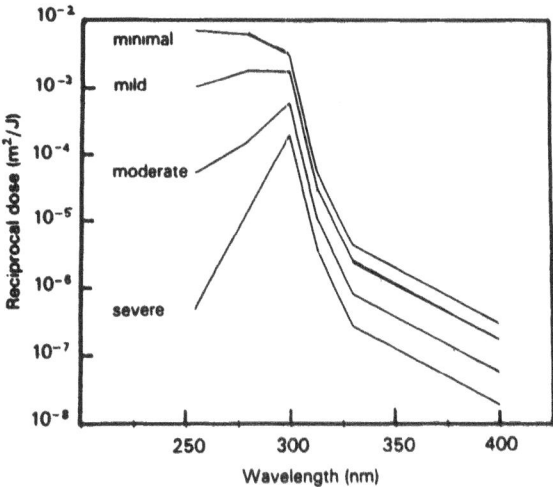

Fig. 3. Action spectra for different degrees of erythema at 24 h (15).

dose required to clear psoriasis was similar to the minimal erythema
dose (MED). It is possible to calculate the LEDD for a given lamp by
combining the spectral power distribution of the lamp with the action
spectrum for clearance of psoriasis:

$$LEDD = \int E(\lambda)d\lambda / \int E(\lambda)p(\lambda)d\lambda \qquad (1)$$

where $E(\lambda)$ is the spectral emission of the lamps at wavelength λ nm, and
$p(\lambda)$ is the reciprocal of the minimum daily dose required for clearance
of psoriasis (Fig. 2) at wavelength λ nm.

Spectrum of erythemal response of human skin

The erythemal response of human skin to UVR is highly wavelength-
dependent and is described by the erythema action spectrum in which the
reciprocal of the dose required for a given degree of erythema is
plotted against wavelength. As the slope of the dose-response curve for
UVR-induced erythema is also wavelength-dependent(10), the shape of the
erythema action spectrum will depend upon the degree of erythema chosen
as the end point (Fig. 3).

By combining the spectral power distribution of a lamp with the
action spectrum for a given degree of erythema (mild, moderate, etc), it
is possible to calculate the erythemal dose (EryD) needed to produce the
given degree of erythema from a particular lamp:

$$EryD = \int E(\lambda)d\lambda / \int E(\lambda)e(\lambda)d\lambda \qquad (2)$$

where $E(\lambda)$ is the spectral emission of the lamp at wavelength λ nm; and
$e(\lambda)$ is the reciprocal of the dose required for the given degree of
erythema (Fig. 3) at wavelength λ nm.

The Phototherapy Index

A phototherapy index which describes the therapeutic effectiveness of a particular lamp in the treatment of psoriasis at each degree of erythema chosen as an acceptable end point (mild, moderate, etc) may be defined as(15)

$$\text{Phototherapy index} = \text{EryD/LEDD} \qquad (3)$$

When this index is less than one, the degree of erythema chosen as the end point will be reached before the patient has received the lowest effective daily dose required for clearance of psoriasis. Clearly, therefore, it is important to induce a degree of erythema such that the index is equal to or greater than one, at which point clearance of psoriasis can be expected at the chosen degree of erythema. So, the higher the index, the more effective the treatment.

ULTRAVIOLET LAMPS FOR PHOTOTHERAPY

It will be apparent that ultraviolet lamps are of two broad types:

i) medium and high pressure mercury arc lamps with or without metal halide additives;

ii) fluorescent tubes with a variety of spectral emissions depending upon the chemical nature of the phosphor lining the lamp envelope

The lamps in category (i) are point sources of bright light and may be used singly or mounted in front of reflectors in a vertical column. A single column houses five or six lamps, and from one to four columns may be used to treat patients depending on factors such as cost, floor space, and so on. Fluorescent lamps, on the other hand, are normally mounted vertically around the inner walls of a cylindrical or hexagonal cabinet, or horizontally in a sunbed and sun canopy. A whole body treatment unit may contain from 20 to 60 tubes the length of 2 metres.

In terms of suitability for phototherapy it is possible to classify lamp systems into one of five categories:

Type A: A single medium-pressure mercury arc or metal halide lamp mounted on a vertical pole; this type of phototherapy unit has been available for many years.

Type B: A vertical column containing five or six optically-filtered high-pressure metal halide lamps.

Type C: A canopy or cubicle containing UV-B fluorescent lamps which emit significant amounts of UV-C (< 290 nm) radiation. Lamps falling into this category are the Westinghouse FS sunlamp, Philips TL12, and Sylvania UV21 lamps.

Type D: A canopy, sunbed or cubicle incorporating UV-B fluorescent lamps which emit negligible amounts of UV-C radiation. A lamp falling into this category includes the Wolff Helarium.

Type E: A newly-developed UV-B fluorescent lamp (Philips TL01) emitting a narrow band of radiation around 311 - 312 nm.

Spectral power distributions

The spectral power distributions measured by the author on lamps from each of the five categories are shown in Fig. 4. Note that for lamps types A and C, the UV-B emission is only tenfold or less than that of the UV-C emission, but with lamp types B, D and E the UV-B power is several hundred times that of the UV-C power (Table 1).

Phototherapy Indices

By combining the spectral power distributions shown in Fig 4 with the action spectra shown in Figs 2 and 3, it is possible to calculate phototherapy indices for each lamp type and for different degrees of delayed erythema(Table 2). It may be seen that with lamps types A and C (those with a relatively high UV-C component) at least a mild to moderate erythema must be achieved for satisfactory clearing of psoriatic lesions. On the other hand, only a minimal erythema is required for satisfactory clearing with lamp types B, D, and particularly E. These theoretical predictions have been borne out by results from experimental studies(16-19).

Treatment Times

Treatment times depend not only upon the spectrum of the radiation, but also upon factors such as electrical power, number of lamps, lamp to skin distance and differences in patient susceptibility to sunburn. Very roughly, initial treatment times with lamp types A and C are about 0.5 to 1 minute, with slightly longer initial treatment times of 1 to 3 minutes for lamp types B, D and E. Treatment times need to be increased throughout a course of phototherapy in order to maintain an erythema on increasingly-acclimatized skin. Regimens for phototherapy have been reviewed by van der Leun and van Weelden(20).

Table 1 The UV-B and UV-C components of Phototherapy Lamps

Type of Lamp	Ratio UV-B : UV-C
A	2 - 10
B	> 500
C	~10
D	> 250
E	800

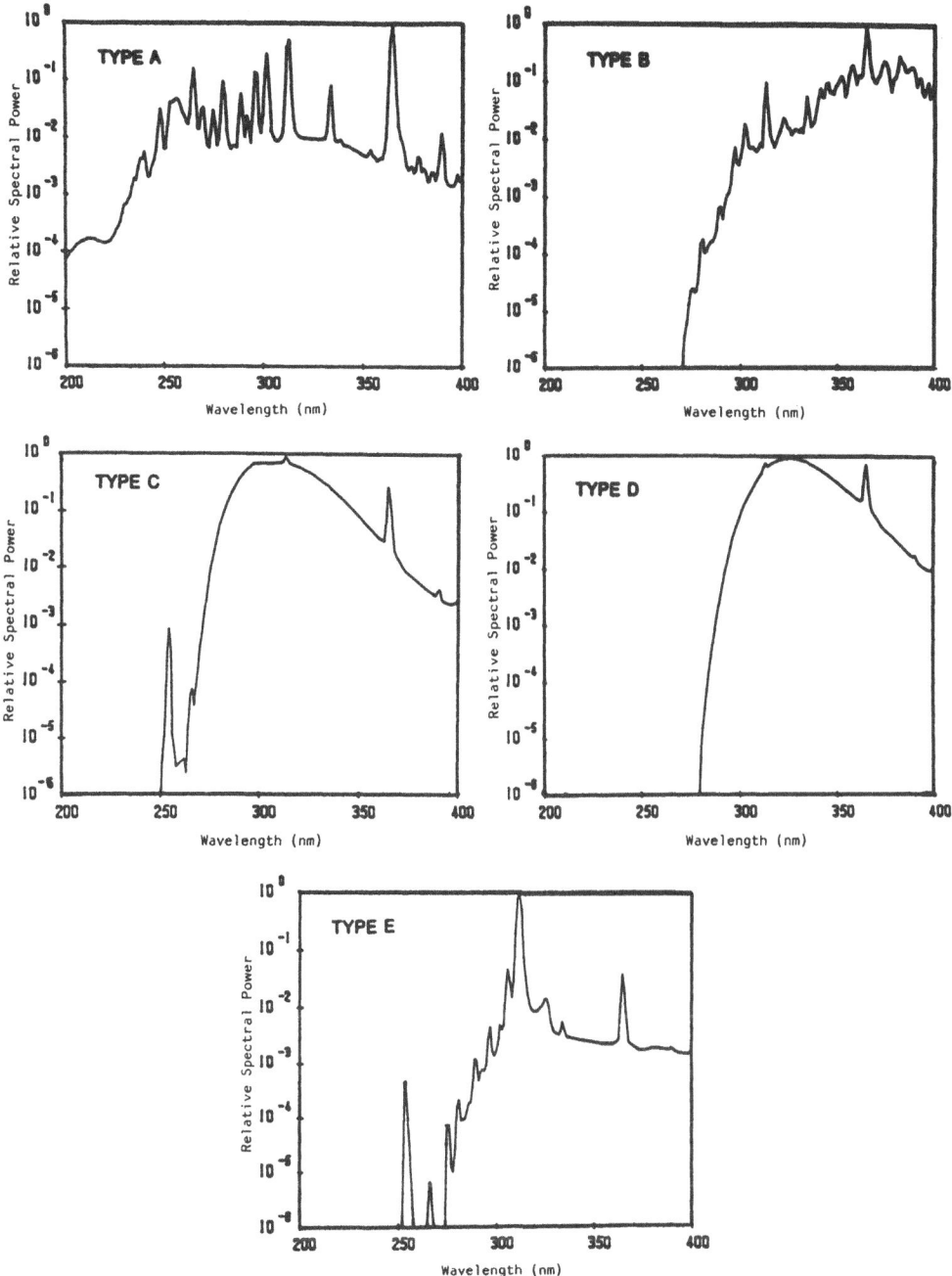

Fig. 4. Spectral power distributions of different types of phototherapy lamp.

Type A: unfiltered medium pressure mercury arc lamp
Tupe B: optically filtered iron iodide lamp
Type C: fluorescent sunlamp (Philips TL12)
Type D: Wolff Helarium lamp
Type E: narrow-band UV-B flurescent lamp (Philips TL01)

Table 2 Phototherapy Indices

Type of Lamp	Phototherapy Index for degree of delayed (24h) erythema		
	Minimal	Mild	Moderate
A	0.3	0.9	5
B	1.2	2.1	7
C	0.6	1.3	5
D	1.3	2.2	7
E	1.5	2.8	8

Uniformity of Irradiation

Lamp type A limits radiation to regions such as the chest or back. In order to achieve whole body irradiation, the lamp to patient distance would be so great that the treatment times become unacceptably long. All other lamp types are designed for partial or whole body irradiation. Studies have shown(21-24) that the vertical distribution of ultraviolet radiation in phototherapy cabinets is non-uniform when fluorescent lamps are used, with a reduction in intensity of up to 20-50% near the ends of the tubes compared with the middle, resulting in significantly lower radiation doses at the extremities(24). In contrast when columns incorporating five or six high pressure metal halide lamps are used the vertical variation of radiation intensity is normally no more than 10%(24).

Photoirradiation systems of type B have the advantage that not all the five or six lamps need to be switched on so that partial body irradiation is possible. This is not the case with fluorescent lamp systems (types C, D and E), although, of course, fluorescent lamps are available in a variety of lengths ranging from 30cm to 2m so that units designed for treating small areas, such as the hands or feet, are available.

In addition to the geometrical problems discussed above, it is apparent that the variation in irradiance over a patient's skin will depend also on the topology and self-shielding of the patient's body. Measurements of the ultraviolet dose received at different body sites from irradiation in phototherapy cabinets have shown(25,26) that a large fraction of the body surface area receives more than 70% of the maximum exposure, while areas such as the groin and the axillae receive a smaller fraction, as expected.

Installation Requirements

Lamp type A requires no special electrical supply or room modification. With fluorescent lamp systems (C,D and E) it may be necessary to install a high current mains electrical supply for whole body units incorporating large numbers of lamps. Consideration should also be given to maintaining a satisfactory environmental temperature by installing air conditioning units. The high power lamps used in system type B nearly always require the installation of high current mains supply and some means of room temperature stabilization.

In the UK there are no specific guidelines published on the installation of phototherapy machines. This type of apparatus falls under the umbrella of Medical Electrical Equipment and as such it should be installed in compliance with the British Standard BS5724.

Servicing

There is very little required in the way of servicing phototherapy units. Lamps surfaces should be cleaned regularly to remove dust and skin, both of which will attenuate the radiation. Lamps should be replaced when the irradiance has dropped sufficiently low that treatment times become unacceptably long. When this point occurs depends very much on local circumstances and workload. Changing lamps, both high pressure and fluorescent, is a tedious but straightforward process and could be carried out by hospital electronics technicians. Maintenance Contracts on phototherapy equipment which are offered by some suppliers may be unnecessary and uneconomical if adequate technical support is available locally.

SAFETY

Considerations of safety apply to both patients and staff.

Patient Safety

It goes without saying that there must be adequate protection against electrical hazards. The patient (or operator) should not be able to touch any live electrical parts and all metal components, such as handrails and safety grids, must satisfy national electrical safety standards and codes of practice.

The patient should not be able to come in contact with bare lamps. In high pressure (type B) units this is achieved by interposing glass filter(s) between the patient and the lamps. Manufacturers of fluorescent lamp units have adopted a variety of approaches(27): wire mesh between the lamps and the patient; ultraviolet-transmitting perspex screen; teflon sleeves around individual lamps; U-shaped reflectors around each lamp such that the straight arms of the 'U' project beyond the lamp towards the patient, so inhibiting contact between the tube and patient. However protective devices which do not fully enclose the lamps will not protect the patient against flying glass if a fluorescent lamp implodes.

Other features which relate to patient safety include hand-rails to support patients during treatment, a cord within the cabinet that can be pulled by the patient to summon help, doors that can be opened easily by the patient from inside the irradiation cabinet, non-skid floor in the cabinet, and adequate air flow to maintain patient comfort during the irradiation period.

Finally, there is one potential hazard associated with the type B phototherapy units which incorporate optical filters to allow either UV-A or UV-A plus UV-B irradiation. If only UV-A irradiation is intended but the operator fails to ensure that the correct filter is in place, the patient may be exposed to high doses of UV-B (depending on the treatment times) which can lead to severe, painful erythema. A similar hazard exists with combination units incorporating both UV-A and UV-B fluorescent lamps.

Staff Safety

It is well known that exposure to ultraviolet radiation can produce harmful effects in the eyes and the skin and measurements have indicated(28) that an ultraviolet exposure hazard exists in the vicinity of many lamps used for dermatological phototherapy. At 1 m from many of the unenclosed lamps used for phototherapy, particularly the type A lamps, the maximum permissible exposure for 8 hour working periods recommended by national regulatory authorities(29,30) can be exceeded in less than 2 minutes(28) For this reason operators should always wear protective eyewear and keep away from the primary beam as much as is practicable. Nevertheless it is inevitable that staff will be exposed to low-level scattered and reflected UVR. It is quite feasible for staff to be exposed to sub-clinical irradiation, so that no acute effects such as erythema are manifest, but to receive annual occupational doses that can result in a risk of non-melanoma skin cancer during a working life of 40 years comparable with that of outdoor workers(33). Measures which can be taken to minimize the unnecessary exposure of staff to ultraviolet radiation include(32): proper engineering design of UVR-emitting installations and suitable enclosures; wearing of appropriate goggles or face shields accompanied, if necessary, by suitable UVR-opaque clothing; limiting access to persons directly concerned with the work; and ensuring that staff are aware of the potential hazards associated with exposure to ultraviolet radiation sources.

REFERENCES

1. Bach H. Ultraviolet light. New York, 1916

2. Scott BO. Clinical uses of ultraviolet radiation. In: Therapeutic Electricity and Ultraviolet Radiation (Licht S, ed.). Connecticut: Elizabeth Licht, 1967; 325-78

3. Phillips R. Sources and Applications of Ultraviolet Radiation. London: Academic Press, 1983; 265

4. Diffey BL. The spectral emissions from ultraviolet radiation lamps used in dermatology. Photodermatol 1986; 3: 179-85

5. McKinlay AF, Diffey BL. A reference action spectrum for

ultraviolet induced erythema in human skin. CIE Journal 1987; 6:
17-22

6.   Hartmann KM. Action Spectroscopy. In 'Biophysics' (Hoppe W,
     Lohmann W, Markl H, Ziegler H, eds) Berlin: Springer-Verlag, 1983;
     136

7.   Henderson ST, Marsden AM. 'Lamps and Lighting', London: Edward
     Arnold 1972; 213-269

8.   Diffey BL. The stability of light sources : implications for
     photobiological studies. Photochem Photobiol 1988; 47 : 317-320

9.   Parrish JA. Phototherapy of skin diseases. In: 'The Science of
     Photomedicine' (Regan JD, Parrish JA, eds). New York: Plenum Press,
     1982; 511-31

10. Farr PM, Diffey BL. The erythemal response of human skin to
     ultraviolet radiation. Brit J Dermatol 1985; 113: 65-76

11. Parrish JA, Jaenicke KF. Action spectrum for phototherapy of
     psoriasis. J Invest Dermatol 1981; 76 : 359-62

12. Parrish JA. The treatment of psoriasis with longwave ultraviolet
     light (UV-A). Arch Dermatol 1977; 113 : 1525-8

13. Fischer T. UV-light treatment of psoriasis. Acta Derm Venereol
     1976; 56 : 473-9

14. Young E, van der Leun JC. Treatment of psoriasis with longwave
     ultraviolet light. Dermatologica 1975; 150 : 352-4

15. Diffey BL, Farr, PM. An appraisal of ultraviolet lamps used for
     the phototherapy of psoriasis. Brit J Dermatol 1987; 117 : 49-56

16. Schothorst AA, Boer J, Suurmond D, Kenter CA. Application of
     controlled high dose rates in UV-B phototherapy for psoriasis.
     Brit J Dermatol 1984; 110: 81-7

17. Farr PM, Diffey BL, Marks J. Phototherapy and anthralin treatment
     of psoriasis: new lamps for old. Brit Med J 1986; 294: 205-207

18. Van Weelden H, Baart de la Failla H, Young E, van der Leun JC. A
     new development in UVB phototherapy of psoriasis. Brit J Dermatol
     1988; 119: 11-19

19. Green C, Ferguson J, Lakshmipathi T, Johnson BE. 311 nm UVB
     phototherapy - an effective treatment for psoriasis. Br J Dermatol
     1988; 119: 691-6

20. Van der Leun JC, van Weelden H. UVB Phototherapy: principles,
     radiaton sources, regimens. In: 'Therapeutic Photomedicine'
     (Honigsmann H, Stingl G, eds) Basel: Karger, 1986; 39-51

21. Horwitz SN, Frost P. A phototherapy cabinet for ultraviolet
     radiation therapy. Arch Dermatol 1981; 117: 469-73

22. Petrozzi JW, Kaidbey, KM, Kligman AM. Topical methoxsalen black light in the treatment of psoriasis. Arcb Dermatol 1977; 113: 292-5

23. Heub CC, Krause W. Die relative Bestrahlungsintensitat auf der menschlichen Hautoberflache in UV - Bestrahlumgikabea. Hautarzt 1985; 36: 142-5

24. Chue B, Borok M, Lowe NJ. Phototherapy units: comparison of fluorescent ultraviolet B and ultraviolet A units with a high-pressure mercury system. J Am Acad Dermatol 1988; 18: 641-5

25. Diffey BL, Harrington TR, Challoner AVJ. A comparison of the anatomical uniformity of irradiaton in two photochemotherapy units. Br J Dermatol 1978; 99: 361-3

26. Fanselow D, Crone M, Dahl MV. Dosimetry in phototherapy cabinets. J Am Acad Dermatol 1987; 17: 74-7

27. Mountford PJ. Phototherapy and photochemotherapy ultraviolet irradiation equipment. Photodermatology 1986; 3: 83-91

28. Diffey BL, Langley FC 'Evaluation of Ultraviolet Radiation Hazards', Report 49 London: IPSM, 1986

29. National Institute for Occupational Safety and Health 'Criteria for a Recommended Standard... Occupational Exposure to Ultraviolet Radiation', Washington, DC: DHEW, 1972

30. National Radiological Protection Board. 'Protection against Ultraviolet Radiation in the Workplace' London: HMSO, 1977

31. Diffey BL. Ultraviolet radiation and skin cancer: are physiotherapist at risk? Physiotherapy 1989; 75: 615-616

32. Diffey BL. Ultraviolet Radiaton Safety. In: 'Handbook of Laboratory Health and Safety Measures' (Pal SB, ed) Lancaster: MTP Press Ltd, 1985: 255-97

PHOTOTHERAPY AND PHOTOCHEMOTHERAPY OF PSORIASIS

Thomas B. Fitzpatrick

Department of Dermatology, Massachusetts General
Hospital
Boston, Massachusetts, 02114, U.S.A.

INTRODUCTION

Two discoveries in the last two decades markedly improved the use of ultraviolet radiation for the treatment of disease or disorders of the skin, including psoriasis:

(a) the introduction of high-intensity UVA sources (Sylvania) which are used in a therapeutic regimen in which topically administered UVA interacts in the skin with orally administered photoactive drugs; this interaction functions as a potent anti-inflammatory modality for the treatment of psoriasis and several other inflammatory disorders of the skin;

(b) new protocols using newly developed high-intensity total-body UVB irradiators, but substituting emollients for crude tar, with therapeutic results equal to the Goeckerman treatment.

PHOTOCHEMOTHERAPY WITH PSORALENS

The first high-intensity source of UVA was made available by U.S. Sylvania in 1972; in January 1974, the first patient with psoriasis was treated at the Massachusetts General Hospital in Boston with an oral photoactive drug (psoralen) followed by exposure to UVA, using the newly developed Sylvania irradiator. Following successful treatment of 21 patients with this method of oral psoralens and UVA, a new general pharmacologic principle was proposed called photochemotherapy[1]: the combination of light + drugs as a therapeutic modality. This new method was, for convenience, termed PUVA (Psoralen + UVA) and the treatment, PUVA photochemotherapy. The questions of efficacy and safety (i.e., acute side effects) of PUVA photochemotherapy were resolved by multicenter cooperative controlled trials. These large clinical trials were completed in over 35 centers in the United States and in Europe during the late 1970s in over 5000 patients. PUVA became an FDA-approved treatment for psoriasis in 1982. Treatment programs based on these highly successful multicenter trials were then available. Most important was an NIH-sponsored PUVA Follow-up Study comprising the original population of 1380 patients and was organized first in 1975; this prospective study, now in its fifteenth year, helped define the long-term efficacy and toxicity of PUVA treatment for psoriasis.

Optronic Techniques in Diagnostic and Therapeutic Medicine
Edited by R. Pratesi, Plenum Press, New York, 1991

In the 16 years since PUVA photochemotherapy was introduced, there has been substantial progress toward optimizing this treatment's usefulness in the treatment of psoriasis. This progress has been a result of clinical and laboratory investigations that have helped to define optimal dosimetry, criteria for patient selection, and the usefulness of protocols that combine PUVA with other agents, including methotrexate, UVB, or retinoids. After a decade of thorough study, PUVA's efficacy for the treatment of psoriasis is unquestioned. The two principal unanswered questions about PUVA are its mechanism of action and its long-term toxicity.

NEW TYPE OF UVB PHOTOTHERAPY

The technology used in the development of the UVA irradiators was then applied to the development of new irradiators for UVB; in 1983, the first computerized, total-body, high-intensity UVB irradiators were made available. Using these newly developed UVB irradiators, new protocols were established for the treatment of psoriasis[2]. Adrian et al.[3] using the paired-comparison method proved that an emollient (e.g., mineral oil) + UVB was as effective as tar + UVB. Inasmuch as crude coal tar used in the Goeckerman program was no longer necessary, UVB phototherapy then became an ambulatory treatment--in the physician's office or by home UVB units. UVB phototherapy, using emollients and administered in ambulatory settings is now widely used throughout the world for the treatment of psoriasis.

THE ERA OF PHOTOMEDICINE

The introduction of photochemotherapy in the early 1970s inaugurated a renaissance of photobiology applied to the treatment of a heterogeneous group of diseases; this field was called photomedicine. Because nonionizing ultraviolet radiant energy has such low penetration compared to ionizing radiation, this new technology has been most beneficial in treating diseases of the skin. Therapeutic photomedicine is the use of nonionizing electromagnetic radiation that interacts with endogenous or exogenous chromophores for the treatment of diverse diseases. UVA is divided into UVA_1 (340-400 nm) and UVA_{11} (320-340 nm) each of which appear to have different effects on the skin. UV phototherapy uses nonionizing electromagnetic radiation (NEMR) alone (as in UVB phototherapy); photochemotherapy is the interaction of NEMR with an exogenous and artificial chromophore (as in PUVA).

MECHANISMS OF ACTION OF PUVA

The therapeutic action of PUVA in psoriasis is not known. We do know that psoralens interact with nucleic acids and intercalate between base pairs. Exposure of cells containing psoralen to long-wave UV radiation (320-400 nm, UVA) results in the formation of cyclobutane adducts in deoxyribonucleic acid (DNA). Initially, monofunctional adducts of psoralen with pyrimidine bases are formed[4-6]. Further radiation leads to the formation of bifunctional adducts and cross-links between DNA strands[7]. Photoconjugation of psoralens to DNA is well established from electron microscopic evidence in vitro and in vivo[8]. Further evidence for PUVA's effects on epidermal DNA comes from the observed dosage-related increase in sister chromatid exchanges in human epidermal cells exposed to PUVA in vitro[9]. PUVA's therapeutic effects on psoriasis may be due to psoralen-DNA interaction which results in a decrease in the rate of epidermal proliferation. The relative importance of monofunctional and bifunctional adducts for PUVA's therapeutic and toxic effects has not been established. PUVA's therapeutic effect in psoriasis may be due to factors other than

its interaction with nuclear DNA. Psoralen adducts may damage cell organelles, such as mitochondria, which are essential for proliferation. Alternatively, PUVA's effect on the immune system may be responsible for part
of its therapeutic effect[10]. The Nobel Laureate in Medicine, Dr. Charles
Huggins, has said, "The only thing worse than not knowing how a drug works
is not having a treatment at all."

LARGE CLINICAL TRIALS IN THE UNITED STATES AND EUROPE ESTABLISH EFFICACY IN REMISSION INDUCTION OF PSORIASIS

PUVA's efficacy in the treatment of psoriasis was first demonstrated
in paired-comparison studies: only half the patient's body surface was
exposed to PUVA[1]. Soon after publication of the initial trial groups in
the United States, European centers began trials of PUVA for psoriasis.
As detailed below, treatment protocols and patient populations varied among
groups. In spite of these differences, nearly all trials shared high success rates in clearing psoriasis. Whether PUVA photochemotherapy can induce
long-term remission is less clear.

Initial dosage was determined from an individual's Skin Phototype, with
increments depending on skin type and degree of erythema from the previous
treatment. European centers tended to utilize more frequent treatments
(four per week) and to determine the minimum phototoxic dose (MPD) of the
patient; this permitted the use of higher dosages of UVA[11,12]. Examination by a physician prior to each treatment permitted the administration
of maximal dosages. Advocates of this more aggressive European regimen
assert that this approach permits a therapeutic effect before melanization
and stratum corneum thickening occur which reduce the penetration of UVA.

Cooperative clinical trials in the United States studied 1380 patients
at 16 centers and formed the nucleus for what became the PUVA Follow-up
Study[13]. The initial clinical trials in the United States administered
treatment two or three times per week. Eighty-eight percent of patients
treated cleared. The type of psoriasis influenced the success rate. While
2% of patients with guttate or plaque-type psoriasis experienced treatment
failure, among patients with erythrodermic psoriasis 16% failed to clear.
Patients with guttate psoriasis cleared most rapidly and those with
erythrodermic psoriasis required the largest number of treatments to clear.
The most frequent acute side effects were phototoxic erythema (10% of
treatments), nausea (3%), pruritus (14%), headache (2%), and dizziness (2%).
This study also demonstrated that patients remaining on maintenance therapy
remained free of significant disease more often than patients who discontinued treatment.

An 18-center European PUVA Study (EPS) of 3175 patients reported a
success rate nearly identical to that reported by the first U.S. multicenter
trial (89% vs 88%)[12] in a protocol calling for four treatments per week
(the U.S. trial used two different protocols, 2x and 3x weekly). In the EPS
trial patients with the uncomplicated forms of psoriasis cleared in a
comparable number of treatments to that in the U.S. trial, but were cleared
in a shorter time and with a far lower cumulative dosage of PUVA. The
failure rate for erythrodermic psoriasis was also similar in the U.S. and
European study (14% in the EPS and 16% in the U.S. multicenter trial).

Additional studies that document PUVA efficacy in clearing psoriasis
have been published by investigators in several countries including
Scandinavia, France, and Britain. These prospective trials all confirm
the results of the earliest investigations, and of the large multicenter
trials; PUVA is effective in rapidly clearing psoriasis and there is little
short-term risk, if patient selection and dosimetry are carefully monitored.

EFFICACY OF MAINTENANCE FOLLOWING REMISSION INDUCTION

An Austrian study[14] has clearly demonstrated that maintenance treatment is important to keep the patient in remission after clearing. This maintenance regimen is used in many centers: after clearing, all patients receive maintenance treatment for two months. Maintenance therapy is then stopped in those patients who are still in remission at this time. Follow-up observations have shown that the chance of these patients remaining in remission for at least 6 to 12 months is excellent. A Dutch study noted that 58% of patients who received maintenance treatment stayed in remission for one year, and 40% for 2 years[15]. At the study's conclusion, the majority of patients were receiving treatment every other week or less frequently.

The PUVA Follow-up Study investigated factors related to continued reliance on PUVA therapy. After an average of 1.8 years, patients with the largest number of previous hospitalizations, and those with histories of adverse reactions to sunlight were most likely to discontinue PUVA[16]. After initial clearing, patients averaged 30 treatments per year. Factors associated with increased rates of PUVA utilization included a history of isomorphic response to injury (Koebner phenomenon) and multiple hospitalizations for psoriasis.

ACUTE SIDE EFFECTS OF PUVA PHOTOCHEMOTHERAPY

While photochemotherapy using 8-methoxypsoralen (8-MOP) is now widely accepted as the standard treatment for severe forms of psoriasis, not infrequently its use is limited by drug intolerance (nausea and/or vomiting) or rendered more difficult because of its relatively narrow therapeutic range which requires a very cautious UVA dosimetry, especially in Skin Phototype I and II individuals.

5-Methoxypsoralen (5-MOP), a less erythemogenic compound, has been reported to be effective in clearing, but with less risk of phototoxic reactions. Recently, a new liquid preparation of 5-MOP has been developed and tested in a large-scale clinical trial. 5-MOP, when given in the same dosage as 8-MOP, clearly was not as effective as 8-MOP in inducing remission of psoriasis. However, by doubling the dose of 5-MOP the therapeutic results came close to those obtained with 8-MOP. In particular, there was no significant difference in the most important parameter, namely, the cumulative UVA dose required for clearing. Likewise, no treatment failures were found in the high-dose 5-MOP group. Even with higher doses of 5-MOP there was a low incidence of phototoxicity and no intolerance (nausea, dizziness, etc.). There was a high patient acceptance in the 5-MOP-treated groups. As with 8-MOP PUVA, the combination with retinoids also increases the efficacy of 5-MOP PUVA.

MODIFICATIONS OF THE THERAPEUTIC PUVA PHOTOCHEMOTHERAPY PROTOCOL

Other Psoralens

The majority of trials with psoralens for psoriasis have utilized 8-MOP, and more recently 5-MOP. Several psoralen derivatives are being tested in psoriasis to find the compound with the least risk of long-term side effects, especially skin carcinogenesis. One of these compounds is 7-methylpyridopsoralen (7-MPP), a monofunctional furocoumarin[17]. In vitro studies have demonstrated that it has a high binding affinity for DNA and, upon irradiation with UVA, forms only monoadducts. 7-MPP is less mutagenic and three to four times less carcinogenic in the albino mouse than the bifunctional psoralens 8-MOP and 5-MOP.

168

4,6,4'-Trimethylangelicin (4,6,4'-TMA) is another monofunctional furocoumarin, which has been developed as a potential photochemotherapeutic agent for the treatment of psoriasis[18]. It has been shown in vitro and in preliminary clinical studies that 4,6,4'-TMA has a strong antiproliferative activity and is not phototoxic.

Combination of PUVA with Other Modalities

Retinoids (Chemophotochemotherapy). The combination of PUVA with oral retinoids[19] has been a major improvement. Most of the studies have been performed with etretinate, an aromatic retinoid. Though of only limited value in the treatment of psoriasis if used as monotherapy, retinoids appear to be potent accelerators of PUVA therapy. When the retinoid (etretinate or isotretinoin) is given five to seven days prior to PUVA therapy in a dosage of 1 mg/kg body weight and is continued until clearing, the response rate is accelerated, the number of treatments is reduced by one-third, and most importantly, the total PUVA dosage is reduced by more than half when compared to conventional PUVA. After complete remission has been induced, regular PUVA maintenance therapy without retinoid is continued. Follow-up studies have indicated that the tendency for relapses is not greater than when PUVA is given alone[12]. A particularly important advantage of chemophotochemotherapy is that the final clearing dosage of UVA, which is usually employed as a constant treatment dosage during maintenance therapy, is much lower than in standard PUVA. Thus, the cumulative UVA dosage during maintenance can also be kept at lower levels[19,20]. In addition, "poor PUVA responders" whose disease cannot be cleared completely by PUVA alone are frequently controlled by this regimen. The high efficacy of this combined therapy has been confirmed by other large-scale studies[20-23]. Chemophotochemotherapy has now been adopted as routine in many European centers.

UVB and PUVA. PUVA therapy has also been combined with UVB[24]. UVB has also been used in combination with methotrexate and oral retinoids. These combinations all proved to be effective in clearing psoriasis. With this treatment less than half the number of exposures were required for clearing psoriasis than with the use of either modality alone. The patients received UVA and UVB simultaneously after ingestion of 8-MOP. The total UVA dosage was less than half that necessary for PUVA alone, and the total UVB dosage was about 20% of that required for successful UVB phototherapy. The lowest dosages of both PUVA and UVB required in this protocol may result in less cumulative damage to the skin.

Methotrexate. The combination of methotrexate and PUVA has been used with success in patients with generalized pustular psoriasis and psoriatic erythroderma in uncontrolled studies. Morison et al.[25] treated 30 patients with a three-week course of methotrexate (5 mg orally every 12 hours for three doses) followed by a combination of PUVA and methotrexate. After the psoriasis cleared, methotrexate was stopped and PUVA therapy was used alone as maintenance therapy. Clearing of psoriasis was achieved in 28 patients, one-half of whom had had treatment failures with PUVA or UVB alone. Duration of treatment, number of exposures, and total UVA dosage were significantly reduced. Since methotrexate was discontinued at the end of the clearing phase, the total dosage remained well below the minimum reported for hepatotoxicity. The only significant adverse effect observed was unexplained prolonged phototoxicity in eight patients. Nevertheless, long-term treatment with concomitant methotrexate may remain a source of concern because of its synergistic carcinogenic potentials.

Abundant evidence indicates that PUVA is mutagenic in a variety of test systems. Schenley[26] showed that PUVA causes mutation in mammalian cells in a dosage-dependent fashion. Increased DNA synthesis and an elevated frequency of presumed gene mutations have been detected in the circulating lymphocytes of patients treated with PUVA.

The induction of cutaneous tumors in animals exposed to very high doses of intraperitoneal methoxsalen and very high doses of UVA radiation was reported in the 1950s[27]. Although oral administration in animals appeared to be less carcinogenic, this may have been due to the poor absorption of 8-MOP in rodents.

Squamous Cell Carcinoma in Humans

The 16-center PUVA Follow-up Study documented nearly a threefold increase in the risk of well-differentiated squamous cell carcinoma (SCC) in patients who were treated for an average of 2.1 years[28]. At that time, a total of 2.0% of patients (30/1380) had developed SCCs. The proportion of SCC was greater than expected in this group and there were SCCs in areas habitually exposed to the sun (e.g., legs). After five years of follow-up, a dosage-dependent increase in the risk of SCC has been noted, with patients receiving high-dosage therapy exhibiting a more than 10-fold increase in the risk of SCC over members of the cohort exposed to low dosages of PUVA.

The most recent report from the PUVA-48 Group[29] that, in general, received relatively low cumulative doses yields similar findings. There was a trend of increasing number of SCCs in patients with higher doses. Most SCCs in the PUVA-48 study occurred on body areas not normally exposed to sunlight but exposed to PUVA (e.g., the lower extremities).

However, the extent of the risk of nonmelanoma skin cancer from long-term PUVA photochemotherapy is still controversial because of the lack of concordance of U.S. and European reports. While the first U.S. 16-center study[30] showed a definite correlation between total UVA, which was independent of prior exposure, and other risk factors, in the large European experience[31-34] there was no increase in skin cancers except in patients exposed to other risk factors (arsenic and ionizing radiation). Honigsmann concludes, "At present there exists no explanation for this discrepancy as the mean irradiation in the 16-center study comes close to that used in our own and the European trial. Current discussions focus on the question of whether few but large phototoxic irradiations are more harmful than frequent small irradiations and whether an aggressive regimen with rest periods is safer or carries more risks than a continuous nonaggressive treatment. One can, of course, compare the various regimens in use and the incidence of tumors but too many variables are involved to permit meaningful conclusions."

After 16 years of extensive use of PUVA in thousands of patients throughout the world it is still not clear what are the limits of radiation dosage within which PUVA can be considered safe for the treatment of psoriasis.

Professor Malcolm Greaves of London has said, "For bad psoriasis we have methotrexate or PUVA--it is easier to remove a squamous cell carcinoma than to remove the liver"[35].

Combination therapies, such as that with retinoids, reduce total cumulative radiation load, and thus, perhaps, the risk of oncogenic effects[19].

Melanoma

In the PUVA Follow-up Study[30], four of the 1380 patients have developed malignant melanoma. This is slightly more than twice the risk for melanoma in the general population, but this observed increase is not statistically significant. Other large studies have not found an increase in the incidence of melanoma.

Noncutaneous Carcinomas

Changes in immunologic function that occur with PUVA may alter the risk of noncutaneous malignancy. After 10 years of follow-up, no overall excess of deaths due to malignancy compared with other causes of death has been noted. Four cases of leukemia or lymphoma were noted, an incidence not significantly different from that expected in a cohort of this size[36]. Other investigators have reported cases of acute myeloid leukemia and pre-leukemic syndromes.

OTHER TOXIC EFFECTS OF PUVA

Ophthalmologic Risks

In five years of prospective study[37] there was no significant difference in the prevalence of cataracts among patients aged 53 to 72 in the PUVA cohort and the Framingham Study.

Hepatotoxicity

Analysis of more than 5000 serial laboratory examinations of PUVA-treated patients has not revealed any abnormalities in hepatic or renal function[30]. There is a widespread belief that psoralens are hepatotoxic; there are no scientific data that 8-MOP is hepatotoxic.

RISK-BENEFIT CONSIDERATIONS

There are only a few therapeutic options for patients with generalized psoriasis (Table I). For the present, PUVA photochemotherapy can be regarded as a safe, highly effective therapy for the control of psoriasis and the technique, if used according to protocol, will result in a method of treating generalized psoriasis that is an acceptable alternative to methotrexate, the only other available therapy with comparable efficacy[38]. Therapy of psoriasis with methotrexate also has problems, especially liver toxicity; it is the last alternative for generalized psoriasis in persons under 50 years of age. UVB phototherapy is a first choice for young patients, but UVB phototherapy has only a limited efficacy, especially in thick, plaque psoriasis, and maintenance schedules have not been as well worked out as in PUVA photochemotherapy. Unlike PUVA, UVB maintenance cannot be reduced below twice, or rarely once a week. At less frequency phototoxic burning results, and suboptimal dose leads to flare-up of psoriasis.

It is to be hoped that in the future there will not only be new methods of photochemotherapy but also new adjunctive treatments, such as new retinoids. At present retinoids are not an effective long-term monotherapy for common plaque-type psoriasis because of the bone toxicity; retinoids are, when used for short periods in combination therapy with PUVA, able to effect a significant reduction of the total cumulative doses of UVA. Moreover, patients who do not respond to PUVA or to UVB treatment as monotherapies, will clear or be maintained when combination treatments are used: retinoids + PUVA or retinoids + UVB. Combinations of methotrexate and PUVA are also highly effective in reducing the total number of treatments.

Table I. Treatment Options for Generalized (>10%) Psoriasis

	Efficacy*	Toxicity[+]
Phototherapy (UVB + emollients)	+++	A
PUVA Photochemotherapy	+++	A,B
Methotrexate	+++	C
Retinoids	+	D
Methotrexate + PUVA	++++	A,B,E
Retinoids + PUVA	++++	A,B,E

* ++++, very effective; +, limited efficacy
[+] Toxicity:
   A - Phototoxicity, induction of new lesions, lentigines
     (rare with UVB, 30% with PUVA)
   B - Dermatoheliosis ("photoaging")
     Nonmelanoma skin cancer in high-risk populations:
      Skin Phototypes I and II
       - previous Rx with ionizing radiation, including
        grenz rays
       - ?prolonged treatment sessions
   C - Hepatic cirrhosis: related to total cumulative dose,
     alcohol
     Bone marrow depression
   D - Hyperostosis after prolonged therapy, hypertri-
     glyceridemia
   E - Combination therapy reduced toxicity of monotherapy
     with PUVA or retinoids or methotrexate

Table II. Diseases Responsive to Oral Psoralen Photochemotherapy (PUVA)*

Therapeutic Response	Prophylactic Response
Psoriasis	Polymorphous light eruption
Palmoplantar pustulosis	Hydroa vacciniforme[+]
Mycosis fungoides (stages 1A, 1B)	Solar urticaria[+]
Vitiligo	Persistent light reaction[+]
Atopic dermatitis	Actinic reticuloid[+]
Generalized lichen planus	Erythropoietic protoporphyria[+]
Urticaria pigmentosa	
Pityroasis lichenoides[+]	
Lymphomatoid papulosis[+]	
Generalized granuloma annulare	
Lymphomatoid papulosis	
Prurigo nodularis	
Scleromyxedema[+]	
Graft-vs.-host reaction	
Transient acantholytic dermatosis	
Urticaria pigmentosa	

* Although all of the dermatoses listed have been reported to respond to
  PUVA therapy, not all represent a true indication for this treatment.
[+] Experience limited to small number of patients.

Therapies have a life span and, over time, certain problems emerge that are not evident until there has been extensive experience over a period of years. The problem with PUVA photochemotherapy, after 16 years of usage, is that psoralen binds to DNA. Therefore, abnormal mutations can occur and there have been a small percentage of patients (2.0%) who have developed "low grade" squamous cell carcinomas, especially on the lower extremities. This has been noted particularly in patients who have had a large number of PUVA treatments, or who have a previous history of exposure to therapeutic ionizing radiation (including grenz rays) for psoriasis, or who have ingested oral arsenical compounds (inorganic trivalent arsenic) which were used years ago in the treatment of psoriasis.

Oral PUVA photochemotherapy has been successfully used to treat more than 20 diverse disorders (see Table II). In many of these (e.g., vitiligo, polymorphous light eruption, solar urticaria, and actinic reticuloid) the length of therapy is finite and toxicity of long-term therapy is, therefore, not a consideration in treating these disorders.

Photochemotherapy using agents that will inhibit cell proliferation but which do not bind to DNA are not yet available but there is promise of such an approach, e.g., parenterally administered tin-protoporphyrin + light.[39] This new photochemotherapy proposed from Sweden has one clear and important advantage over PUVA photochemotherapy--its action is most probably not on the DNA of the epidermal cells (although these authors did not speculate on the mechanism of action). Porphyrins interacting with light have their pharmacologic action principally on cell membranes. Thus it would not alter DNA and there would not be a risk of skin cancer as a result of tin-protoporphyrin + UVA.

REFERENCES

1. J. A. Parrish, T. B. Fitzpatrick, L. Tanenbaum, M. A. Pathak, Photochemotherapy of psoriasis with oral methoxsalen and long wave ultraviolet light, N. Engl. J. Med. 291:1207-1212 (1974).
2. J. A. Parrish, Ultraviolet phototherapy of psoriasis, Pharmacol. Ther. 15:313-320 (1982).
3. R. M. Adrian, et al., Outpatient phototherapy for psoriasis, Arch. Dermatol. 117:623 (1981).
4. L. Musajo and G. Rodighiero, Studies on the photo-C4-cycloaddition reactions between skin photosensitizing furocoumarins and nucleic acid, Photochem. Photobiol. 11:27 (1970).
5. M. A. Pathak and D. M. Kramer, Photosensitization of skin in vivo by furocoumarins (psoralens), Biochim. Biophys. Acta 195:197 (1969).
6. L. Musajo, et al., Photoreaction at 3655 A linking a 3-4 double bond of furocoumarins with pyrimidine bases, Photochem. Photobiol. 6:927 (1967).
7. R. S. Cole, Light-induced cross-linking of DNA in the presence of furocoumarin (Psoralen). Studies with phage lambda, Escherichia coli, and mouse leukemia cells, Biochim. Biophys. Acta 217:30 (1970).
8. A. Lerche, J. Sondergaard, S. Wadskov, V. Leick, and V. Bohr, DNA interstrand crosslinks visualized by electron microscopy in PUVA-treated psoriasis, Acta Derm. Venereol. (Stockh.) 59:15 (1979).
9. M. R. West, M. Johansen, and M. J. Faed, Sister chromatid exchange frequency in human epidermal cells in culture treated with 8-methoxypsoralen and long-wave UV radiation, J. Invest. Dermatol. 78:67 (1982).
10. W. L. Morison and J. A. Parrish, The in vivo effect of PUVA on lympho-

cyte function (abstr.), J. Invest. Dermatol. 72:204 (1979).

11. K. Wolff, H. Honigsmann, F. Gschnait, et al., Photochemotherapie bei Psoriasis. Klinische Erfahrungen bei 152 Patienten, Dtsch. Med. Wochenschr. 100:2471 (1975).

12. T. Henseler, K. Wolff, H. Honigsmann, et al., Oral 8-methoxypsoralen photochemotherapy of psoriasis. The European PUVA Study: A cooperative study among 18 European centres, Lancet 1:853 (1981).

13. J. W. Melski, L. Tanenbaum, J. A. Parrish, et al., Oral methoxsalen photochemotherapy for the treatment of psoriasis: A cooperative clinical trial, J. Invest. Dermatol. 68:328 (1977).

14. K. Wolff and H. Honigsmann, Clinical aspects of photochemotherapy, Pharmacol. Ther. 12:381 (1981).

15. B. A. Gilchrest, J. A. Parrish, L. Tanenbaum, et al., Oral methoxsalen photochemotherapy of mycosis fungoides, Cancer 38:683 (1976).

16. R. S. Stern and J. W. Melski, Long-term continuation of psoralen and ultraviolet-A treatment of psoriasis, Arch. Dermatol. 118:400 (1982).

17. L. Dubertret, D. Averbeck, E. Bisagni, J. Moron, E. Moustacchi, C. Billardon, D. Papadopoulo, S. Nocentini, P. Vigny, J. Blais, R. V. Bensasson, J. C. Bonfard-Haret, E. J. Land, F. Zajdela, and R. Latarjet, Photochemotherapie using pyridopsoralens, Biochimie. 67:417 (1985)

18. M. Cristofolini, A. Guiotto, P. Rodighiero, G. Pastorini, P. Manzine, F. Bordin, F. Baccichetti, F. Cavlassare, G. Recchia, D. Vedaldi, F. Dall'Aqua, and G. Rodighiero, Synthesis of new 6-methylangelicins as potential agents for the photochemotherapy of psoriasis, Acta Derm. Venereol. (Stockh.) [Suppl.] 113:170 (1984).

19. P. O. Fritsch, H. Honigsmann, E. Jaschke, et al., Augmentation of oral methoxsalen-photochemotherapy with an oral retinoid acid derivative, J. Invest. Dermatol. 70:178 (1978).

20. H. Honigsmann and K. Wolff, Isotretinoin-PUVA for psoriasis, Lancet 1:236 (1983).

21. G. Heidbreder and E. Christophers, Therapy of psoriasis with retinoid plus PUVA: Clinical and hostologic data, Arch. Dermatol. Res. 264:331 (1979).

22. C. Grupper and B. Berretti, Treatment of psoriasis by oral PUVA therapy combined with aromatic retinoid (Ro 10-9359; Tigason^R), Dermatologica 162:404 (1981).

23. J. Laurahanta, et al., A clinical evaluation of the effects of an aromatic retinoid (Tigason), combination of retinoid and PUVA, and PUVA alone in severe psoriasis, Br. J. Dermatol. 103:325 (1982).

24. K. Momtaz and J. A. Parrish, Combination of UVB and PUVA in the treatment of psoriasis, J. Invest. Dermatol. 76:303 (1981).

25. W. L. Morison, et al., Combined methotrexate-PUVA in the treatment of psoriasis, J. Am. Acad. Dermatol. 6:46 (1982).

26. R. L. Schenley and A. W. Hsie, Interaction of 8-methoxypsoralen and near-UV light causes mutation and cytotoxicity in mammalian cells, Photochem. Photobiol. 33:179 (1981).

27. A. C. Griffin, Methoxsalen in ultraviolet carcinogenesis in the mouse, J. Invest. Dermatol. 32:367 (1959).

28. R. S. Stern, et al., Risk of cutaneous carcinoma in patients treated with oral methoxsalen photochemotherapy for psoriasis, N. Engl. J. Med. 300:809 (1979).

29. H. H. Roenigk, Jr., and W. A. Caro, Skin cancer in the PUVA-48 cooperative study, J. Am. Acad. Dermatol. 4:319 (1981).

30. R. S. Stern, N. Laird, J. Melski, et al., Cutaneous squamous cell carcinoma in patients treated with PUVA, N. Engl. J. Med. 310:1156 (1984).

31. H. Honigsmann, et al., Keratoses and non-melanoma skin tumors in long-term photochemotherapy (PUVA), J. Am. Acad. Dermatol. 3:406 (1980).

32. A. Lassus, et al., PUVA treatment and skin cancer. A follow-up study, Acta Derm. Venereol. (Stockh.) 61:141 (1981).

33. A.-M. Ros, et al., Long-term photochemotherapy for psoriasis: A histo-pathological and clinical follow-up study with special emphasis on tumour incidence and behaviour of pigmented lesions, Acta Derm. Venereol. (Stockh.) 63:215 (1983).

34. A. Tanew, H. Honigsmann, B. Ortel, et al., Nonmelanoma skin tumors in long-term photochemotherapy treatment of psoriasis. An 8-year follow-up study, J. Am. Acad. Dermatol. 15:960 (1986).

35. H. Honigsmann and G. Stingl, Therapeutic Photomedicine: Current Problems in Dermatology, vol. 15. S. Karger, Basel (1986).

36. N. E. Hansen, Development of acute myeloid leukemia in a patient with psoriasis treated with oral 8-methoxypsoralen and long-wave ultraviolet light, Scand. J. Haematol. 22:57 (1979).

37. R. S. Stern, J. A. Parrish, and T. B. Fitzpatrick, Ocular findings in patients treated with PUVA, J. Invest. Dermatol. 85:269 (1985).

38. R. S. Stern, Long-term use of psoralens and ultraviolet-A for psoriasis: Evidence for efficacy and cost savings, J. Am. Acad. Dermatol. 14:520 (1986).

39. L. Emtestam, L. Berglund, B. Angelin, G. S. Drummond, and A. Kappas, Tin-protoporphyrin + UVA: a new photochemotherapy, Lancet 1:1231 (1989).

LIGHT THERAPY FOR NEONATAL JAUNDICE

John F. Ennever

Department of Pediatrics
Case Western Reserve University
Cleveland, Ohio 44106 U.S.A.

INTRODUCTION

The use of light in the treatment of neonatal jaundice is no doubt the most common therapeutic use of light in medical practice today. It has been used on millions of infants worldwide since its serendipitous discovery more than 30 years ago.[1,2] There has been little change over this period in how this phototherapy is administered and essentially no improvement in its efficacy. What has changed is our understanding how light affects bilirubin, the pigment which produces the yellowish discoloration of skin called jaundice.

Bilirubin is the end product of heme catabolism in most animals, including all mammals.[3] The catabolism of heme involves two enzymatic steps; the first and rate limiting step is catalyzed by heme oxygenase, an enzyme present in the reticuloendothelial system, largely in the spleen. This enzyme catalyzes the oxidative cleavage of the closed protoporphyrin IX ring of heme at the alpha carbon, yielding three products: CO, Fe^{3+}, and the blue-green pigment biliverdin IXα. The CO is excreted by the lungs, the Fe^{3+} is salvaged for reuse in subsequent heme synthesis, and the biliverdin IXα is reduced in a second step by the enzyme biliverdin reductase to form the yellow pigment bilirubin IXα.[4]

Bilirubin IXα is a linear tetrapyrrole containing two exocyclic double bonds, between carbons 4 and 5 and carbons 15 and 16 (Figure 1). Because of the asymmetrical substitution around these two double bonds, there are two possible *configurations*, designated either Z or *E*. The configurations of these two exocyclic double bonds in the native bilirubin molecule is determined by the configuration of these double bonds in the parent molecule, heme. In the native configuration, both exocyclic double bonds are in the Z configuration. 4Z,15Z-Bilirubin IXα is a highly lipophilic molecule and is only sparingly soluble in water at physiological pH.[5] The reason for this unexpected lipophilicity is the three dimensional structure assumed by the molecule (Figure 2). Bilirubin assumes a "ridge tile" conformation in which all of the polar functional groups of the molecule are involved in intramolecular hydrogen bonds.[6] Under normal physiological conditions, the bilirubin molecule is transported in the plasma firmly bound to albumin[7,8] from its site of formation to the liver where it is taken up by the hepatocyte, and rendered more polar by the enzymatic esterification of a glucuronic acid moiety to one or both of the propionic acid side chains of the molecule. This chemically modified form of bilirubin (called "conjugated bilirubin") is much more water soluble than the parent molecule and is readily excreted, primarily into the bile.[4]

During fetal life, bilirubin formed in the developing organism is transported across the placenta and taken up by the maternal liver where the chemical modification and excretion into the bile occurs. The glucuronic acid conjugates of bilirubin are not able to pass through the placenta, so that any bilirubin conjugates formed by the fetus would be trapped on the

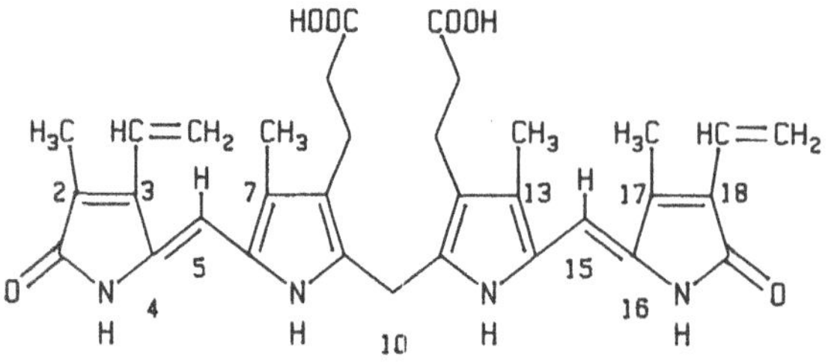

Figure 1. Linear representation of 4Z,15Z-bilirubin IXα.

fetal side. Therefore, the hepatic machinery necessary for the normal physiological clearance of bilirubin, including uptake into the hepatocyte, intracellular transport, conjugation with glucuronic acid, and biliary excretion, are all suppressed during fetal life. At birth, the newborn infant must assume total responsibility for clearance of his or her bilirubin; however, essentially none are up to the task immediately at birth.[9] The production of bilirubin by the newborn infant almost always exceeds his/her ability to eliminate it, and thus, bilirubin typically accumulates in the infant. This accumulation is usually monitored by measuring levels in the serum. By three to five days of life, however, the hepatic functions necessary to chemically solubilize and excrete the bilirubin are able to meet the demand, and bilirubin levels begin to decline. In approximately half of all newborn infants, the level of bilirubin rises sufficiently high to produce visible discoloration of the infant, termed jaundice. This is so common that it is termed "physiological jaundice" by pediatricians. [Although not intended, the term physiological jaundice could imply a benefit from having jaundice. Recent *in vitro* work by Stocker and his co-workers suggest a potential beneficial role for bilirubin as an antioxidant, effectively scavenging peroxyl radicals.[10,11] Perhaps the jaundice which is so common in newborn infants is indeed physiological, helping to protect the newborn infant as he or she adjusts to the relatively hyperoxic extrauterine environment.]

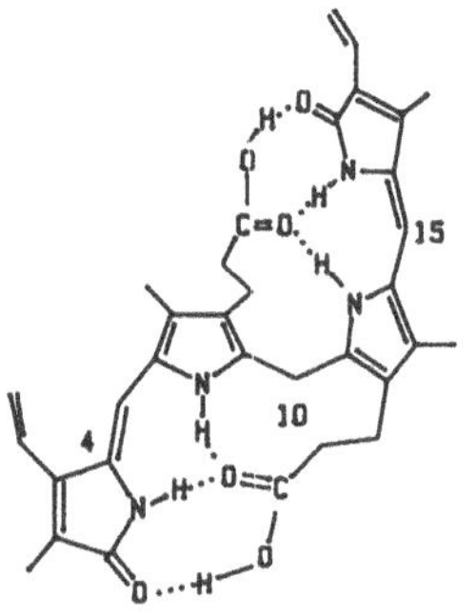

Figure 2. Three dimensional representation of 4Z,15Z-bilirubin IXα. Internal hydrogen bonds are represented with dotted lines.

The moderate elevation of bilirubin in the first few days of life is of little concern, other than perhaps to anxious new parents. High levels of bilirubin, however, are of concern because of the association with a clinical syndrome termed kernicterus, which consists of irreversible brain damage that is often severe enough to cause death in the newborn period.[12] Those infants who survive are left with profound psychomotor retardation. Before effective therapy to lower bilirubin levels was introduced in the late 1940's kernicterus was a leading cause of death in the newborn period. The initial therapy was simple mechanical removal of the bilirubin, by exchanging the jaundiced infants blood for non-jaundiced donor blood. This exchange transfusion therapy, which is still used occasionally today, is extremely time consuming, costly and carries significant morbidity and mortality. This therapy has been largely replaced by phototherapy in which bilirubin *in situ* is transformed photochemically into forms which can be eliminated by the newborn infant, bypassing the temporary metabolic block in the liver. The remainder of this chapter will review the photochemistry of bilirubin, the fate of the bilirubin photoproducts in babies, and prospects for improving the way light is used to treat neonatal jaundice.

BILIRUBIN PHOTOCHEMISTRY

The primary event in phototherapy is the absorption of a photon by bilirubin. The absorption spectrum of bilirubin bound to human albumin is shown in Figure 3. This rather simple beginning, however, is somewhat complicated in bilirubin because the molecule is not one, but rather two chromophores (pyrromethenones) joined by a saturated carbon bridge at C-10 (Figure 1). The two chromophores are quite similar, but are not identical: the pyrromethenone on the left half of the molecule has the vinyl group positioned *endo*, whereas the pyrromethenone on the right has the vinyl group positioned *exo*. Although the two chromophores are separate, they do not function independently. When bilirubin absorbs a photon, the excitation energy is localized in one or the other pyrromethenone chromophore; however, this energy can rapidly move from one chromophore to another.[13-16] As a result, the absorption of light by one half of the bilirubin molecule can lead to photochemistry in the other half. The rate at which this energy transfer takes place depends on the relative orientation of the two chromophores in space, which in turn can be influenced *in vivo* by the way in which the bilirubin is bound to albumin. This ability to transfer energy from one half of the molecule to the other has the effect of complicating (or making more interesting) the photochemistry of bilirubin. Bilirubin has been shown to undergo three photochemical reactions *in vivo* during phototherapy: configurational isomerization, structural isomerization, and oxidation.

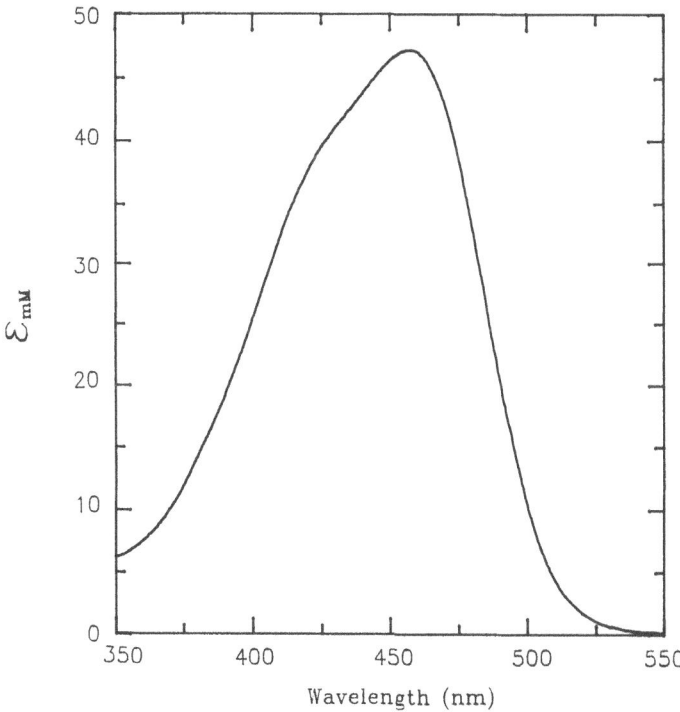

Figure 3. Absorption spectrum for bilirubin bound to human serum albumin, pH 7.4.

The most quantum efficient photochemical reaction of bilirubin is a $Z \to E$ configurational isomerization. Since bilirubin IXα contains two non-identical exocyclic double bonds, one between carbons 4 and 5 and another between carbons 15 and 16, there are four possible isomers: 4Z,15Z, 4Z,15E, 4E,15Z, and 4E,15E. Irradiation of native 4Z,15Z-bilirubin in organic solvents leads to the accumulation of the 4Z,15E and 4E,15Z isomers. These photoproducts can absorb a second photon and undergo a second photochemical reaction forming the fourth isomer, 4E,15E. The absorption spectra of these *E*-isomers of bilirubin are similar to and overlap with the absorption spectrum of the naturally-occurring 4Z,15Z-isomer. Because the $Z \to E$ isomerization is photochemically reversible, irradiation in a closed system results in the formation of a photoequilibrium mixture containing all four isomers.

The composition of the photoequilibrium mixture depends upon the solvent and the wavelength of light used. In all solvents, the native 4Z,15Z-isomer is the most abundant, and the two photon product, the 4E,15E-isomer, is the least abundant. Under most conditions, the single photon photoproducts, 4Z,15E and 4E,15Z, are formed in comparable amounts. The exception is for bilirubin bound to human albumin, where the configurational isomerization is highly regioselective for the double bond at position 15.[17] With irradiation at the absorption maximum ($\approx$ 450 nm), the ratio of 4Z,15E to 4E,15Z at photoequilibrium is nearly 100:1. In contrast, for bilirubin bound to rat or bovine albumin, this photoproduct ratio is approximately 2:1. McDonagh *et al.* have recently shown using laser irradiation that the regioselectivity of the configurational isomerization is wavelength dependent for bilirubin bound to human albumin.[16] Irradiation at 457.9 nm produces a 4E,15Z:4Z,15E ratio of less than 0.015, which increases nearly tenfold with irradiation at the extreme long wavelength edge of the absorption band (Figure 4). The basis for this wavelength dependence is thought to be energy transfer processes in the excited state (for a detailed discussion, see reference 16).

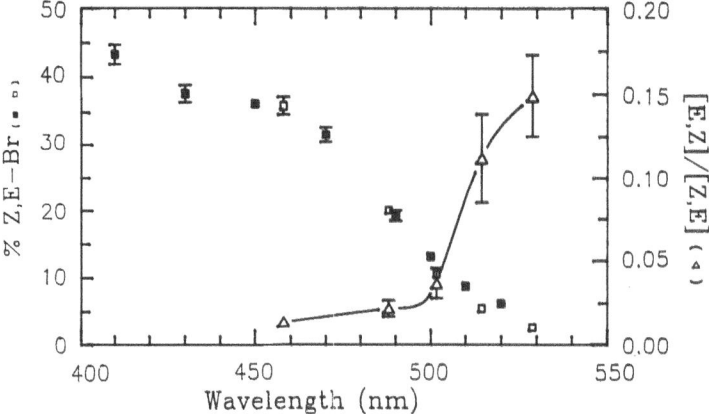

Figure 4. Dependence of the photoequilibrium concentration and isomeric composition of configurational isomers on excitation wavelength. Data with open symbols are from McDonagh *et al.*[16]; data with solid symbols are unpublished data from Ennever.

The photoequilibrium concentration of the principal configurational isomer, 4Z,15E, produced by the irradiation of bilirubin bound to human albumin is highly dependent upon the excitation wavelength.[18] The highest concentration is produced with irradiation at the short wavelength edge of the bilirubin absorption band, e.g., more than 40% with 410 nm light, and the least with irradiation at the long wavelength edge, e.g., less than 3% with 528.7 nm light (Figure 4). For polychromatic irradiation, the photoequilibrium mixture depends upon the relative emission of the source within the bilirubin absorption band.[19] This wavelength dependence in the amount of 4Z,15E-isomer formed is also seen *in vivo* in the serum of infants treated with phototherapy (*vide infra*).

The bilirubin isomers containing an *E* configuration at one or the other double bonds are significantly more polar than the native all *Z* molecule. The reason for this is illustrated for the 4Z,15E-isomer in Figure 5. In this configuration, the half of the molecule containing the *E* double bond is prevented from forming the internal hydrogen bonds with the propionic acid side chain from the opposite half. As a result of the polar side chain is exposed to the solvent, making the molecule less lipophilic. The *E*-isomers of bilirubin are thermodynamically unstable and they readily revert to the more stable 4Z,15Z form in the dark. However, when these photoproducts are bound to human albumin they are stable for many hours at physiological pH and temperature.

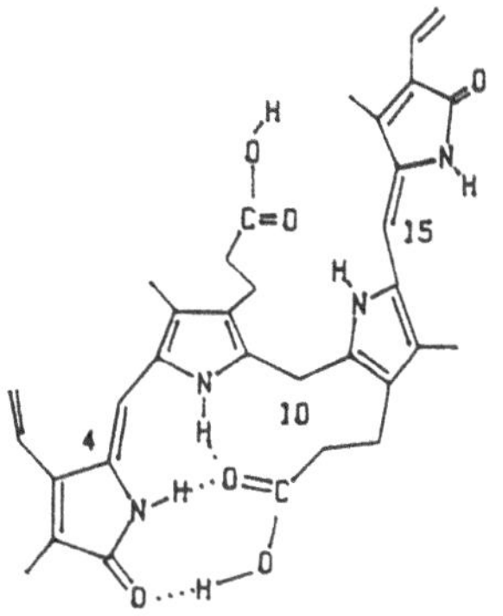

Figure 5. Three dimensional representation of 4Z,15E-bilirubin IXα.

The quantum yield for the configurational isomerization of bilirubin bound to human albumin is quite high. In 1982, Lamola et al.[20] and Sloper and Truscott[21] independently reported quantum yield values of 0.2 for irradiation near the absorption maximum. More recent work suggests a somewhat lower value of 0.109 at 457.9 nm which decreases with longer wavelength irradiation (Agati et al., unpublished). Despite this downward revision, the formation of the 4Z,15E-isomer remains the most quantum efficient photochemical reaction of bilirubin bound to human albumin.

Structural isomerization

Bilirubin can undergo a second class of isomerization reaction, called structural isomerization, to form a cyclized product shown in Figure 6. This product has been given a variety of names, including lumirubin,[17] photobilirubin II,[22] and (*EZ*)-cyclobilirubin.[23] Two changes have occurred in going from native bilirubin to lumirubin: the configuration of the double bond between carbons 4 and 5 has changed from *Z* to *E*, and the vinyl group attached to carbon 3 has undergone addition to the adjacent pyrrole ring. Unlike configurational isomerization, structural isomerization can occur only on the left half of the bilirubin molecule, because the requisite vinyl group on the right half of the molecule is attached to the outside of the pyrrole ring (Figure 1). Another dissimilarity is that the structural isomerization to form lumirubin is *essentially* irreversible; lumirubin does not readily photoisomerize back to native bilirubin. Lumirubin does undergo a further reversible photochemical reaction, forming an *E*-isomer at the 15 position. Lumirubin contains two chiral centers, denoted with asterisks in Figure 6. Because of these two chiral centers, there is not one lumirubin, but rather four diastereoisomers, consisting of a pair of enantiomers. McDonagh et al. have shown that lumirubin formed from bilirubin bound to human albumin is optically active, whereas lumirubin formed by irradiation of bilirubin in an achiral organic solvent is not optically active.[24] Since it appears that the photochemistry of bilirubin during phototherapy occurs on albumin-bound bilirubin, it is likely that lumirubin formed *in vivo* is also optically active. Because lumirubin is unstable and undergoes autoxidation in the dark. It has not yet been crystallized so that little is known of its binding to albumin or other properties. In fact, it structure is not known with certainty.[25]

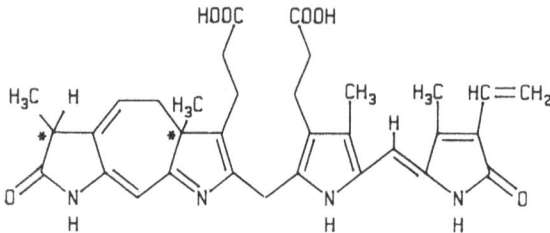

Figure 6. Structure of lumirubin. Asterisks indicate location of chiral centers.

The formation of lumirubin is much less quantum efficient than configurational isomerization. Greenberg *et al.* reported the quantum yield as 0.0015 at 450 nm, rising to 0.0032 at 510 nm.[26] McDonagh *et al.* have reported similar measurements,[16] but with substantially lower quantum yields. The majority of the discrepancy of these two sets of data can be accounted for by differences in the correction factors used in analyzing the HPLC data. Figure 7 shows the recent data of McDonagh *et al.* plotted along with the earlier data of Greenberg *et al.* which has been recalculated using the same HPLC correction factors.

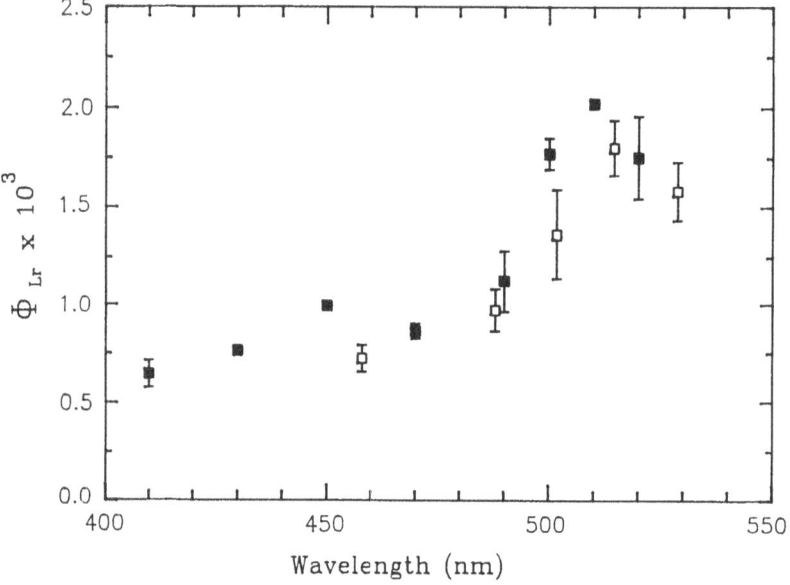

Figure 7. Wavelength dependence of the quantum yield for formation of lumirubin. Solid symbols from McDonagh *et al.*;[16] open symbols from Greenberg *et al.*[26] which were recalculated using HPLC correction factors from McDonagh *et al.*[16] Error bars are one standard deviation. Data points with no error bars indicated had standard deviations which were smaller than the symbol used.

In theory, the formation of lumirubin could occur one of two ways, either directly in a concerted single photon reaction from native bilirubin, or via a sequential two photon route in which native bilirubin first undergoes configurational isomerization to form the 4E,15Z-isomer which in a second photochemical reaction cyclizes to form lumirubin. Itoh and Onishi have used kinetic data to argue[27] that only the second, two photon, pathway exists. However, lumirubin formation occurs readily under conditions where little 4E,15Z-isomer is formed, i.e., bilirubin bound to human albumin irradiated with blue light. This observation means that if lumirubin were formed *only* with 4E,15Z as an intermediate, either the extinction coefficient of the 4E,15Z-isomer must be much greater than that of the other configurational isomers, or the quantum yield for the cyclization of the 4E,15Z-isomer must be very high.

Ennever and Dresing have demonstrated the latter to be true.[28] These investigators have determined that the quantum yield for the cyclization of the 4E,15Z-isomer bound to human albumin to form lumirubin is 0.12 at 450 nm and 0.19 at 510 nm. The *predicted* minimum quantum yield necessary for the cyclization of 4E,15Z to account for all of the formation of lumirubin starting with native bilirubin was 0.12 at 450 nm and 0.20 at 510 nm. The agreement between this predicted minimum and the measured quantum yields suggest that the major pathway for the formation of lumirubin from bilirubin occurs via a sequential two photon pathway with the 4E,15Z-isomer as an intermediate. These results do not disprove the existence of a single photon pathway from native bilirubin to lumirubin, but indicate that such a pathway is at most minor contributor to the overall formation of lumirubin from bilirubin bound to human albumin.

Lumirubin is more polar than native bilirubin for the same reason that the configurational isomers are more polar, namely the disruption of intramolecular hydrogen bonding. On reversed-phase HPLC, lumirubin is more polar than either of the two single-photon configurational isomers.

Photooxidation

Configurational and structural isomerization involve only light and bilirubin; the third photochemical reaction of bilirubin which occurs during phototherapy involves a third reagent, molecular oxygen. This third reaction is photooxidation which yields a number of different products of lower molecular weight which are mostly colorless and soluble in water.[29] The structures of the principal photooxidation products are shown in Figure 8.

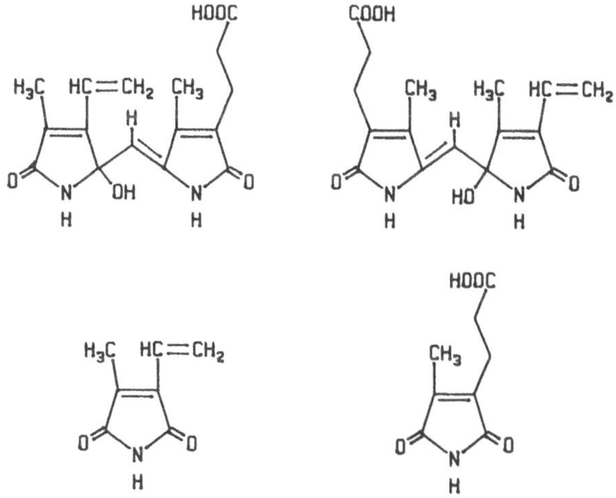

Figure 8. Photooxidation products of bilirubin. The compounds, clockwise from top left, are: endovinyl-B-water propentdyopent, exovinyl-A-water propentdyopent, methylvinylmaleimide, and hematinic acid imide. The major products found in the urine of infants undergoing phototherapy are the two water-propentdyopent, hematinic acid imide, and hydrolysis products of methylvinylmaleimide.[29]

Photooxidation was the first recognized photochemical reaction of bilirubin and a key to the development of phototherapy. One of the chance events which led to the development of phototherapy was the loss of measurable bilirubin in a sample of serum left on a window sill for an hour exposed to bright summer sunshine. The sample was from a severely jaundiced infant and the bilirubin level measured in the clinical laboratory was far lower than suspected clinically from the degree of jaundice in the infant. Measurement of bilirubin in a second serum sample not exposed to sunlight gave the expected value, whereas a repeated measurement on the initial sample left exposed for an additional hour yielded an even lower bilirubin value.[2]

The photochemical mechanism for bilirubin oxidation was initially thought to be a self-sensitized formation of singlet oxygen via energy transfer from an excited state of bilirubin.[30] Although bilirubin reacts with singlet oxygen, it is a poor sensitizer for the formation of singlet oxygen because of its low intersystem quantum yield and its extremely short-lived triplet lifetime.[31] More recent evidence suggests that the reaction is an electron transfer from excited state bilirubin with ground state oxygen which is followed by free radical rearrangements to give the various products.

MECHANISM OF PHOTOTHERAPY IN VIVO

The Gunn Rat

Much of our understanding of the mechanism by which phototherapy works to lower serum bilirubin levels comes from work using an animal model, the Gunn rat. The homozygous Gunn rat has a congenital absence of the hepatic enzyme responsible for esterifying glucuronic acid to bilirubin, and therefore is unable to form excretable bilirubin conjugates. The animals have life-long jaundice and respond to phototherapy with a decline in serum bilirubin levels. Early work by Ostrow[32] using Gunn rats with bile duct fistula and radiolabeled bilirubin demonstrated that the major site of bilirubin elimination during phototherapy was into the bile, with a minor amount appearing in the urine. In addition, the failure of phototherapy to lower significantly bilirubin levels in Gunn rats in whom biliary excretion is blocked by ligation of the bile duct is further evidence that bile is the principal route of photoproduct elimination in the rat.[33]

A key to the understanding of phototherapy was the development of HPLC methodology to identify and quantitate the various bilirubin isomers both *in vitro* and in biological fluids.[34,35] McDonagh has used his HPLC method to demonstrate that bile from a Gunn rat kept in the dark contains no bilirubin photoisomers and within minutes of initiating phototherapy, isomers begin appearing in the bile.[36] Within 90 minutes of starting phototherapy, steady-state amounts of bilirubin photoproducts are found in the bile, although none are detectable in the serum. All of the isomers of bilirubin are found in the bile: 51% 4Z,15E; 18% 4E,15Z; 2% 4E,15E; 14% Z-lumirubin; 7% E-lumirubin. The remaining 8% is 4Z,15Z, presumably formed from the thermal re-isomerization of some of the configurational photoisomers.[37] Photooxidation products have not been unambiguously identified in serum or urine of Gunn rats exposed to phototherapy, and are probably quantitatively unimportant. Thus, in the Gunn rat, configurational isomers are the major pigment excreted into the bile during phototherapy. The contribution of configurational isomers to net elimination of bilirubin in the intact rat may be somewhat lower (and structural isomers higher), because of the ready re-isomerization of the E-isomers in the bile back to native 4Z,15Z-bilirubin which can be reabsorbed into the circulation in the distal gastrointestinal tract.

The Human Infant

The basic photochemistry of bilirubin in the human infant treated with phototherapy is similar to that in the Gunn rat. However, significant differences do exist. One difference is that human albumin imparts a regioselectivity on the configurational isomerization reaction so that the principal isomer formed is 4Z,15E. The major difference between the Gunn rat and the human infant is in the kinetics of excretion. Whereas in the rat the various isomers are essentially absent from the serum, in the human infant treated with phototherapy, the configurational and structural isomers are easily detectable.

Figure 9 shows a typical HPLC analysis of serum from a jaundiced infant undergoing phototherapy. The major bilirubin species in the serum is the native 4Z,15Z-isomer. (The HPLC system is reversed phase so that the least polar components come off the column last.) The principal photoproduct in the serum is the 4Z,15E-isomer with relatively little of the other single photon configurational isomer, 4E,15Z, which is not completely resolved in this HPLC system. In addition to these two configurational isomers, Z-lumirubin is also found in the serum. Typically, the configurational isomers, principally 4Z,15E, represent 15-20% of the total pigment in the serum and lumirubin is 2-5% of the total. Photooxidation also occurs in infants treated with phototherapy. Various mono- and dipyrrolic fragments of bilirubin have been identified in the urine of irradiated infants; however, their quantitative contribution appears to be small.[29] Thus in the human infant, as in the Gunn rat, the principal photoproducts are those formed fastest *in vitro*, i.e., configurational isomers. However, in the human infant, the rate of excretion of the photoproducts is much slower than in the rat and determining the relative contributions of the various products to net bilirubin elimination during phototherapy has been more difficult.

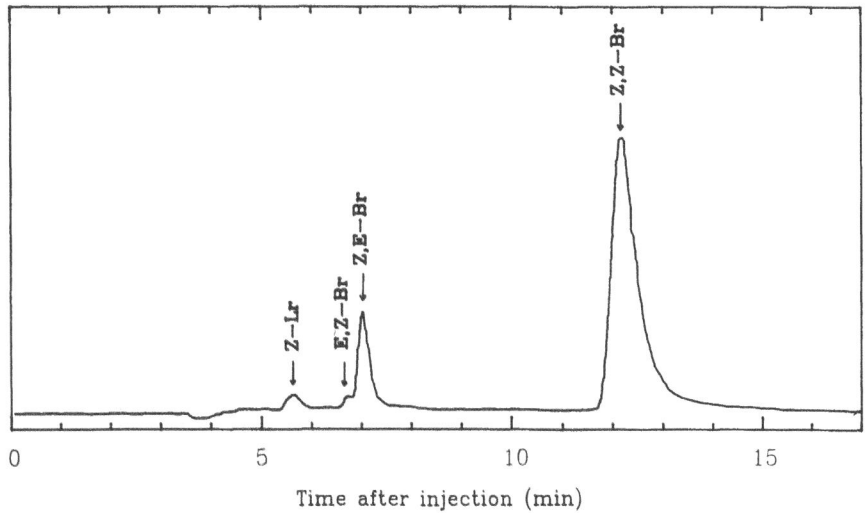

Time after injection (min)

Figure 9. HPLC analysis of serum from a jaundiced infant receiving phototherapy. The y-axis is absorbance monitored at 450 nm.

Several lines of evidence suggest that it is excretion of the structural isomer lumirubin which accounts for the majority of the net excretion of bilirubin in human infants during phototherapy. All of the evidence, however, is indirect. For example, measurement of the disappearance of the isomers from the serum after phototherapy is stopped has yielded the *apparent* half-life of the 4Z,15E-isomer and lumirubin in the serum. The half-life of the 4Z,15E is approximately 15 hours,[38] whereas the half-life of lumirubin is less than 2 hours.[39] In addition, analysis of bile from the duodenum of jaundiced infants receiving phototherapy has shown that the principal pigment is lumirubin, and that it is absent before beginning or several hours after stopping phototherapy.[39,40] While these results are consistent with lumirubin playing a major role in bilirubin excretion during phototherapy, limitations inherent in investigations on human infants preclude the type of definitive quantitative analysis of nascent bile necessary to prove this. For example, little is known about the formation and excretion of the 4E,15Z or 4E,15E-isomers in human infants. It is also possible that photooxidation products which are found in low concentrations in the urine are excreted in high concentrations in the bile. Thus available evidence is consistent with, but does not *prove*, that formation and excretion of lumirubin is the major pathway for bilirubin elimination during phototherapy.

The use of light in the treatment of neonatal jaundice has changed little in the 35 years since its discovery. Improvement in phototherapy could involve either increased efficacy or safety. There is little doubt that phototherapy is safe. It has been used on millions of infants over the past three and a half decades and no significant toxicity has been identified. The side effects which have been attributed to phototherapy, including increased insensible water loss, diarrhea, hypocalcemia, riboflavin deficiency and rashes, are minor and resolve upon after stopping treatment. Concern about potential mutagenicity of the treatment has been raised because of the observation that visible light in the 400 to 450 nm region (violet and blue) has been shown to be genotoxic in a number of *in vitro* systems.[41,42] The applicability of *in vitro* genotoxicity in the prediction of human risk has been questioned.[43] However, even if this wavelength region is mutagenic, the amount of this irradiation received by an infant during phototherapy is only a minute fraction of the total irradiation he or she will receive from natural sources during the first year of life.

While phototherapy is no doubt effective in lowering bilirubin levels, much room exists for improving its efficacy. As currently administered phototherapy requires continuous treatment, typically for 2 to 4 days. In otherwise health full-term infant, the need for phototherapy is the most common reason for extended hospitalization, adding more than $50 million to annual hospital costs in the United States.[44] Premature infants, who are at risk for neurological damage from relatively low bilirubin levels, often receive continuous phototherapy for a week or more.

The first phototherapy unit consisted of eight fluorescent lamps housed in a reflective canopy positioned about 50 cm above the infant. This same configuration remains the most commonly used today. Most of the effort at improving phototherapy has focused on what color light to use. Figure 10 shows emission spectra of three types of lamps which have been used for phototherapy: daylight (white), Special Blue and green. The most commonly used is the broad-spectrum daylight lamp. In the early 1970's the narrow spectrum Special Blue lamp was introduced for use in phototherapy. This lamp has a narrow spectral output with a peak emission at ≈ 450 nm which closely matches the *in vitro* absorption spectrum of bilirubin bound to human albumin. Because this lamp has essentially all of its output in the bilirubin absorption band, it is more effective in the treatment of jaundice than broad-spectrum daylight lamps. Its use clinically, however, has been limited because of resistance from those who have to work around the nauseating light produced by these lamps. In addition, a clinically-important parameter, the infant's skin color (i.e., cyanosis) is impossible to judge when this intense blue light is used. As a result, the most commonly-used lamps for phototherapy are broad-spectrum white fluorescent lamps.

In 1983, Vecchi *et al.* published[45] the results of a small clinical trial comparing the efficacy of narrow spectrum green fluorescent lamps, which have a peak emission at ≈ 525 nm (Figure 10) with broad-spectrum white lamps. They found a statistically-significant greater decrease in serum bilirubin levels in those infants treated with green light phototherapy than in those treated with broad-spectrum white light. A later study by the same authors reported an equivalent clinical response to narrow-spectrum blue and green lights.[46] Subsequent studies by others have provided contradictory results with some supporting the equivalence of blue and green[47,48] and another reporting superior efficacy of blue.[49] A part of the contradictory results can be attributed to the lack of a true *control* group (i.e., no therapy) in any study. Furthermore, none of the three groups of investigators use the same blue light. Vecchi *et al.*[46] used a Westinghouse Special Blue (F20T12/BB) which have an emission spectrum identical to that shown in Figure 10; Ayyash *et al.*[47,48] used Sylvania "regular" blue lamps (F20T12/B) which have a very broad emission (see Figure 1 in reference 19); Tan used a Philips "special" blue lamp (TL20W/52) which has a narrow emission spectrum which is shifted 5 to 10 nm towards the green relative to that of the Westinghouse Special Blue lamp.[50]

Despite the controversy as to whether or not blue and green lamps are equivalent it is clear that green lamps are far better than one would predict based upon the overlap of the emission spectrum of these lamps with the *in vitro* absorption spectrum of bilirubin bound to human albumin. A part of this unexpected efficacy can be ascribed to wavelength-dependent photochemistry with enhanced formation of lumirubin with green light.[16,23,26] These effects,

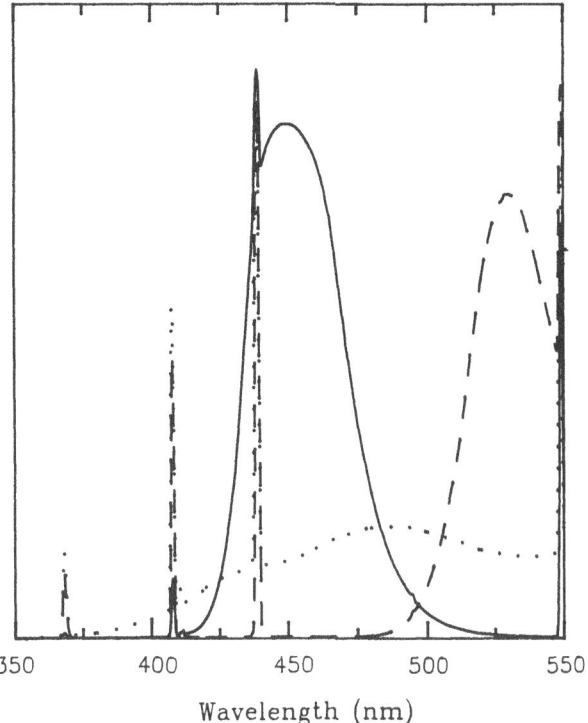

Wavelength (nm)

Figure 10. Emission spectra of three lamps used for phototherapy: (————————)
Westinghouse "special" blue (F20T12/BB); (·········) General Electric daylight
(F20T12/D); (– – – – – –) Sylvania green (F20T12/G). Y-axis is an arbitrary linear scale
of intensity which is the same for each lamp.

however, are probably small. More important is that the transmission of light through skin
increases with increasing wavelength[51] over the bilirubin absorption band. This physical
phenomenon will have the effect of shifting the peak of the action spectrum to a wavelength
longer than the peak of the *in vitro* absorption spectrum. In addition, the binding of long-
chain fatty acids to albumin shift the absorbance spectrum of the albumin bound bilirubin
towards the green.[52] The net result of all of these factors is that a lamp whose output is
closely matched to the *in vitro* absorption spectrum of bilirubin (i.e., the Special Blue lamp)
will be less effective clinically than one would expect; and a lamp whose output overlaps only
the long wavelength edge of the *in vitro* absorption spectrum of bilirubin (i.e, the green lamp)
will be more effective clinically than one would expect. Precisely which wavelength light will
be most effective in the treatment of neonatal hyperbilirubinemia is not known, although the
best data suggest that it would be in the 480-510 nm range.[16]

Unfortunately, no commercially-available lamp has a narrow spectral output in the
theoretically best range of 480-510 nm. [Even if one were available, it would no doubt suffer
from the same practical limitations that plague the blue and green narrow spectrum lamps -
complaints from nursing personnel who have to provide the continuous care to eerily-colored
infants.[49]] There is no convincing theoretical basis for filtering broad-spectrum light to
provide only the therapeutically "best" light, since this would remove useful, albeit perhaps
less effective, light.

Most of the effort at improving the effectiveness of phototherapy has focused on
defining what is the "best" light to use. As a result, a important parameter in the use of light in
the treatment of jaundice, intensity, has been largely ignored. Tan has demonstrated a clear
proportionality between the intensity of light used and the rate of decline in serum bilirubin

187

levels, up to relatively high light intensities ($\approx$ 40 μwatt·cm^{-2}·nm^{-1} measured in the 425 to 475 nm range).[50] The typical intensity of phototherapy in use today is one-fourth or less of this saturating dose. A reason that higher dose phototherapy is not in routine use is the practical difficulty in delivering such high intensity light with currently available fluorescent units without overheating of the infant. Thus under conditions where one is only able to deliver far fewer photons than is optimal it makes sense to use only the best. A new technology is under development and clinical testing which employs a woven fiber optic pad to deliver high intensity light (> 50 μwatt·cm^{-2}·nm^{-1}) which has no ultraviolet or infrared irradiation. (For comparison, the intensity in the 425 to 475 nm range on a summer day is $\approx$ 40 μwatt·cm^{-2}·nm^{-1} and more than 120 μwatt·cm^{-2}·nm^{-1} in direct sunlight.) If this high-intensity phototherapy is more clinically effective, as expected, then when the optimal number of photons are delivered, it may then be the right time to revisit the unanswered questions of the "best" color light to use.

ACKNOWLEDGEMENT

This work was supported by grant number DK-38575 from the United States Public Health Service.

REFERENCES

1. R. J. Cremer, P. W. Perryman, and D. H. Richards, Influence of light on the hyperbilirubinaemia of infants, *Lancet* 1:1094 (1958).
2. R. H. Dobbs and R. J. Cremer, Phototherapy (Looking back), *Arch. Dis. Child.* 50:833 (1975).
3. J. D. Ostrow, J. H. Jandl, and R. Schmid, The formation of bilirubin from hemoglobin in vivo, *J. Clin. Invest.* 41:1628 (1962).
4. R. Schmid and A. F. McDonagh, Formation and metabolism of bile pigments in vivo, *in* "The Porphyrins" (volume VI), D. Dolphin, ed., Academic Press, New York (1979). 5. R. Brodersen, Bilirubin. Solubility and interaction with albumin and phospholipid, *J. Biol. Chem.* 254:2364 (1979).
6. R. Bonnett, J. E. Davies, and M. B. Hursthouse, Structure of bilirubin, *Nature* 262:326 (1976).
7. J. D. Ostrow and R. Schmid, The protein-binding of ^{14}C-bilirubin in human and murine serum, *J. Clin. Invest.* 42:1286 (1963).
8. R. Brodersen, Binding of bilirubin to albumin, *Crit. Rev. Clin. Lab. Sci.* 11:305 (1980).
9. M. J. Maisels, Neonatal Jaundice, *Semin. Liver Dis.* 8:148 (1988).
10. R. Stocker, Y. Yamamoto, A. F. McDonagh, A. N. Glazer, and B. N. Ames, Bilirubin is an antioxidant of possible physiological importance, *Science* 235:1043 (1987).
11. R. Stocker, A. N. Glazer, and B. N. Ames, Antioxidant activity of albumin-bound bilirubin, *Proc. Natl. Acad. Sci. USA* 84:5918 (1987).
12. G. B. Odell and H. S. Schutta, Bilirubin encephalopathy, *in:* , "Cerebral Energy Metabolism and Metabolic Encephalopathy," D. McCandless, ed., Plenum Press, New York (1985).
13. A. A. Lamola, Effects of environment on photophysical processes of bilirubin, *in:* "Optical Properties and Structure of Tetrapyrroles," G. Blauer and H. Sund, eds., Walter de Gruyter, Berlin, (1895).
14. A. F. McDonagh and D. A. Lightner, Intramolecular energy transfer in bilirubins, *in:* "NATO Advanced Study Institute on Primary Photo-processes in Biology and Medicine," R. Bensasson, G. Jori, E. Land, and T. Truscott, eds., Plenum Press, New York, (1985).
15. D. A. Lightner, J. K. Gawronski, and W. M. D. Wijekoon, Complementarity and chiral recognition: enantioselective complexation of bilirubin, *J. Am. Chem. Soc.* 109:6354 (1987).
16. A. F. McDonagh, G. Agati, F. Fusi, and R. Pratesi, Quantum yields for laser photocyclization of bilirubin in the presence of human serum albumin. Dependence of quantum yield on excitation wavelength, *Photochem. Photobiol.* 50:305 (1989).

17. A. F. McDonagh, L. A. Palma, and D. A. Lightner, Phototherapy for neonatal jaundice: stereospecific and regioselective photoisomerization of bilirubin bound to human serum albumin and NMR characterization of intramolecularly cyclized photoproducts, *J. Am. Chem. Soc.* 104:6867 (1982).

18. R. Pratesi, G. Agati, and F. Fusi, Configurational photoisomerization of bilirubin in vitro - I. Quenching of Z to E isomerization by two-wavelength irradiation, *Photochem. Photobiol.* 40:41 (1984).

19. J. F. Ennever, M. Sobel, A. F. McDonagh, and W. T. Speck, Phototherapy for neonatal jaundice: in vitro comparison of light sources, *Pediatr. Res.* 18:667 (1984).

20. A. A. Lamola, J. Flores, and F. H. Doleiden, Quantum yield and equilibrium position of the configurational photoisomerization of bilirubin bound to human serum albumin, *Photochem. Photobiol.* 35:649 (1982).

21. R. W. Sloper and T. G. Truscott, The quantum yield for bilirubin photoisomerisation, *Photochem. Photobiol.* 35:743 (1982).

22. M. S. Stoll, N. Vicker, C. H. Gray, and R. Bonnett, Concerning the structure of photobilirubin II, *Biochem. J.* 201:179 (1982).

23. S. Onishi, S. Itoh, and K. Isobe, Wavelength-dependence of the relative rate constants for the main geometric and structural photoisomerization of bilirubin IXα bound to human albumin, *Biochem. J.* 236:23 (1986).

24. A. F. McDonagh, D. A. Lightner, M. Reisinger, and L. A. Palma, Human serum albumin as a chiral template. Stereoselective photocyclization of bilirubin, *J. Chem. Soc. Chem. Commun.*:249 (1986).

25. A. F. McDonagh, A. F. and D. A. Lightner, Phototherapy and the photobiology of bilirubin, *Semin. Liver Dis.* 8:272 (1988).

26. J. M. Greenberg, V. Malhotra, and J. F. Ennever, Wavelength dependence of the quantum yield for the structural isomerization of bilirubin, *Photochem. Photobiol.* 46:453 (1987).

27. S. Itoh, and S. Onishi, Kinetic study of the photochemical changes of (ZZ)-bilirubin IXα bound to human serum albumin, *Biochem. J.* 226:251 (1985).

28. J. F. Ennever and T. J. Dresing, Quantum yields for the cyclization and configurational isomerization of 4E,15Z-bilirubin, *Photochem. Photobiol.*, in press.

29. D. A. Lightner, W. P. I. Linnane, and C. E. Ahlfors, Bilirubin photooxidation products in the urine of jaundiced infants receiving phototherapy, *Pediatr. Res.* 18:696 (1984).

30. A. F. McDonagh, The role of singlet oxygen in bilirubin photo-oxidation, *Biochem. Biophys. Res. Commum.* 44:1306 (1971).

31. D. A. Lightner, Structure, photochemistry, and organic chemistry of bilirubin, *in*: , "Bilirubin" (volume 1), K. Heirwegh and S. Brown, eds., CRC Press, Boca Raton, Florida (1982).

32. J.D. Ostrow, Photocatabolism of labeled bilirubin in the congenitally jaundiced (Gunn) rat, *J. Clin. Invest.* 50:707 (1971).

33. D. R. Davis, R. A. Yeary, and K. Lee, The failure of phototherapy to reduce plasma bilirubin levels in the bile duct-ligated rat., *J. Pediatr.* 99:965 (1981).

34. A. F. McDonagh, L. A. Palma, F. R. Trull, and D. A. Lightner, Phototherapy for neonatal jaundice: configurational isomers of bilirubin, *J. Am. Chem. Soc.* 104:6865 (1982).

35. S. Onishi, N. Kawade, S. Itoh, K. Isobe, and S. Sugiyama, High-pressure liquid chromatographic analysis of anaerobic photoproducts of bilirubin-IXα in vitro and its comparison with photoproducts in vivo, *Biochem. J.* 190:527 (1980).

36. A. F. McDonagh, Molecular mechanisms of phototherapy for neonatal jaundice, *in*: "Neonatal Jaundice," F. Rubaltelli and G. Jori, eds., Plenum Press, New York, (1984).

37. A. F. McDonagh and L. A. Palma, Bilirubin photoisomer excretion in Gunn rats: Effects of green and blue light. (Abstr.), *Photochem. Photobiol.* 45S:96S (1987).

38. J. F. Ennever, I. Knox, S. C. Denne, and W. T. Speck, Phototherapy for neonatal jaundice: In vivo clearance of bilirubin photoproducts, *Pediatr. Res.* 19:205 (1985).

39. J. F. Ennever, A. T. Costarino, R. A. Polin, and W. T. Speck, Rapid clearance of a structural isomer of bilirubin during phototherapy, *J. Clin. Invest.* 79:1674 (1987).

40. S. Onishi, K. Isobe, S. Itoh, M. Manabe, K. Sasaki, R. Fukuzaki, and T. Yamakawa, Metabolism of bilirubin and its photoisomers in newborn infants during phototherapy, *J. Biochem.* (Tokyo) 100:789 (1986).

41. M. O. Bradley and N. A. Sharkey, Mutagenicity and toxicity of visible fluorescent light to cultured mammalian cells, *Nature* 266:724 (1977).

42. B. S. Rosenstein and J. M. Ducore, Induction of DNA strand breaks in normal human fibroblast exposed to monochromatic ultraviolet and visible wavelengths in the 240-546 nm range, *Photochem. Photobiol.* 38:51 (1983).

43. R. W. Tennant, B. H. Margolin, M. D. Shelby, E. Zeiger, J. K. Haseman, J. Spalding, and R. Minor, Prediction of chemical carcinogenicity in rodents from in vitro genetic toxicity assays, *Science* 236:933 (1987).

44. H. A. Hein, Why do we keep using phototherapy in healthy newborns, *Pediatrics* 73:881 (1984).

45. C. Vecchi, G. P. Donzelli, M. G. Migliorini, and G. Sbrana, Green light in phototherapy, *Pediatr. Res.* 17:461 (1983).

46. C. Vecchi, G. P. Donzelli, G. Sbrana, and R. Pratesi, Phototherapy for neonatal jaundice: clinical equivalence of fluorescent green and 'special' blue lamps, *J. Pediatr.* 108:452 (1986).

47. H. Ayyash, E. Hadjigeorgiou, J. Sofatzis, A. Chatziioannou, D. Nicolopoulos, and E. Sideris, Green light phototherapy in newborn infants with ABO hemolytic disease, *J. Pediatr.* 111:882 (1987).

48. H. Ayyash, E. Hadjigeorgiou, I. Sofatzis, H. Dellagrammaticas, and E. Sideris, Green or blue light phototherapy for neonates with hyperbilirubinaemia, *Arch. Dis. Child.* 62:843 (1987).

49. K. L. Tan, Efficacy of fluorescent daylight, blue, and green lamps in the management of nonhemolytic hyperbilirubinemia, *J. Pediatr.* 114:132 (1989).

50. K. L. Tan, The pattern of bilirubin response to phototherapy for neonatal hyperbilirubinemia, *Pediatr. Res.* 16:670 (1982).

52. R. R. Anderson and J. A. Parrish, The optics of human skin, *J. Invest. Dermatol.* 77:13 (1981).

53. V. Malhotra, J. W. Greenberg, L. L. Dunn, and J. F. Ennever, Fatty acid enhancement of the quantum yield for the formation of lumirubin from bilirubin bound to human albumin, *Pediatr. Res.* 21:530 (1987).

THE HAZARDS OF COSMETIC TANNING WITH UVA RADIATION

Antony R. Young

Photobiology Unit
Institute of Dermatology
St Thomas's Hospital, London

INTRODUCTION

The use of UVA (315-400 nm) tanning technology has increased markedly over the last 5 years in Western societies. The UVA tanning industry tends to claim, or imply, that a UVA tan is a safe tan when compared with a natural suntan. This paper proposes to examine the validity of this claim. Before any assessment on the safety of ultraviolet radiation (UVR) on the skin can be made it is necessary to distinguish between short-term effects and long-term consequences.

SHORT-TERM EFFECTS

The most obvious short term effects of solar UVR (290-400 nm) are erythema, which at 24 h post-exposure may range from a mild to a severe sunburn, and tanning. Microscopic examination of mammalian skin fixed within hours to days after the exposure may reveal changes in the epidermis such as individual keratinocyte death (sunburn cell formation) (Young, 1987), depletion of the antigen presenting cells (Langerhans cells) (Aberer et al, 1981; Obata et al, 1985), melanogenesis, hyperplasia and stratum corneum thickening (Johnson, 1984). The dermis may show an inflammatory infiltrate (Johnson, 1984). From a practical point of view, these effects are due to the UVB (290-315 nm) component of solar UVR. This is because UVB is more potent than UVA (per unit dose) by about 3 orders of magnitude (Parrish et al, 1982). Whereas, by and large, the UVA content of solar UVR is greater than its UVB content by only 1-2 orders of magnitude.

The comparative sensitivity of the skin to UVB compared with UVA means that relatively small miscalculations of exposure time with UVB emitting sources (including the sun) can result in severe sunburn. This problem does not arise with pure UVA (> 315 nm) sources. It is quite difficult to accidently obtain a dose of 3-4 minimal erythema doses (MEDs) UVA, even with high intensity sources. Furthermore, the mechanisms of UVB and UVA erythemas are different, the latter being oxgygen dependent (Auletta et al, 1986). UVA does not result in the sunburn reaction typical of high dose UVB. Thus, with resect to short term effects, UVA tanning is safer than UVB tanning, whether by the sun or a lamp. A recent study did however show that UVA sunbed use was associated with side effects such as skin itching and dryness and polymorphic light eruption (Rivers et al, 1989).

Any short-term advantages of avoiding a sunburn type reaction with UVA exposure may be lost if the sunbed user is taking photosensitizing medication or is wearing perfumes/ cosmetics/sunscreens with bergamot oil which may contain the photosensitizing agent 5-methoxypsoralen (5-MOP).

LONG-TERM CONSEQUENCES

The long term effects of solar radiation are skin cancer (MacKie et al, 1987) and accelerated skin ageing which may be termed dermatoheliosis (Gilchrest, 1984) or photoageing. These effects have been determined in human skin by their association with solar exposure especially in people who sunburn easily and tan poorly if at all (skin types I and II). In humans, it is not possible to determine whether these effects are due to UVB or UVA. The relative importance of each waveband can only be studied under more controlled conditions by the use of animal models. Until relatively recently, skin cancer was widely believed to be a UVB phenomenon. It has now been convincingly shown that UVA is a carcinogen in mouse skin (Sterenborg et al, 1987). This is not suprising as it has been shown that UVA is mutagenic in mammalian cells (Wells and Han, 1984; Jones et al, 1987; Enninga et al, 1986; Lundgren and Wulf, 1988) and damages DNA in human cells in vitro (Enninga et al, 1986; Rosenstein et al, 1983) and in human skin in vivo (Freeman et al, 1987). DNA repair mechanisms are important in protecting cells against damage caused by UVB radiation. However, the endogenous radical scavenger glutathione is believed to be more important than DNA repair mechanims in protecting cells from UVA damage (Tyrrell, 1989). This and other defence mechanisms may be overstretched by high intensity UVA exposure (Tyrrell, 1989).   Unlike UVB, UVA does not induce stratum corneum thickening. A thickened stratum corneum is believed to afford protection against further UVR induced damage. Comparable UVB and UVA induced tans in human subjects gave comparable protection against UVB-induced epidermal DNA damage but a UVB-induced tan gave much better protection against UVB-induced erythema (Gange et al, 1985).

It is important to recognize that many so-called UVA sources also contain up to 3% UVB (280-315 nm). The 315 nm boundry between UVA and UVB (Commission Internationale de l'Eclairage, 1970) and the 320 nm boundry that is also used by many authors were fixed long before current knowledge of the important mechanistic differences of biological effects of these two wavebands. Broadly speaking, UVA effects are oxygen dependent. It is now apparent, from many studies, that "UVB type" mechanisms also extend out to beyond 330 nm. Genetic effects, at least, are likely to be a consequence of direct, but poor, UVR absorption by DNA in this region (DNA shows an absorption maxiumum at 260 nm with exponentially decreasing absorbance towards the longer wavelengths (Sutherland and

Griffin, 1981)). Thus, unlike those of UVB, the effects of UVA (defined mechanistically as opposed to physically) on DNA are believed to be due to the generation of active oxygen species after absorption by endogenous chromophores (Tyrrell and Keyse, 1990). In practice, many cosmetic UVA sources are also UVB sources from a mechanistic point of view.

It has been recently reported that the action spectra for tanning, by a single exposure, in human skin and photocarcinogenesis in mouse skin have similar shapes in the UVB-UVA range (Roza et al, 1989). On the assumption that mouse data are predictive for relative rather than absolute risk; the aquisition of a tan (with the mimimum doses necessary) whether obtained by UVB or UVA, and independent of mechanisims, is likely to be associated with a skin cancer risk of the same order of magnitude. UVR induced immunosuppression in an important factor in mouse skin photocarcinogenesis (Krutmann and Elmets, 1988) but the role of such immunosuppression in human skin has yet to be clarified. However, it has been clearly demonstrated that UVA (with a small UVB content) and pure UVA exposure may deplete Langerhans cell markers (Rivers et al, 1989) and depress human immune function (Hersey et al, 1983; Mork et al, 1987; Baadsgard et al, 1987). Thus, the claim that a UVA tan is a safe tan is not be valid. It should be stated that in practice the long-term risk of skin cancer by UVA exposure is not know. One study indicated increased risk of skin cancer with sunlamp use (Swerdlow et al, 1988) but without differentiating between the different types of UVR source, some of which almost certainly emitted substantial quantities of UVB. The only way to establish the risk of skin cancer with UVA sunbed use is to conduct large scale long-term epidemiological studies.

The hairless albino mouse has been used as a model for photoageing (Kligman et al, 1985, 1987; Bisset et al, 1987). Histologically, solar elastosis may be seen in the dermis and biochemical changes of collagen may be observed (Plastow et al, 1987, 1988). It has been shown that both UVB and UVA induce photoageing. In human skin, UVA penetrates the dermis much more than UVB which may mean that the relative effects of UVA may be much more important.

Recent reports have shown that skin fragility and blistering may result from long-term sunbed exposure in fair-skinned people (Farr et al, 1988; Murphy et al, 1989).

CONCLUSIONS

Any comparison of the safety of long-term effects of UVA sunbed use and solar exposure must take into account human behavioural aspects and the skin type of the person who is tanning. As discussed above, exposure to minimum tanning doses, whether of UVB or UVA, is likely to carry a risk of the same order of magnitude. With sunbed use, it is relatively easy to monitor and control exposure so that it is kept as low as possible to achieve a tan, especially if the tanning is done in a reputable salon. Many people, especially in sun-starved Northern Europe, seem to find it quite difficult to control their solar exposure both at home and on holiday.

Thus, they may receive very much more than their minimum tanning UVB dose. On the other hand, people may expose themselves to UVA sunbeds throughout the year but to sun for only relatively short periods.

Skin types III and IV tan easily whereas skin types I and II do not tan or tan poorly. If the latter expose themselves to the sun or sunbeds they are likely to be accumulating risk of skin cancer and photoageing, often with little or no cosmetic benefit, whereas the former tan with relatively low doses of UVR.

In conclusion, there is probably no such thing as a safe tan. The best advice is not to tan especially in the case of skin types I and II. However, if an individual is determined to tan, the risk may be reduced by recognizing the genetic contraints of ones skin type and achieving ones tan with the minimum amount of UVR exposure whether from the sun or artificial sources.

REFERENCES

Aberer, W. Schuler, G. Stingl G., Hönigsmann, H. and Wolff, K., 1981, Ultraviolet light depletes surface markers of langerhans cells, J. Invest. Dermatol., 76:202.

Auletta, M., Gange, R.W., Tan, T.O. and Matzinger, E., 1986, Effect of cutaneous hypoxia upon erythema and pigment responses to UVA, UVB, and PUVA (8-MOP + UVA) in human skin, J. Invest. Dermatol., 86:649.

Baadsgard, O., Cooper, K.D., Lisbys, S., Wulf, H.C. and Wantzin, G.L., 1987, Dose response and time course for induction of T6- DR+ human antigen presenting cells by in vivo ultraviolet A, B and C irradiation, J. Amer. Acad. Dermatol., 17:792.

Bisset, D.L., Hannon, D.P. and Orr, T.V., 1987, An animal model of solar aged skin: Histological, physical and visisble changes in UV-irradiated hairless mouse skin, Photochem. Photobiol., 46:376.

Commission Internationale de l'Eclairage, 1970. International Lighting Vocabulary, 3rd Edition, Publication CIE No 17 (E-1.1.), Paris.

Enninga, I.C., Groenendijk, R.T.L., Filon, A.R., van Zeeland, A.A. and Simons, J.W.I.M., 1986, The wavelength dependence of uv-induced pyrimidine dimer formation, cell killing and mutation induction in human diploid skin fibroblasts, Carcinogenesis, 7:1829.

Farr, P.M., Marks, J.M., Diffey, B.L. and Ince, P., 1988, Skin fragility and blistering due to use of sunbeds, Brit. Med. J., 296:1708.

Freeman, S.E., Gange, R.W., Sutherland, J.C., Matzinger, E.A. and Sutherland, B.M., 1987, Production of pyrimidine dimers in DNA of human skin exposed in situ to UVA radiation, J. Invest. Dermatol., 88: 430.

Gange, R.W., Blackett, A.D., Matzinger, E.A., Sutherland, B.M. and Kochevar, I.E., 1985, Comparative protection efficiency of UVA- and UVB-induced tans against erythema and the formation of endonuclease-sensitive sites in DNA by UVB in human skin, J. Invest. Dermatol., 85:362.

Gilchrest, B.A., 1984 Dermatoheliosis (Sun-Induced Aging), in: "Skin and Aging Processes", CRC Press, Boca Ratan, Florida.

194

Hersey, P., Bradley, M., Hasic, E. Haran, G., Edwards, A. and McCarthy, W.H., 1983, Immunological effects of solarium exposure, <u>Lancet</u>, i:545.

Johnson, B.E., 1984, Reactions of Normal Skin to Solar Radiation, <u>in</u>: "The Physiology and Pathophysiology of The Skin", vol 8, A. Jarret, ed., Academic Press, London.

Jones, C.A., Huberman, E., Cunningham, M. L. and Peak, M.J., 1987, Mutagenesis and cytotoxicity in human epithelial cells by far-and near-ultraviolet radiations: Action spectra, <u>Radation Research</u>, 110:244.

Krutmann, J. and Elmets, C.A., 1988, Recents studies on mechanims in photoimmunology, <u>Photochem Photobiol.</u>, 48:787.

Kligman, L.H., Akin, F.J. and Kligman, A.M., 1985, The contributions of UVA and UVB to connective tissue damage in hairless mice, <u>J. Invest. Dermatol.</u>, 84:272.

Kligman, L.H., Kaidbey, K.H., Hitchens, V.M. and Miller, S.A., 1987, Long wavelength (>340 nm) ultraviolet-A induced skin damage in hairless mice is dose dependent, <u>in</u>: "Human Exposure to Ultraviolet Radiation: Risks and Regulations", W.F. Passchier and B.F.M. Bosnjakovic, eds., Elsevier, Amsterdam.

Lundgren, K. and Wulf, H.C., 1988, Cytotoxicity and geno-toxicity of UVA irradiation in Chinese hamster ovary cells measured by specific locus mutations, sister chromatid exchanges and chromosome aberrations, <u>Photochem. Photobiol.</u>, 47:559.

MacKie, R.M., Elwood, J.M. and Hawk, J.L.M., 1987, Links between exposure to ultraviolet radiation and skin cancer, <u>J. Royal Coll. Phys.</u>, 21:91.

Mork, N.J., Gaudernack, L.R. and Braathen, L.R., 1987, Effect of UVA and PUVA on alloactivationg and antigen-presenting capacity of human epidermal Langerhans cells, <u>Photodermatol.</u>, 4:66.

Murphy, G.M., Wright, J., Nicholls, D.S.H., McKee, P.H., Messenger, A.G., Hawk, J.L.M. and Levene, G.M., 1989, Sun-bed induced psuedoporphyria, <u>Brit. J. Dermatol.</u>, 120: 555.

Obata, M. and Tagami, H., 1985, Alteration in murine epidermal langerhans cell population by various UV irradiations: Quantitative and morphologic studies on the effects of various wavelengths of monochromatic radiation on Ia-bearing cells, <u>J. Invest. Dermatol.</u>, 84:139.

Parrish, J.A., Jaenicke, K.F. and Anderson, R.R., 1982, Erythema and melanogenesis action spectra of normal human skin, <u>Photochem. Photobiol.</u>, 36:187.

Plastow, S.R., Lovell, C.R. and Young, A.R., 1987, UVB-induced collagen changes in the skin of the hairless albino mouse, <u>J. Invest. Dermatol.</u>, 88:145.

Plastow, S.R., Harrison J.A. and Young, A.R., 1988, Early changes in dermal collagen of mice exposed to chronic UVB irradiation and the effects of a UVB sunscreen, <u>J. Invest. Dermatol.</u>, 91:590.

Rivers, J.K., Norris, P.G., Murphy, G.M., Chu, A.C., Midgeley, G., Morris, J., Morris, R.W., Young, A.R. and Hawk, J.L.M. UVA Sunbeds: tanning, photoprotection, accute adverse effects and immunological changes, <u>Brit. J. Dermatol.</u>, 120:767

Rosenstein, B.S. and Ducore, J.M., 1983, Induction of DNA strand breaks in normal human fibroblasts exposed to monochromatic ultraviolet and visible wavelengths in the 250-546 nm range, <u>Photochem. Photobiol.</u>, 38:51.

Roza, L., Baan, A.R., van der Leun, J.C. and Kligman, L., 1989, UVA hazards in skin associated with the use of tanning equipment, <u>J. Photochem. Photobiol. B Biol.</u>, 3: 281.

Sterenborg, H.J.C.M. and van der Leun, J.C., 1987, Action Spectra for tumorigenesis by ultraviolet radiation, <u>in</u>: "Human Exposure to Ultraviolet Radiation: Risks and Regulations", W.F. Passchier and B.F.M. Bosnjakovic, eds., Elsevier, Amsterdam.

Sutherland, J.C. and Griffin, K.P., 1981, Absorption spectrum of DNA for wavelengths longer than 300 nm, <u>Rad. Res.</u>, 86:399.

Swerdlow, A.J., English, J.S.C., Mackie, R.M., O'Doherty, C.J., Hunter, J.A.A., Clark, J. and Hole, D.J., 1988, Fluorescent lights, ultraviolet lamps, and risk of cutaneous melanoma, <u>Brit. Med. J.</u>, 297:647.

Tyrrell, R.M., 1989, UVA hazards in skin associated with the use of tanning equipment - a comment, <u>J. Photochem. Photobiol. B Biol.</u>, 4:227.

Tyrrell, R.M. and Keyse, S.M., 1990, The interaction of UVA radiation with cultured cells, <u>J. Photochem. Photobiol. B Biol.</u>, 4:349.

Wells, R.L. and Han, A., 1984, Action spectra for killing and mutation of Chinese hamster cells exposed to mid-and near-ultraviolet monochromatic light, <u>Mutation Research</u>, 129:251.

Young, A.R., 1987, The sunburn cell, <u>Photodermatol.</u>, 4:127.

THE ROLE OF THE NEODYMIUM YTTRIUM ALUMINIUM GARNET

(Nd:YAG) LASER IN MEDICINE

H.Barr and S.G. Bown

The Department of Surgery, John Radcliffe Hospital, Oxford
and National Medical Laser Centre, University College London

Although this laser was originally developed for the military as a tank aiming device, the Nd:YAG laser has become one of the work horses of medical laser systems being used in a variety of areas of the body to treat a range of conditions. The medium excited to produce the laser beam is a synthetic crystal rod of yttrium aluminium garnet (YAG) which has been doped with a very small amount of neodymium (Nd) atoms. It is excited by a powerful krypton lamp focused onto the rod by elliptical reflectors. The neodymium atoms are excited to a high energy level and decay (non-radiatively) back to a long lived metastable condition. Transition from these states to intermediate and subsequently non-radiatively to the ground state produces infrared light of 1064nm wavelength. It is a four level laser with the population inversion between the long lived high energy and the short lived low energy intermediate states. Mirrors placed at either end of the rod produce a laser cavity.

The Nd:YAG laser is efficient by laser standards, the output being 1-2% of the input power. For clinical use laser powers up to 100W are required. Most of the power is lost as heat so continuous water cooling is necessary. The infrared beam is invisible so a coaxial beam of helium neon laser red light is used as an aiming device.

Prior to the consideration of this specific laser, it is vital that the reasons for the clinical application of laser therapy are understood. Too often the adoption of new techniques has been impeded by their empirical application with little thought being given to the basic biological reactions and without sound scientific rationale.

THE PROPERTIES OF LASER LIGHT RELEVANT TO MEDICAL APPLICATIONS

The light produced from lasers exhibits several special properties, only some of these are at present directly relevant for medical applications.

Coherence and Collimation

In light produced by a laser the photons are travelling in the same direction and are identical being in step with each other in both time (temporal coherence) and space (spatial coherence). Temporal coherence implies that the relative phases between two points in time remain constant; and spatial coherence implies that the relative phases between two points in space remains constant. Only lasers produce temporally and spatially coherent light. However, even with a laser the light waves are not truly identical and gradually drift out of phase as they travel. The high spatial coherence of laser light means that the divergence of a laser beam is very small. Thus the full power of a laser beam can be focused onto a very small spot and it is this property of a laser that enables almost all the power to be coupled into and transmitted through a small diameter fibreoptic light guide. Optical fibres transmit light using the principle of total internal reflection. If a light ray is travelling in a medium of high refractive index and encounters a medium of low refractive index, the light is reflected back at the interface if its angle of incidence to the interface is greater than the critical angle. The angle of convergence of the focused beam must not exceed the acceptance angle of the optical fibre. The output of the Nd:YAG laser is easily focused to a $200\mu m$ spot and is usually coupled to a $400\mu m$ optical fibre. To reduce power losses the ends of the fibre must be perpendicular to the longitudinal axis and clean.

These optical fibres can in turn be passed down the instrumentation channel of flexible endoscopes. Endoscopes allow hollow parts of the body to be inspected through normal bodily orifices or very tiny incisions with minimum disturbance to the patient (Bown, 1990). In addition laser energy can be delivered along fibres passed through the operating channels of flexible and rigid endoscopes to inaccessible parts of the body and along blood vessels. The energy is then used to cut, coagulate or destroy tissue.

Monochromaticity, Output Power and Polarisation

Laser light is also monochromatic or quasi-monochromatic being of one or just a few distinct wavelengths, this being a property of the temporal coherence of the light.The output of the laser is centred on one wavelength of the light and since the beam of light from a laser is highly collimated, the irradiance (power per unit cross/sectional area) of the beam is very high, and can be focused to very tiny spots producing enormous irradiance and localised power. It is virtually impossible to achieve the same levels with conventional light and heat sources. The wavelength of light can also be matched to specific endogenous or exogenous chromophores to produce precise non-specific thermal absorption or non-thermal photochemical reactions in tissue. The wavelength of the laser is an important factor in the interaction and penetration into the tissue.

THE INTERACTION OF LASER LIGHT WITH TISSUE

An understanding of the interactions of laser light with tissue is fundamental to the rational use of lasers as instruments in medicine, and also determines which laser is chosen for which task. These are best considered by examining the fate of a laser beam externally irradiating a block of tissue (Figure 1). On striking an air or fluid/tissue interface some of the photons are reflected back from the tissue surface (specular reflection). On penetrating the tissue some photons are 'backscattered' and escape from the tissue back into the air (diffuse reflection). When a photon is scattered its energy is not significantly altered (elastic scattering) but the direction of travel is changed. Most light reflected from tissue is diffuse reflection as a result of backscattering from tissue components. Scattering is caused by the inhomogeneous nature of tissue and is determined by the variations in the refractive index between parts of cells and also by the difference in optical properties between cells and their surrounding media. Light that has penetrated deeper than a few fractions of a millimetre will have been subject to multiple scattering. A laser beam is found to be scattered into an approximately isotropic distribution (spherical distribution) after travelling less than 1mm into tissue. It is important to note that scattered light does not produce any biological effect. Similarly, light that is transmitted through tissue exerts no biological effect.

In order for a biological reaction to occur light must be absorbed by the tissue. Non-specific absorption occurs when a variety of tissue components absorb the light. In general absorption in tissue is determined by tissue chromophores such as melanin, haemoglobin and myoglobin and also by tissue water. The absorption characteristics of individual tissues vary enormously and are also highly dependent on the wavelength of the light. The carbon dioxide laser beam (wavelength 10,600nm, in the far infrared end of the spectrum) is strongly absorbed by water; whereas the argon ion laser beam is strongly absorbed by haemoglobin

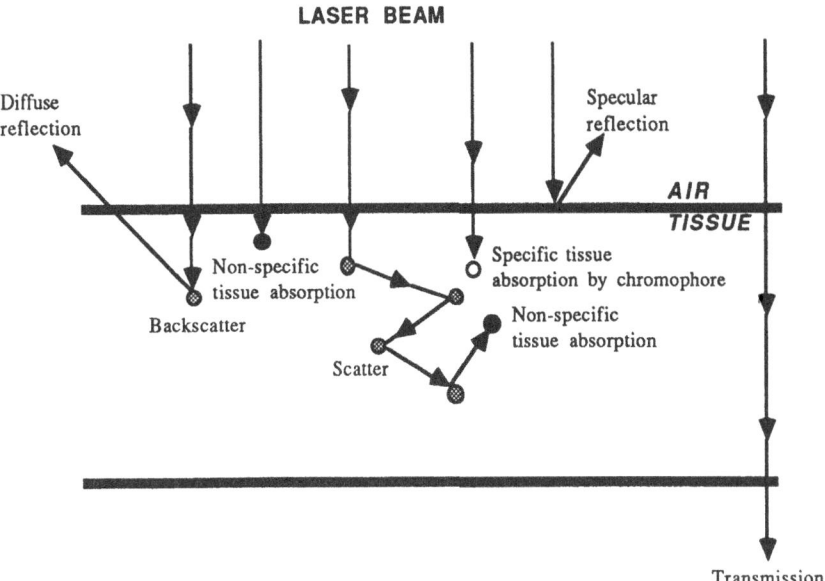

LASER BEAM

Diffuse reflection

Specular reflection

AIR

TISSUE

Non-specific tissue absorption

Specific tissue absorption by chromophore

Backscatter

Non-specific tissue absorption

Scatter

Transmission

Figure 1.   Interactions of light with tissue.

molecules and other pigments. In certain circumstances it may be desirable to administer an exogenous chromophore, to produce specific adsorption of light in certain tissues such as malignant tumours. This allows photochemical reactions to be produced and is the basis of photodynamic therapy.

Thermal Effects

Absorption of light in a non-specific manner may produce thermal changes in the tissue. These are at present the most widely used and surgically useful biological effects produced by a laser beam.

The thermal properties of tissue are important and are governed by three different mechanisms: the ability of the tissue to transport heat by thermal conduction and diffusion, the ability to store heat and finally the ability to transfer heat by blood flow. Local heating of malignant tissue to the temperature region of 41-45ºC may produce a degree of selective hyperthermic destruction of malignant cells, since they are slightly more sensitive to this temperature range than normal tissue. This differential cell kill is lost above 45ºC and all cells are rapidly killed above 50ºC. Further heating causes thermal contraction and coagulation of proteins, and as the tissue shrinks small vessels can be sealed, arresting haemorrhage. Thrombosis of the occluded vessel seems to occur as a secondary event. This haemostatic effect of laser light is best when the volume of tissue heated is relatively large (5mm diameter). The Nd:YAG laser is able to seal vessels up to 1mm in diameter. If more energy is used tissue necrosis with vaporisation, laser ablation and burning occurs. Vaporisation occurs at 100ºC when water boils.

For short exposure times, the distribution of effects in tissue depends on the thermal relaxation time, which is a measure of the time required for the tissue to cool by dissipation of heat. If an area of tissue can be destroyed before the heat transfer damages adjacent tissue more localised tissue destruction can be produced. Thermal relaxation time is more precisely defined as the time for cooling to 1/e of the excess temperature. This is important when considering delivery of the laser energy in pulses rather than continuously. The pulse length must be short compared with the thermal relaxation time and the gap between pulses must   be

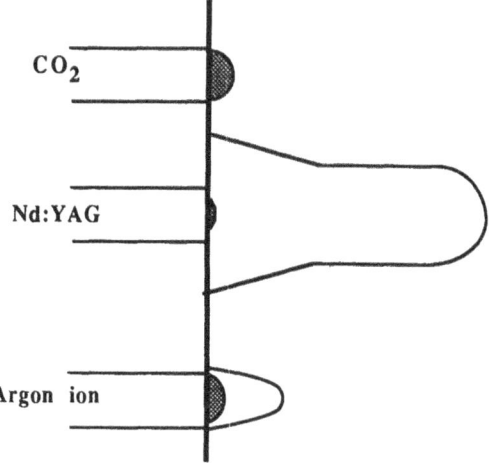

Figure 2. Relative extents of vapourisation [▨]
and coagulation [☐] for the CO_2, Nd:YAG and
Argon lasers. The CO_2 beam is so rapidly absorbed
that all the energy produces vapourisation.

comparatively large. This effect has assumed importance in the treatment of vascular malformations in the skin and also with the use of the pulsed Nd:YAG laser for laser angioplasty. The thermal relaxation time relates to the size of the irradiated tissue and the spot diameter.

Three lasers are at present used for the thermal effects they produce; the carbon dioxide (CO_2), argon ion and Nd:YAG. All produce different thermal effects because of the wavelength at which they operate. The CO_2 laser beam is in the far infra-red (10,600nm), being strongly absorbed by water it is very effectively absorbed when it enters biological tissue, so produces a very localised effect. The laser energy is rapidly absorbed by tissue, causing intracellular water to boil, disrupting cells, and cutting through tissue. There is very little scattering of the CO_2 beam, thus there is only a small area (0.1mm) of coagulation beyond the vaporisation crater. The thermal effects of the CO_2 laser make it very suitable as a non-contact laser scalpel.

The Nd:YAG beam can produce coagulation up to 6mm into tissue, but superficial vaporisation will occur if a high enough power is used. At lower powers only coagulation will occur. This contrasts to the CO_2 laser. The reason is that a photon of light from the Nd:YAG laser is ten times more likely to be scattered than absorbed, thus it will travel further into the tissue before it is absorbed. Thus if a more diffuse thermal effect is required the Nd:YAG laser should be used, since it will penetrate tissue further before it is absorbed (Figure 2). This shows the iso-intensity lines for ninety percent extinction for the three different thermal lasers. That is these lines connect all the points in the tissue at which the light intensity is one tenth of the light intensity impinging on the surface. The difference between the profiles is strictly related to the wavelength. It is clear that the energy delivered by the Nd:YAG laser will be distributed to a larger volume of tissue than that of the others.

The blue/green argon ion laser beam is absorbed in blood and pigmented tissue, and produces thermal coagulation 1-2mm below its vaporisation crater. The argon ion laser is most suitable for producing specific thermal destruction of vascular areas such as port-wine birth marks under the skin. The frequency doubled Nd:YAG laser will produce light at a wavelength of 532 nm and can be used in this role also.

Non-Thermal Effects

In certain circumstances the thermal effects of the laser beam are not required, but photons are required to drive photochemical reactions in tissue. The most promising technique involves the administration of photosensitising agents, that are retained with some selectivity in malignant tissue. When activated by light of a wavelength that is absorbed mainly by the photosensitiser and less by non-specific tissue components, a cytotoxic substance (singlet oxygen) is produced and tissue destruction occurs. The higher concentration of photosensitiser in malignant tumours offers the possibility of selective tumour destruction. Most photosensitisers absorb at red and near infra-red (600-800nm) wavelengths to allow adequate tissue penetration. Ideally peak absorption near $1\mu m$ would allow optimum tissue penetration and there are some photosensitiser that would absorb from 800-1100nm and could be activated using the Nd:YAG laser. However, the generation of singlet oxygen becomes energetically unfavourable since the triplet state energies of these photosensitisers are too low. The use of two photon activation would overcome this problem, and there are indeed very reliable pulsed Nd:YAG lasers that can be used for this purpose. However, the non-linear intensity dependence of multi-photon absorption is a weak process (MacRobert et al, 1989).

Other non-thermal laser effects are the 'non-linear' reactions that occur when tissue is exposed to pulsed laser light. The excimer (ultraviolet-wavelength 193-351nm) lasers produce a laser beam in which the individual photon energy is very high and which is highly absorbed by most biological tissues. This combination means that the light beam is capable of breaking interatomic bonds and a chemical photoablation of tissue occurs. For example the carbon to carbon bond in organic molecules has an energy of 3.6eV, this is exceeded by photons with a wavelength of less than 340nm. The important biological feature of photoablation is the very sharp cut-off (of the order of a few microns) between normal cells and ablated tissue. There is no charred zone as with thermal laser ablation. The Nd:YAG laser can be made to emit very short laser pulses, which if focused can produced very high energies in a very small area. These high energies strip electrons from atoms and produce a rapidly expanding plasma. This can generate powerful mechanical forces that can disrupt tissues. The Q-switched Nd:YAG laser is used by ophthalomolgists to punch holes in the opacified posterior lens capsule that can occur some months after cataract surgery. The laser beam diverges steeply after being focussed on this membrane, so that by the time it reaches the retina the light intensity is at a safe level. A further 'non-linear' effect is utilised for laser lithotripsy in the endoscopic fragmentation of renal and biliary calculi. Pulsed Nd:YAG and dye laser beams are transmitted down flexible optical fibres, the end of which is placed just touching the stone. The laser pulses produce a localised shock wave that can shatter stones.

SPECIFIC APPLICATIONS OF THE Nd:YAG LASER IN MEDICINE

Ophthalmology

It is not surprising that the largest application of lasers in medicine is in the field of ophthalmology since ophthalmic surgeons were the first to recognise the potential. The Q-switched Nd:YAG laser is predominantly used as described in the previous paragraph for disruption of the lens capsule that may have opacified following cataract surgery. In addition, glaucoma can be treated by creating a hole in the iris to improve drainage. The area in the eye to be treated can be selected by careful use of the eyes own optics, and abnormal strands of tissue in the posterior chamber very near the retina can be precisely disrupted without damage to surrounding areas. Most of these procedures can be performed as an office procedure with the laser incorporated into a slit lamp or microscope.

Cardiovascular

Cardiovascular disease, secondary to atherosclerotic occlusion of blood vessels is responsible for over 100,000 deaths and 5,000 limb amputations in the United Kingdom alone every year. Many patients are treated by bypass surgery or with intravascular balloons which can be passed over a guide wire and used to dilate narrowed or blocked vessels

(Gruntzig & Hopf, 1974). The potential of laser angioplasty lies in the possibility of re-boring the vessel to create a channel for a guide wire and so make it accessible to balloon dilatation. Both the pulsed and continuous wave Nd:YAG laser are being investigated for this purpose. It has become clear that to avoid vessel perforation it is essential to have the laser beam transmitted through a device on the tip of the laser fibre. Various systems are now available involving metal, sapphires or ceramic tips. One of the simplest methods is to melt the end using energy passed down the fibre and mould a ball or bulbous end.

The thermal relaxation time of arterial tissue for a 100μm spot size is in the order of 25ms. If the pulse length is short compared with this then surrounding arterial damage may be limited (Selzer et al, 1985). The theoretical danger of diffuse thermal damage is that the wall may be weakened and prone to aneurysm formation and delayed rupture (Lee et al, 1984).

Initial clinical results, using a pulsed Nd:YAG laser, in 40 patients with severely ischaemic limbs unsuitable for other more conventional procedures produced a recanalisation rate of 63% with one late occlusion with a maximum follow up of 18 months (Michaels et al, 1989).

A major goal is the production of a laser angioplasty system for use in coronary artery disease. Potentially, patients would only require insertion of a catheter into an artery in the groin under local anaesthetic and recanalisation under X-Ray screening, avoiding the need for open heart surgery . However, the dangers of perforation with a fatal outcome are substantial in the arteries to the heart and this technique is still in the experimental stage.

The Nd:YAG laser although being investigated primarily for laser angioplasty is also under study for ablation of areas of the heart lining which are producing abnormal electrical activity, and also for valvuloplasty (clearing of damaged heart valves).

The role of the Nd:YAG laser as a thermal welding devices to perform microvascular anastomosis (sealing the joint between two cut vessels) is still under investigation and remains controversial.

Other types of laser are probably more appropriate (Flemming et al, 1988).

Gastroenterology

Upper gastrointestinal tract haemorrhage (UGI) is responsible for 50-100 acute hospital admissions per 100,000 of the population per annum, and most of these are due to peptic ulcer haemorrhage. The mortality from this disease has changed little over the past 3 decades, despite advances in diagnostic endoscopy and intensive care. The majority of 'avoidable deaths' in patients with UGI bleeds are due to post-operative complications of the surgery required to arrest haemorrhage. Therefore, although most episodes of acute upper gastrointestinal haemorrhage cease spontaneously, there is a need for an endoscopic haemostatic device to treat those lesions that are likely to continue bleeding in order to avoid surgery with its associated morbidity and mortality. Lasers were the first non-surgical endoscopic method to be investigated in any detail. The aim is to cause thermal contraction of the feeding vessel, not to directly hit the visible vessel in the ulcer crater which often precipitates bleeding (generally stopped by further laser therapy in the correct area). Controlled trials have confirmed the efficacy of this treatment, of 39 ulcers with a high risk of rebleeding only 6 rebleed after laser therapy, as compared with 23 of 43 control untreated ulcers ($p<0.005$) (Swain et al, 1986) There was also a 1% mortality in the laser treated group and a 12% mortality in the control patients. The Nd:YAG laser is an effective endoscopic method for haemostasis, although it is likely that cheaper methods such as injection sclerotherapy will prove equally effective.

Endoscopic therapy for the relief of obstruction to swallowing from advanced cancers of the gullet primarily involves the use of Nd:YAG laser therapy. Treatment is entirely local and the laser is used to coagulate and vaporise the tumour. In general therapy is started at the lower end of the tumour and continues upwards removing tissue gradually using the laser set to deliver 70-80W in 1 second pulses. However, in 25% of patients with total obstruction to the oesophagus this method is not possible and treatment has to be commenced at the upper end. This method is slightly more hazardous since the direction of the lumen has not been identified. Overall there is a 5% perforation rate but most of these can be managed conservatively without surgical intervention (Bown et al, 1987). The laser is most suitable for the treatment of totally obstructing tumours and those tumours high in the oesophagus unsuitable for endoscopic prosthetic intubation. It is unsuitable if the tumour is predominantly extrinsic to the lumen or if a tracheo-oesophageal fistula is present. In these circumstances intubation is the preferred treatment. A problem with laser therapy is that it must be repeated at monthly intervals in order to maintain the oesophageal lumen, but recent studies suggest that a combination of laser therapy and palliative radiotherapy may help this problem.

The vast majority of colorectal cancers are best treated by surgical resection both for cure and palliation. However, there remain some 5% of patients with advanced metastatic spread or severe concomittant disease who are unsuitable for operative intervention. Palliative colostomy will only bypass the obstruction and does little to relieve the local problems of discharge, bleeding, tenesmus and incontinence. Local fulguration or cryotherapy has provided some relief for these patients, but fulguration requires a general anaesthetic and both are restricted to the management of lower rectal cancers. The endoscopic management of these tumours is possible using the Nd:YAG laser. As with upper gastrointestinal tumours the aim of treatment is to remove all the exophytic tumour. The symptoms that are predominantly improved are bleeding, discharge and obstructive symptoms (Bown et al, 1986). Incontinence is little improved and invasive pain is not helped at all. Quantitative quality of life measurement of those patients treated with palliative laser therapy has shown that little improvement occurs if the patient has pain as the predominant symptom or has only a short time to live (<10 weeks).

Endoscopic Nd:YAG laser therapy is also possible for benign colonic tumours. Most polyps can be very effectively treated by snare diathermy, however some large villous adenomas may be best treated with laser ablation. A large study reported complete eradication of 42 of 56 villous adenomas with the laser and followed for 3-24 months (Brunetaud et al, 1985). These results are comparable to those that would be expected after surgery.

Laser therapy has also proved useful for the treatment of angiodysplasias (areas of fragile blood vessels that are prone to bleed) of the colon and stomach. If the lesion can be identified it can be treated with the Nd:YAG laser and completely destroyed. The major problem is identifying the lesion that has bleed. The principle of treatment is slightly different than that for bleeding peptic ulcers. First a circumferential ring of tissue around the lesion is treated to produce thermal contraction of any feeding vessels. Finally the lesion itself is treated. Oozing is often induced but rapidly ceases with continued laser application.

The use of the Nd:YAG laser as a thermal knife (with or without a sapphire probe attached) is not yet established. In certain circumstances it may be useful, for instance to reduce the blood loss during major liver resection, but many more claims have been made for the technique than can be justified scientifically. It is clear that to cut the skin with the laser offers little advantage, and may be detrimental.

Urology and Gynaecology

The Nd:YAG laser may have a considerable impact in urology in the treatment of bladder cancer. It is important to remember that a relatively safe and efficient method of treatment for superficial bladder cancer has been present for a number of years. Transurethral resection of tumours with a diathermy loop not only removes the growth but provides histological staging of the disease. This is not available when the tumour is totally ablated with the laser. The advantages claimed for the laser in the treatment of urothelial cancer are that it is contact less producing necrosis without loss of mechanical stability. There is no liberation of free cells to re-implant, the lymphatics are sealed preventing cancer dissemination and it is a more predictable thermal effect. After transurethral resection, some 50-70% of lesions will be expected to recur at some point in the patients lifetime. It is possible that one of the explanations of this high recurrence rate may be due to implantation of cells dislodged at the time of resection. Laser treatment is generally carried out with the Nd:YAG laser. It is coupled into a flexible optical fibre and passed through a rigid or flexible endoscope. In some instances laser therapy can be performed through a flexible cystoscope under local anaesthesia as an outpatient, whereas transurethral resection and cystodiathermy is uncomfortable in the conscious patient. The bladder is distended with saline and the laser fibre is held 0.5cm from the tumour, which is vaporised and coagulated with a laser power of 40-50W. One year after laser therapy a 16-18% recurrence rate has been reported. A controlled trial comparing transurethral resection (TUR) with Nd:YAG laser therapy has been performed. It demonstrated some benefit for laser therapy combined with transurethral resection over transurethral resection alone (Beisland & Sealand, 1986).

Recently the Nd:YAG laser has been used to treat menorrhagia. The endometrium (the lining of the uterus) is examined and treated at hysteroscopy, the laser being used to completely ablate all the endometrium. A series of 300 patients have been treated by this method for dysfunctional uterine bleeding (heavy periods in women who have completed their families). Complete success was reported in 90%. If these initial findings are confirmed then the hysterectomy rate in the UK may be drastically reduced (Goldrath, 1986)

Respiratory Medicine

The development of endoscopic laser treatment of obstructing cancers of the major airways has paralleled the endoscopic recanalisation of advanced cancers of the gullet. Unfortunately most patients with lung cancer are inoperable at the time of diagnosis. 50-60% of these patients will have haemoptysis (coughing up blood) or airways obstruction causing shortness of breath (dyspnoea) at some stage. Chemotherapy (anti-cancer drugs) and radiotherapy are effective in some patients, however are associated with considerable toxicity and complications. The principle of laser therapy is to remove the bulk of tumour causing the blockage to relieve airways obstruction or to coagulate tumours that are bleeding. Treatment can be performed under sedation with a flexible bronchoscope, or under general anaesthesia (which makes the technique safer and easier) with a combination of rigid and flexible bronchoscopy. The laser is used to vaporise and coagulate the tumour. If coagulated dead tumour sloughs some time after treatment, it may be inhaled and occlude a major airway, therefore all dead tissue is debrided with endoscopic forceps immediately after laser coagulation. This provides instant improvement in the airway and reduces the risk of temporary obstruction due to swelling or sloughed dead tumour tissue. These large pieces of dead tumour can be removed most easily through a rigid bronchoscope (Hetzel et al, 1985).

Good palliation of dyspnoea and haemoptysis has been reported, although tumour is likely to recur and the mean length of time between treatments is 70 days. Measurement of peak rate of air flow before and after treatment has demonstrated that 63 % of patients have at least a 25% improvement immediately after treatment. The most dangerous complication is that of major haemorrhage which can be rapidly fatal. This is easier to control under general anaesthesia with rigid bronchoscopy than through the flexible instrument under sedation. At present laser therapy is most appropriate for intraluminal tumour. However, some 40% of patients with dyspnoea are breathless because of extrinsic compression of the airway. It may be possible to treat these patients by inserting a laser fibre through the wall of the airway and coagulating the tumour with the laser in the hope that it will shrink. This is at present being investigated, but no results are yet available.

Laser Lithotripsy (Stone Fragmentation)

Laser lithotripsy was originally developed to deal with kidney stones impacted in the ureter (the tube joining the kidney to the bladder), that could not be removed endoscopically or shattered using extracorporeal shock wave lithotripsy (ESWL). Lasers have proved to be less traumatic to the ureter itself when compared with ultrasound (which requires a rigid probe and is relatively slow) and electrohydraulic lithotripsy (which can cause local tissue heating). Continuous wave Nd:YAG lasers were originally investigated for laser lithotripsy, but as they depended on breaking the stone by a thermal processes, there was considerable risk of damaging the surrounding ureter (Lux et al, 1986). At present the most used system is a pulsed tuneable dye laser (tuned to 450-500nm) coupled into a flexible optical fibre (Watson & Wickham, 1986). However, it has recently become apparent that the Q-switched Nd:YAG (20Hz, pulse energy 30-40mJ) laser is a very effective lithotripsy device particularly for stones in the bile duct. It reduces stones to smaller fragments than do the other lasers, most of the fragments measuring less than 2mm, which can then be passed spontaneously. Approximately 500 hundred pulses are required to completely shatter a 30mm stone of any composition. If inadvertent damage occurs to surrounding tissue it is restricted to the superficial layers and rapid healing occurs without dire consequences. However, it is yet to be fully established which is the ideal instrument.

The technique of treatment involves the laser fibre being passed down a flexible ureteroscope or duodenoscope and placed just touching the stone, which is shattered by repeated laser pulses. Laser lithotripsy by the 'free floating' technique under radiological control appear safe but not as reliable as lithotripsy under direct vision. ESWL is the preferred method for stone destruction in the kidney and upper ureter, but laser lithotripsy is the best method of endoscopic management for impacted stones in the rest of the ureter and biliary tree.

Interstitial Hyperthermia

The continuous wave Nd:YAG laser appears to be the ideal instrument for interstitial hyperthermia. Recently there has been an increasing interest in the use of this technique for the treatment of solid cancers particularily in the pancreas and liver, that are not easily resectable. Interstitial hyperthermia involves reducing the Nd:YAG laser power from 50-80W for 0.5-1 second (as used for the advanced tumours of the gullet and lungs) to between 1-2 W delivered over a longer time (up to 1000 seconds) and inserting the fibre directly into tissue. The aim of using low power is to avoid vaporisation but produce destruction of the tissue by slow thermal coagulation. For clinical treatment, multiple fibres are inserted into the tumour under ultrasound control. The development of damage in the tumour can be monitored in real time by ultrasound so once the area to treated is totally destroyed, the laser fibres are removed. The procedure can be repeated as often as necessary and has the potential for revolutionising minimally invasive surgery. So far the technique has been applied to small numbers of tumours of the liver, pancreas and breast, but in principle could be applied to tumours in the centre of any solid organ, as long as there is sufficient surrounding normal tissue to ensure safe healing (Steger et al 1989).

REFERENCES

Bown,S.G., Barr,H., Matthewson,K., Hawes,R., Swain,C.P., Clark,C.G., Boulos,P.B.,1986, Endoscopic treatment of inoperable colorectal cancers with the Nd:YAG laser, Br. J.Surg. 73:949-52.

Bown,S.G., Hawes,R., Matthewson,K., Swain,C.P., Barr,H., Boulos,P.B. and Clark, C.G., 1987, Endoscopic laser palliation for advanced malignant dysphagia, Gut, 28:799-807.

Bown,S.G., 1990, Lasers-the minimally invasive surgeons of the future, Sci.Publ.Affairs, 4:41-63.

Beisland, H.O. and Seland, P.,1986, A prospective randomised study on Nd:YAG laser irradiation versus TUR in the treatment of urinary bladder cancer, Scand. J. Urol. Nephrol. 20:209-212.

Brunetaud,J.M., Mosquet,L.and Houcke,M.,1985, Villous adenoma of the rectum-results of endoscopic treatment with argon and Nd:YAG lasers, Gastroenterology, 89:832-7.

Goldrath, M.H., 1986, Hysteroscopic laser ablation of the endometrium, in: "Gynaecological laser surgery," J.A.Jordan and F.Sharp, ed. Perinatology Press, New York.

Michaels, J.A., Cross,F.W., Shaw,P., Raphael,M., Bowker,T.J., Bown,S.G., Adiseshiah, M. and Marston, A., 1989, Laser angioplasty with a pulsed Nd:YAG laser: early clinical experience, Br.J.Surg., 76:921-924.

Gruntzig,A.R.and Hopf, H.,1974,Perkutane rekanalisation chronischer arterieller verschlusse mit einem neuen dilationskatheter. Modifikation der Dotter-technik. Dtsch. Med. Wochenschr., 99:2502.

Hetzel, M.R., Nixon,C., Edmonstone,W.M. and Mitchell,D.M.,1985, Laser therapy in 100 tracheobronchial tumours, Thorax, 40:341-345.

Flemming, A.F.S., Colles, M.J., Guillianotti, R., Brough, M.D. and Bown, S.G.,1988, Laser assisted microvascular anastomosis of arteries and veins: laser tissue welding. Br.J.Plas.Surg., 41:378-388.

Lee, G., Ikeda, R.M., Theis, J.H., Chan, M.C., Stobbe, D., Ogata, C., Kumagai, A. and Mason, D.T., 1984, Acute and chronic complications of laser angioplasty: vascular wall damage and formation of aneurysms in the atherosclerotic rabbit, Amer.J.Cardiol., 53:290.

Lux, G., Ell, Ch., Hochberger, J., 1986, The first endoscopic retrograde laser lithotripsy of common bile duct stones in man using a pulsed Nd:YAG laser, Endoscopy, 18:144.

MacRobert, A.J., Bown, S.G. and Phillips, D., 1989, What are the ideal properties for a photosensitizer, in "Photosensitizing compounds: their chemistry, biology and clinical use," Ciba Foundation Symposium 146, John Wiley & Sons, Chichester.

Selzer, P.M., Murphy-Chutorian, D., Ginsburg, R. and Wexler, L., 1985, Optimizing strategies for laser angioplasty, Invest Radiol., 20:860.

Steger,A.C., Lees,W.R., Walmsley,K. and Bown,S.G.,1989, Interstitial laser hyperthermia: a new approach to local; destruction of tumours, Brit. Med. J., 299:362-365.

Swain, C.P., Kirkham, J.S., Salmon, P.R., Bown, S.G.,Northfield,T.C.,1986, Controlled trial of Nd:YAG laser photocoagulation in bleeding peptic ulcers, Lancet, i:1113-7.

Watson, G.M., Wickham,J.E.A.,1986, Initial experience with a pulsed dye laser for ureteric calculi, Lancet,i,:1357-8.

DIODE LASER PHOTOCOAGULATION IN OPHTHALMOLOGY

Rosario Brancato, Giovanni Leoni, Giuseppe Trabucchi

Department of Ophthalmology - University of Milano
Scientific Institute H S. Raffaele
Via Olgettina,60 - 20132 MILANO (Italy)

INTRODUCTION

Argon and Krypton laser systems are the laser sources commonly used in ophthalmology for retinal photocoagulation procedures. However, they suffer from several ergonomic and economic drawbacks. In fact, ion lasers are expensive, bulky and relatively inefficient devices: the energy is generated by a relatively bulky gas-filled tube with short operating lifetime, high electrical energy consumption, and low electrical-to-optical conversion.

In recent years advances in the technology of semiconductor crystals has led to a great development in the field of miniaturized diode laser. The new generation of diode lasers has almost a 50% efficiency and output power of several watts. Commercially available lasers have a maximum output power of about 1-1.5 Watts emitted in the far-red/near-infrared (750-811 nm) spectral range. These new laser sources present a number of characteristics which lend themselves to ophthalmic application, including standard voltage requirement, the need for a simple air-cooling system, extremely compact size and long operating lifetime.

In 1984 Pratesi[1] suggested the possibility of using high-power diode lasers in photomedicine.

In this paper we report our experience regarding the applications of diode lasers in the photocoagulation of ocular tissues with particular regard to histopathological results, beginning from first experimental investigations to recent clinical trials.

TRANSPUPILLARY CHORIORETINAL PHOTOCOAGULATION

In 1987 we first reported transpupillary photocoagulations on rabbit chorioretina using a continuous wave (CW) AlGaAs diode laser coupled with a slit-lamp microscope. The retinal semi-conductor laser lesions obtained were ophthalmoscopically and histologically similar to those normally produced by the common ion lasers at similar irradiance levels.[2,3] Diode laser lesions appeared ophthalmoscopically as small white areas more or less round and similar to the lesions produced by argon laser.

Histologically, argon laser resulted in a damage to both the inner and the outer retinal layers while diode laser produced damage to the outer retina and choroid where small vessel occlusions and/or oedemas were induced. Necrosis of Retinal Pigment Epithelium (RPE) was similar for both the radiations employed.

Human transpupillary chorioretinal photocoagulations obtained with a diode laser prototype coupled with a slit-lamp microscope were reported by Brancato et al. and McHugh et al. in 1989.[4,5] Ophthalmoscopic and histological patterns of the diode photocoagulation produced on human chorioretina outside the posterior pole did not differ from those obtained with argon and krypton lasers.

Also for infrared radiations the most important cromophores to obtain photocoagulative effects is melanin in RPE and choroid.[6] Due to the highest density of melanin in RPE 80% of laser energy is absorbed at this level.[7] At 810 nm melanin absorption is lower than for ion laser, but substantially higher than for 1064 nm of Nd:YAG and sufficient to produce photocoagulative effects.[8]

Recent histological studies have demonstrated that outside the macular region, no particular differences were noted in the argon, krypton and dye laser chorioretinal lesions for super-threshold energy values.[10] Furthermore, the studies on the biophysics of beam-tissue interactions indicate that the wavelength of the laser source is not of prime importance, but that thermal effects are better related to exposure times.[11]

Therefore, the idea to selectively photocoagulate precise targets within the chorioretinal layers using different wavelengths has been partially confounded.

In macular photocoagulation the difference between blue-green spectrum (488-514 nm) and longer wavelengths, such as 610 nm of krypton or 800 nm of diode laser, is related to the high absorption coefficient of xanthophyll for shorter wavelengths.[14]

Working in the macular region diode laser lesions were similar to those obtained with Krypton laser.[7]

Fig.1.   Similar histological characteristics of Argon green and Diode chorioretinal photocoagulation produced on human eye, outside the macula region.

The emission wavelength of the Diode laser is not within the absorption spectrum of haemoglobin, which is the second most important cromophore for chorioretinal photocoagulation.[8] This is very important considering that diabetic eyes often have cloudy media, due to hemorrhagic phenomena. The lack of light absorption in vascularized structures may be improved by some specific stains. The 810 nm are absorbed selectively by some stains such as indocyanine green. Several investigators are studying the possibility of using this property to photocoagulate pathological vascular tissues after intravenous injection of specific absorption spectrum dye.

Light at 400-520 nm has a considerable attenuation by both scattering and absorption of the ocular media as compared to longer wavelengths (600-1064 nm).[2] For example, over 50 years of age the lens has a substantial loss of transparency to light, due to age related tissue changes, with a prominent drop of transmission in the blue-green portion of the spectrum.[7]

More than 80% of the photocoagulative treatments have to be performed in eyes with age-related lens changes. In these clinical situations the 810 nm of diode laser minimize the loss of energy through the lens enabling us to obtain good photocoagulative effects with lower energy levels.

The absorption-transmission characteristics and the histological findings described indicate that diode laser can be useful in the treatment of retinal pathologies.

Preliminary clinical results obtained in patients with diabetic retinopathy (proliferative and non-proliferative diabetic retinopathy) and ischemic occlusion of retinal veins are similar to those obtained with argon laser.

A randomized clinical multicenter study is now in progress in Europe to confirm the efficacy of this new laser source in the treatment of diabetic retinopathy and to evaluate possible side effects.

CHORIORETINAL ENDOPHOTOCOAGULATION

In 1987 Puliafito et al.,[12] introducing a 100 mµ optic fiber coupled to Spectra Diode Labs AlGaAs diode laser into a rabbit eye, obtained the first chorioretinal endophotocoagulations with an energy value of 40 mW.

Several authors have recently reported on the use of diode laser to perform retinal endophotocoagulation procedures during vitreoretinal surgery in human eyes.[13]
Endophotocoagulation diode laser units have many advantages over the widely used argon laser: they are extremely compact in size, air-cooled and may be commercially available for cheaper initial costs.

LASER TRABECULOPLASTY IN OPEN ANGLE GLAUCOMA

In 1989 Mc Hugh et al.[16] reported the first clinical results of Laser Trabeculoplasty (LTP) with diode laser in the treatment of Open Angle Glaucoma.

Our clinical results confirm that the behavior of intraocular pressure (IOP) after diode LTP is similar to that reported with Argon LTP. The physiopathological mechanisms of IOP decrease are unclear for diode LTP as well as for Argon LTP. The lower absorption coefficient of melanin at 800 nm produces a deeper penetration of diode radiation through the trabecular meshwork. This fact could result in a wider shrinkage of the trabecular meshwork improving aqueous outflow. Also for diode laser trabeculoplasty a clinical multicenter study is in progress to evaluate its actual effectiveness in the treatment of open angle glaucoma.

TRANSSCLERAL PHOTOCOAGULATION OF CILIARY BODY AND CHORIORETINA

Nd:YAG laser transscleral photocoagulation of ciliary body has been shown to be an effective method to lower IOP.[14]

The 810 nm has a lower scleral transmission as compared to Nd:YAG (1064 nm), while it has a higher absorption coefficient for the pigmented structures of the ciliary body. The transmission-absorption coefficient makes the diode laser suitable to produce transscleral thermal lesions. Applying the optic fiber, coupled to a diode laser, directly in contact with the rabbit sclera (0.5 mm from limbus),we obtained transscleral photocoagulations of the ciliary body with a power of 500 mW for 0.5 s confirmed by other authors.[15]

We have also obtained contact transscleral chorioretinal whitish lesions with an energy of 200 mW for 0.2 s. These results are very encouraging but, in our opinion, prior to the introduction of this laser source in clinical transscleral procedures, further investigations and more powerful lasers are needed.

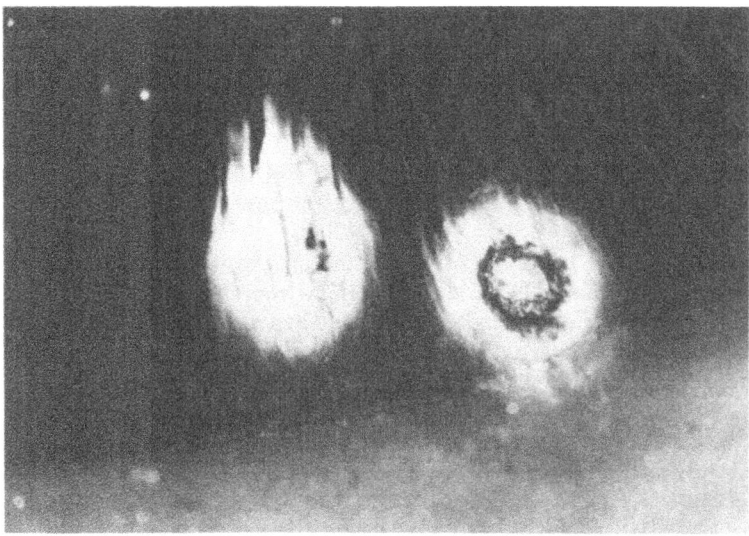

Fig. 2.   Macroscopical view of rabbit ciliary body transcleral photocoagulation produced delivering Diode laser radiation directly in contact on the conjunctiva with a fiber optic system.

CONCLUSIONS

First results obtained indicate an interesting future for these new laser sources in ophthalmology. The compact size, three-phase electrical supplies, water cooling and long operating lifetime make diode lasers suitable to medical applications.

However, further technical shortcomings such as the low power emission, quality of the beam, minimal spot size and the shape of the spot have to be improved for a wider diffusion of diode laser in ophthalmology. Diode lasers with higher energy reserve allow the use of shorter exposure times and consequent safer ocular treatments.

Another feature that has to be improved is the spatial quality of the output beam. The output beam of diode laser always has a wide divergence angle and rectangular pattern. For this reason it is difficult to couple the laser beam to the optics of a slit-lamp and to obtain a convergent cone-angle necessary for retinal photocoagulation. Therefore, the possibility to have laser systems with clear focalization of a round spot on the retina and a well-defined spot size is still lacking. We think that by coupling the diode laser beam to an optic fiber system, some characteristics of the beam profile could be improved. In fact, fiber optics mix together all the beam modes into a quasi-Gaussian beam with smooth and regular slope obtaining a better distribution of energy density.

Pursuant to these results several laser manufacturers are now producing commercial diode laser systems.

Due to their potential low cost and the lack of maintenance requirements for these laser systems a very high diffusion is expected. However, further clinical trials are needed to evaluate the efficacy of diode laser radiation as opposed to Argon and Krypton in retinal photocoagulation before establishing the cost-benefits of this new laser source.

ACKNOWLEDGMENTS

Work supported by the Special CNR-Project "Tecnologie Elettroottiche".

REFERENCES

1. R. Pratesi, Semiconductor lasers in photomedicine, J. Quantum Electronic 20: 1433-1435 (1984).
2. R. Brancato, and R. Pratesi, Applications of diode lasers in ophthalmology, Lasers in Ophthalmol. 1: 119-126 (1987).
3. R. Brancato, R. Pratesi, G. Leoni, G. Trabucchi, L. Giovannoni, and U. Vanni, Retinal photocoagulation with diode laser operating from a slit lamp microscope, Lasers and Light in Ophthalmol. 2: 72-78 (1988).
4. R. Brancato, R. Pratesi, G. Leoni, G. Trabucchi, and U. Vanni, Histopathology of diode and argon laser lesions in rabbit retina, Invest. Ophthalmol. Vis. Sci. 30: 1504-1510 (1989).
5. R. Brancato, R. Pratesi, G. Leoni, G. Trabucchi, and U. Vanni, Semiconductor diode laser photocoagulation of human malignant melanoma, Am. J.Ophthalmol. 107: 296-297 (1989).

6. J.D.A. McHugh, J. Marshall, M. Capon, A. Rothery, and A. Raven, Transpupillary photocoagulation in the eyes of rabbit and human using a diode laser, Lasers and Light in Ophthalmol. 2: 125-143 (1988).

7. J.D.A. McHugh, J. Marshall, T. Ffytche, A. Hamilton, and A. Raven, Macular photocoagulation of human retina with a diode laser: a comparative histopathological study, Lasers and Light in Ophthalmol. 3: 11-28 (1990).

8. J.D.A. McHugh, J. Marshall, T. Ffytche, and A. Hamilton, Initial clinical experience using a diode laser in the treatment of retinal vascular disease, Eye 3: 516-527 (1989).

9. W. Smiddy, A. Quigley, and S. Fine, Comparison of krypton and argon laser photocoagulation, Arch. Ophthalmol. 102: 1087-1092 (1984).

10. H.L. Brooks, R. Eagle, R. Schroeder, W. Annesley, J. Shields, and J. Augsburger, Clinico-pathologic study of organic dye laser in human fundus, Ophthalmology 96: 822-834 (1989).

11. R. Birngruber, P. Gabel, and F. Hillenkamp, Fundus reflectometry: a step towards optimization of retinal photocoagulation, Mod. Prob. Ophthalmol. 18: 383-390 (1977).

12. C.A. Puliafito, T.F. Deutsch, and J. Boll, Semiconductor laser endophotocoagulation of the retina, Arch. Ophthalmol. 105: 424-428 (1987).

13. J. Duker, J. Federman, H. Schubert, and C. Talbot, Semiconductor diode laser endophotocoagulation of the retina, Ophthalmic Surgery 20: 717-719 (1989).

14. R. Brancato, G. Leoni, G. Trabucchi, and C. Pietroni, Contact transscleral cyclophotocoagulation with Nd:YAG laser cw in uncontrolled glaucoma, Ophthalmic Surgery 21: 1-7 (1989).

15. S. Okamoto, H. Takahashi, Y. Fukado, and T. Ozawa, Laser diode application for transscleral photocoagulation, Lasers and light in Ophthalmol. 3: 29-37 (1990).

16. J.D.A. McHugh, J. Marshall, T. Ffytche, A. Hamilton, and A. Raven, Clinical and histopathological results of transpupillary diode laser retinal photocoagulation, Invest. Ophthalmol. Vis. Sci. (suppl.) 30: 371 (1989).

LASERS IN DERMATOLOGY

R. Rox Anderson, and John A. Parrish

Department of Dermatology
Harvard Medical School
Massachusetts General Hospital
Boston, MA 02114

INTRODUCTION

Dermatology has been at the forefront of photomedicine, and has fostered great interest in medical laser applications. In the decade following invention of the ruby laser, this device was used to experimentally treat portwine stains, tattoos, hemangiomas, and melanoma (Goldman and Rockwell, 1971). The ruby laser was then essentially abandoned and replaced during the 1970's by continuous-wave lasers such as argon-ion and CO_2, which offered more predictable, less explosive tissue reactions for coagulation and cutting, respectively. In the 1980's, the major clinical advances were based on selectively-absorbed pulsed lasers, capable of producing histologically-selective thermal injury for treatment of vascular and pigmented lesions, with minimal or absent scarring. This began with using yellow dye laser pulses for treatment of portwine stains, and a concept we called "selective photothermolysis" (Anderson and Parrish, 1981 and 1983). The past decade has seen major advances in basic understanding of laser-induced photothermal, photochemical, and photoacoustic injury of the skin.

Present well-accepted laser applications in dermatology include soft tissue ablation and removal of vascular lesions, tattoos, and benign pigmented lesions. To some extent, lasers have been used to replace other destructive modalities such as electrosurgery. However, physical interactions and treatments unique to lasers are now in widespread use, making it possible for example, to remove some lesions without scarring. New interest in treatment of skin neoplasms with tumor-localizing phototoxic dyes, robotically-controlled laser surgery, and in diagnostic applications of laser spectroscopy are emerging. This chapter is organized by problems and processes in dermatology, with emphasis on basic understanding, correlation of scientific and clinical findings, and problems remaining to be solved.

Optronic Techniques in Diagnostic and Therapeutic Medicine
Edited by R. Pratesi, Plenum Press, New York, 1991

The early experience with ruby laser pulses for removal of tumors showed that irregular, explosive removal of fragments of viable tissue occurred, with living tumor cells being ejected over great distances, and without hemostasis. Such gross "photomechanical" damage proved difficult to control. In contrast, conventional c.w. CO_2 laser surgery (10.6 μm, 1-10 W typical) vaporizes tissue more predictably, does not produce large ablated-tissue fragments, and leaves typically 0.3-1 mm of residual thermally-coagulated skin tissue at the ablation bed (Walsh et al, 1988). The 10.6 μm radiation penetrates about 20 μm (1/e depth) into wet tissue, and the much thicker residual thermal coagulation zone therefore results from heat conduction during the relatively long exposure times used. This coagulum results in hemostasis for most vessels less than approximately 0.5 mm, but also appears to be responsible for the somewhat delayed wound healing and high incidence of scarring associated with many CO_2 laser procedures, even when skin adnexae are preserved. The availability and reliability of clinical CO_2 lasers has led to widespread, appropriate use for ablation of warts and condylomas, many benign tumors, actinic cheilitis, tattoos, and other non-neoplastic lesions (Olbricht et al, 1988). Although keloids can be treated with CO_2 laser ablation, the results are not significantly better than conventional excision (Stern and Lucente, 1989). Use for routine excision of malignancies is problematic because of destruction of histology at the excision margins; this may be less of a limitation for pulsed mid-infrared lasers or appropriately pulsed CO_2 lasers, which leave significantly less residual thermal damage. Such lasers are not yet in clinical use in dermatology.

Minimization of residual thermal damage during laser ablation has significant clinical advantages in dermatology, and may allow new applications. The extent of thermal damage is determined primarily by the distribution of residual energy deposited in tissue after ablation. In general, minimal thermal damage occurs with single-pulse ablation by pulses shorter than the thermal relaxation time associated with the heated layer thickness, given by approximately 1/(absorption coefficient) (Wolbarsht, 1984). In albino guinea pigs, skin wound healing was not significantly delayed by a coagulum thickness of approximately 50-100 μm, which was produced with a TEA CO_2 laser (10.6 μm, 1 μsec pulses, >5 J/cm^2) (Walsh J.T. Jr, Hruza G., Anderson R.R., Deutsch T.F.: unpublished observations). Normal-mode and Q-switched erbium laser pulses (2.9 μm) were also used to ablate guinea pig skin (Walsh et al, 1989). The Q-switched pulses produced approximately 5 μm of residual thermal coagulum, which is insufficient for hemostasis. In pigs, the viability and healing of split-thickness skin grafts was unaffected by residual thermal coagulation of less than approximately 100 μm thickness (Green, H.A., et al unpublished observations). The adherance and viability of keratinocyte cell grafts after laser ablation has not yet been determined.

Approximately 50 to 100 μm of residual thermal coagulum achieves hemostasis without causing delayed wound healing, and therefore may be viewed as a practical goal. At high irradiance (e.g., tightly focussed), conventional c.w. CO_2 lasers and certain "superpulsed" CO_2 lasers may potentially achieve this (Schomacker et al, 1990); however, the rapid scanning necessary for proper surgical control at such high average irradiance is difficult if not impossible to achieve manually. The use of pulsed TEA-CO_2 or scanned infrared laser systems for debridement of major skin burns and ulcers could greatly improve patient care, given proper development of means to control and/or guide rapid laser ablation.

Mechanisms for excimer laser ablation of skin may differ from other visible and infrared pulsed lasers, and appear to involve both photochemical and photothermal mechanisms (Clarke et al, 1987; Marshall et al, 1985; Trokel et al, 1983). Regardless of whether the lack of thermal damage after ablative 193 nm ArF excimer pulses is due to photochemical interactions and/or an explosive thermal tissue removal mechanism or other processes, this laser presents an opportunity for highly precise ablation of tissue surfaces. "Photoacoustic" injury to skin results from ArF laser ablation (Watenabe et al, 1988), with disruption of cells well below the ablation site. Despite this, skin healing after 193 nm ablation into the superficial dermis is apparently rapid. The risk of mutagenesis by 193 nm excimer laser pulses is minimal in comparison to longer UV wavelengths (Kochevar, 1989), although experimental photocarcinogenesis trials have not been performed. Therefore, the 193 nm excimer laser or similar pulses, may indeed have some dermatologic applications. For example, the ability to precisely remove part or all of the outermost stratum corneum (8-15 μm thickness) (Jacques et al, 1987), may in theory allow transcutaneous delivery of large-molecular weight drugs or macromolecules.

The possibility of infective viral particles in laser plume material has appropriately gained recent attention. Intact papilloma virus DNA has been shown in CO_2 laser plume material (Garden et al, 1988), yet the magnitude of risk for infection with papilloma (wart) virus, human immunodeficiency viruses (HIV), or hepatitis viruses remains to be determined. It is important to recognize that the biohazard depends greatly on the laser-tissue interaction and exact procedure. Although it remains unclear whether infective HIV may be aerosolized within CO_2 laser plume, it is essentially certain that even a thermally-sensitive virus such as HIV can survive fragmentary explosive tissue removal with high-energy laser pulses. This includes ruby or pulsed dye laser ablation of soft tissue, in which whole viable cells may be ejected intact.

VASCULAR LESIONS

The treatment of vascular malformations such as portwine stains (PWS, nevus flammeus) or essential telangiectasias, and of benign proliferative vascular lesions such as hemangiomas, is one of the oldest applications of lasers in dermatology. In early work, Goldman treated PWS with some variable success using normal-mode ruby laser pulses (Goldman and Rockwell, 1971). There have since been both great advances and great controversy. It was quickly recognized that argon-ion radiation [488, 514 nm] was well absorbed by hemoglobins, and this laser became the standard treatment during the 1970's and early 1980's, using a variety of exposure techniques. Typically, powers of 1 to 5 watts applied over a 1 mm diameter area, with either freehand scanning, or shuttered exposures of 100-500 msec were used, to the point of causing slate-grey discoloration indicative of coagulation necrosis of the entire epidermis and upper dermis. Early clinical studies indicated an incidence of scarring ranging from approximately 5 to 20%, however more recent studies with somewhat less aggressive treatment have shown a reduced incidence. Argon-ion laser treatment is less effective in children than in adults. CO_2 and other c.w. lasers have also been successfully used to treat portwine lesions, by producing nonselective, superficial coagulation necrosis of the upper dermis, but are probably more prone to produce scarring. The success of treatment with any of these c.w. lasers depends significantly upon the surgical skill of achieving a uniform, superficial coagulation.

Selective Photothermolysis: pulsed dye lasers

In 1981, the concept of selective photothermolysis was developed
for the treatment of portwine stains in younger patients, and pulsed dye
laser use in medicine was initiated (Anderson and Parrish, 1981).
Selective photothermolysis relies on preferential absorption within
microscopic tissue "target" structures such as blood vessels, to produce
selective, localized heating and alteration of the target. A pulse or
exposure duration approximately equal to or less than the thermal
relaxation time associated with cooling of the targets is necessary, in
order to confine thermal damage. The 577 nm absorption band of
oxyhemoglobin was chosen on the basis that the greatest preferential
absorption by cutaneous microvessels is within this band.

The thermal relaxation time of structures varies predominately with
size, being proportional to the square of the target structure
dimension. Thus, within a portwine stain, the typically 20 - 200 μm
diameter abnormal dermal vessels (Barsky et al, 1980) have estimated
thermal relaxation times ranging from tens of microseconds to tens of
milliseconds. A pulse duration in the microsecond domain therefore
allows good thermal confinement to the larger microvessels. Perhaps
more importantly, the smaller-caliber, normal vessels of the upper
dermis appear to escape damage from pulses longer than their thermal
relaxation times, which are about 20 μsec or less. Thus, single 577 nm
pulses shorter than about 20 μsec cause intense microvascular injury
with hemorrhage, compared with longer pulses (Garden et al, 1986). The
hemorrhagic effect of short pulses was also noted in vitro with 577 nm
pulses (Anderson et al, 1983; Gange et al, 1984). Due to radial heat
conduction during the laser pulse, the peak temperature in small vessels
would be less than in larger vessels, even at the same laser pulse
fluence. Because of a shorter thermal relaxation time, small vessels
cool more rapidly. This should further reduce the extent of their
thermal damage, which is a rate-dependent process (Birngruber, 1980).
Many of these theoretical predictions have not been specifically tested
experimentally.

Selective photothermolysis of portwine stains was first reported in
1986 (Morelli et al), and is now the preferred means of treating
portwine stains in children. The incidence of scarring even after the
necessary multiple treatments is on the order of 1%, using 577 nm, 300-
500 μsec, 6-7.5 J/cm^2 pulses delivered in a contiguous array (Garden et
al, 1988). The efficacy in neonates or children is excellent (Tan et
al, 1989a). Unfortunately, multiple painful treatments at approximately
one-month intervals are required. Treatment of superficial or mixed
proliferating hemangiomas in infancy is also possible and appears to be
somewhat effective, despite the fact that these lesions almost always
involute spontaneously. There is therefore controversy over when, if,
and how to treat hemangiomas in early childhood. In general, patients
with lesions near important structures such as the eye, which may lead
to functional impairments, may benefit most from laser or other
treatment.

The dermal penetration of 577-590 nm radiation is limited to
typically several hundred μm (Anderson and Parrish, 1981), with longer
wavelengths penetrating deeper. Histologically, there is somewhat
increased deep vascular injury with 585 vs. 577 nm pulses in pig skin
(Tan et al 1989b), which has led to clinical use of 585 nm pulses.
Despite the lack of clinical comparative trials, it is likely that
somewhat better results can be obtained with adult PWS lesions using
580-600 nm radiation, as compared to 577 nm, because of the increased
individual vessel caliber in adults (Barsky, 1980). In most childhood

216

lesions, however, there is probably little difference. Moreover, the action spectrum for selective photothermolysis of vessels is complex due to laser-induced alterations in hemoglobins, and "ideal" wavelengths for particular lesions remain to be determined. The inability of yellow dye laser pulses to treat deep (e.g., greater than 1 mm) lesions makes this approach of little or no value for deep cavernous hemangiomas, which may be treated with intralesional or systemic steroids and/or Nd:YAG laser coagulation (Apfelberg et al, 1990). Hypertrophic adult portwine lesions frequently appear to respond better to conventional argon-ion or c.w. yellow dye laser treatments, but there are as yet no direct comparative treatment trials.

Other Techniques

Scheibner and coworkers have found that excellent clinical results can be achieved, although with considerable tedium, by manually aiming directly at the ectatic vessels in PWS or telangiecytasia, with a 100-200 μm diameter spot of c.w. argon-ion or 577 nm c.w. yellow dye laser irradiation at low power, typically 0.1-0.3 W (Scheibner and Wheeland, 1989). It is unclear whether this method leads to histologically-selective vascular injury, but it is likely to spare much of the tissue intervening between visible vessels. Treatment times of six or more hours may be necessary for typical facial PWS lesions, and it remains unclear how this putatively nonscarring technique compares with other therapy.

Copper vapor laser treatment of vascular lesions is in widespread use. This laser produces a high-frequency continous train of nanosecond-domain pulses at 511 and/or 578 nm, wavelengths which are preferentially absorbed by hemoglobins. A doubled-Nd:YAG laser known as the "KTP laser", named after the doubling crystal, is also clinically available, producing a continuous pulse train at 532 nm. There is no convincing evidence that selective vascular injury results with these lasers, as they are presently used. The effects of both copper vapor and "KTP" lasers are similar, if not identical, to true c.w. lasers such as argon-ion and c.w. dye lasers at similar wavelength regions and average powers. The recent development of a shuttered delivery device, which produces a hexagonal patterned array of exposure spots, appears to make the clinical results of such c.w. and effectively-c.w. lasers more predictable, perhaps with a reduced risk of scarring (Mordon et al, 1989).

Many fundamental and practical questions regarding laser-induced vascular injury remain. Details of biologic repair are poorly characterized and may offer insight into enhancing efficacy or reducing scarring. The action spectrum for photothermal coagulation of vessels is complex, and remains to be well defined; alterations in hemoglobin species including photodissociation and photocoagulation are probably important factors. The effects of multiple pulses, in which cumulative but highly selective thermal denaturation may occur, remain poorly understood. Photodynamic therapy (PDT) with dihematoporphyrin ether (DHE) causes solid tumor necrosis largely because of vascular injury and thrombosis. The treatment of vascular lesions in skin with DHE or other systemic photosensitizers may have merit based on early in vivo animal studies (Orenstein et al, 1990), and would potentially alleviate some of the problems associated with photothermal laser treatment, especially limitations of tissue depth and scarring. Photodynamic therapy of proliferating hemangiomas may become useful with further development.

There are a variety of medically or cosmetically significant pigmented lesions of the skin, which may be divided into benign vs. malignant vs. pre-malignant, epidermal vs. dermal, and melanotic (i.e., due to deposition of excess melanin by normal melanocytes) vs. nevomelanocytic (i.e., due to abnormal nevus cells which are capable of melanogenesis). The destruction of primary melanomas with lasers is unwarranted, since surgical excision with preservation of histology is necessary to assess prognosis, and is directly related to survival. In contrast, non-scarring treatment of disfiguring but otherwise completely benign lesions is certainly warranted. Laser treatment of pre-malignant or associated lesions, such as dysplastic nevi, may be beneficial, but may also alter or disguise changes potentially leading to melanoma. We expect that both controversy and improved patient care will arise in the near future.

Epidermal Pigmented Lesions

The destruction of benign epidermal pigmented lesions such as senile lentigines ("liver spots", "age spots") is both easily achieved and cosmetically acceptable using a variety of lasers. It is important to differentiate these lesions from lentigo maligna melanoma, which may sometimes have a similar appearance. Low-irradiance c.w. CO_2 laser exposures have been shown to produce lightening or removal of lentigines (Dover et al., 1988a). The 10.6 μm radiation is absorbed by water in the upper epidermis, producing non-selective heating of the pigmented epidermis or cells. However, basal keratinocytes may be preferentially affected by heating, and the clinical results are good, comparable to those with superficial cryotherapy. Clearly, histologically-selective destruction of hyperpigmented epidermis, although possible, is not required for the treatment of epidermal hyperpigmentation. Cryotherapy is easier, cheaper, and more available than any laser.

Studies in our laboratory have generated significant interest in selective photothermolysis of pigmented lesions, cells and organelles, however, which may become clinically useful in the near future. Great control over the level of selectivity, and depth of injury to pigmented cells is possible, by varying laser pulsewidth and wavelength, respectively. It was initially observed in pigmented guinea pigs that 351 nm excimer laser pulses produce preferential, apparently photothermally-driven rupture of cutaneous melanosomes, which are melanized organelles typically 0.5 - 1 μm in diameter (Anderson and Parrish, 1983). This damage leads to necrosis of pigmented basal keratinocytes and melanocytes, causing delayed complete epidermal depigmentation in pigmented guinea pigs, with subsequent repigmentation. Melanosome rupture requires pulses in the submicrosecond domain (Dover et al., 1986), and occurs across a broad UV-visible-near IR spectrum.

Action spectrum studies from 355-1064 nm of melanosome selective photothermolysis (Anderson et al, 1989) show a wavelength dependence for pigment cell injury in accordance with linear absorption by melanin. This was also confirmed using visible tunable-dye laser pulses (Margolis et al., 1989). In the highly penetrating red-near infrared region, melanin is a strong absorber, but oxy- and reduced hemoglobins absorb poorly. Thus, red and near-infrared pulses produce highly selective, deep injury to the cutaneous pigmentary system. Selective injury of pigmented cells in the epidermis, dermis (if present abnormally) and hair follicles is produced, with little or no direct injury of dermis or blood vessels. Q-switched ruby laser pulses (694 nm, 40 nsec) produce

218

melanosome-selective injury (Polla et al., 1987) and epidermal depigmentation in animals, but permanent leukotrichia (white hair) also results (Dover et al.,1988b). It is unclear why epidermal but not follicular pigmentation recovers in this animal model. Of the hundreds of humans treated with Q-switched ruby laser pulses, none have had leukotrichia.

A study in guinea pigs using 630-694 nm pulses ranging from 100 fsec to 300 μsec pulse duration showed that pulses substantially less than 1 μsec produce similar damage thresholds, of approximately 0.4 J/cm^2, with ultrastructurally-selective injury of melanosomes at threshold fluences. The consistent response to 100 fsec pulses with a peak irradiance of over 10^{12} W/cm^2 suggests impressive linearity of absorption by melanin. With pulses of 1 μsec and longer, however, both the damage threshold fluence and cellular extent of thermal injury increase, which is consistent with loss of thermal confinement as the laser pulsewidth exceeds the melanosome thermal relaxation time (Watenabe, Flotte, Anderson unpublished studies).

Clinically, the utility of selective photothermolysis of pigmented cells remains unproven, though promising. The Q-switched ruby laser has recently been approved by FDA for treatment of decorative tattoos (see below). The same laser, at fluences of 2-4 J/cm^2, is capable of removing senile lentigines, with disappearance to normal or slightly hypopigmented skin without scarring (Levin, et al., 1988). Similarly, cafe-au-lait macules, nevus spillus, freckles, labial macules of Peutz-Jegher's syndrome, and other epidermal lesions can often be effectively treated with Q-switched ruby laser pulses in one or two treatments (Taylor, Anderson, Flotte, unpublished observations). This response is approximately equivalent to that of superficial cryotherapy in the hands of an experienced therapist, but may be more predictable. Sherwood, et al.(1989) reported a study in pig skin using 0.3 μsec dye laser pulses, suggesting from gross and histologic observations that 504 nm green laser pulses may be ideal for treating epidermal lesions. Because of poorer penetration into skin, more superficial injury was produced by 504 nm vs. longer wavelength pulses. Fortuitously, this is the same wavelength of a lithotripsy system produced by the dye laser manufacturer involved in the study, and may therefore offer some advantages. In other studies, green pulses produce direct vascular injury in addition to pigmented cell injury, which would make green pulses seen less than ideal (Anderson et al., 1989a, Margolis et al., 1989). There has been no direct comparison of histologically-selective and non-selective laser treatment modalities for the removal of epidermal or dermal pigmented lesions, and no systematic clinical comparison of different wavelengths.

Permanent or prolonged depigmentation of epidermis without scarring, can result from selective photothermolysis using Q-switched ruby laser pulses, and presumably other pulsed lasers, especially with repeated treatments at high fluences. Depigmentation lasting for one year or more occurs in approximately 25-35% of patients treated with high fluence Q-switched ruby laser pulses used for removal of professional tattoos, i.e., 6-10 J/cm^2 in six to eight multiple treatments at monthly intervals (see below). The risk of permanent depigmentation is fluence-dependent. Although the lower fluence limit remains to be determined, the fluences needed for treatment of epidermal lesions rarely induce prolonged depigmentation. Although acceptable in many fair-skinned individuals, depigmentation limits high-fluence Q-switched ruby laser use in dark skin types.

Dermal Lesions

The treatment of dermal pigmentation such as Nevus of Ota or Ito, mongolian spots, post-inflammatory hyperpigmentation, or the dermal pigment component of melasma, has been unpredictable but generally disappointing, using Q-switched ruby laser pulses of 4-8 J/cm^2 (Taylor, Anderson, Flotte, unpublished observations). One must distinguish between dermal melanoses, such as postinflammatory hyperpigmentation in which melanin is phagocyotsed within macrophages and other cells, vs. nevomelanocytoses, in which pigment-producing dermal cells are present. Goldman (1971) has reported success in treatment of nevus of Ota with normal-mode ruby laser. Theoretically, the destruction of dermal nevus cells by selective photothermolysis may offer a nonscarring method of treatment, although clinical response may be slow.

TATTOO REMOVAL

Tattooing is one of the earliest known forms of self-decoration, dating at least to the Egyptians. In the U.S., the prevalence of tattoos is unknown, but appears to be between 5 and 15% of the entire population, and occurs among all socioeconomic classes. Although most tattoos are obtained after some deliberation, they may also be acquired whimsically, through peer pressure, while inebriated, as a means of identification, during sexual abuse, or as a result of trauma. Tattoos often outlive the sentiment or intended message which gave rise to them; there are large numbers of individuals who desire removal of their tattoos, for a great variety of individual reasons. Until recently, all treatment options involved a high risk of disfigurement due to scarring.

Despite major health and potential social risks, tattooing is not regulated and is legal in many U.S. states. Professional tattooing has been associated with transmission of hepatitis, and is a possible route for transmission of AIDS, although no cases have been reported as yet. Tattooing is often performed by marginally-trained individuals using mail-ordered supplies. There is no control by U.S. federal agencies over the identity, purity or content of the injected substances. The widespread practice of tattooing, although deeply rooted in tradition, is now starkly and absurdly juxtaposed against the intense regulation of modern medicines and invasive procedures.

Amateur tattoos consist mainly of amorphous carbon (India ink, cigarette ash) or occasionally graphite, which are deposited into the dermis at greatly variable depth. Professional tattoos consist of denser but more superficial and uniform dermal deposits of a variety of black or colored substances. These are mainly organometallic dye complexes of unknown exact composition, which are applied with an oscillating needle device. Upon tattooing, some of the tattoo ink is desquamated in a crust over several weeks, and some is transported to regional lymph nodes. The remaining ink, which forms the permanent tattoo, is mainly sequestered in phagosomes within macrophages, fibroblasts, mast cells, and occasionally melanocytes, in particles ranging typically from tens of nm to tens of microns is size.

Removal of tattoos is perhaps as old as tattooing itself. Surgical excision remains the treatment of choice for small tattoos in appropriate body sites. Dermabrasion often partially removes tattoo pigments, leaving minimal superficial scarring. CO_2 laser ablation involves vaporization of the upper dermis, well into the level of the tattoo ink. This almost invariably results in scarring to some extent, representing healing after a third-degree partial thickness burn. Application of urea-containing dressings may reduce the need for deep ablation, and decrease scarring (Ruiz-Esparza et al., 1988). However, the use of Q-switched ruby laser pulses is the only method at present which frequently removes the tattoo without scarring.

Goldman was the first investigator to use ruby laser pulses for removal of tattoos, but the normal-mode ruby laser was not well accepted as other lasers became available for surgical procedures. Reid and coworkers (1983) first reported cases of successful removal of amateur tattoos with Q-switched ruby laser pulses. More recently, the fluence-dependence, treatment technique, and mechanisms of action have been examined in our laboratory (Taylor et al., 1990). Despite renewed interest, there remain both fundamental and clinical problems to be solved in order to optimize this treatment.

The 694 nm ruby laser wavelength, as mentioned above, is absorbed almost exclusively by melanin in normal human skin, and penetrates well into the deep dermis (Anderson, 1981). Black, blue and green tattoo inks also absorb well at 694 nm, but most of the red tattoo inks do not (DeCoste and Anderson, unpublished data). After exposure of black, blue or green tattooed skin to Q-switched ruby laser pulses of approximately $2 \ J/cm^2$ or greater, an immediate whitening reaction is observed, often with elevation of the skin and immediate disappearance of the tattoo. Over several minutes, the whitening fades, and a wheal-and-flare reaction ensues, with the tattoo again becoming fully visible. The immediate whitening and elevation are almost certainly due to intradermal gas vacuoles, seen on histology around pigment deposits. Electron microscopy of immediate damage shows gross mechanical disruption of the tattoo-laden dermal cells, with fracturing of individual ink particles, which also become lamellated rather than amorphous when viewed at high magnification (Taylor et al., 1990). The ink particles become smaller on average, and predominately extracellular due to cell disruptions. Epidermal changes are similar to those described above under the discussion of pigmented lesions, and include intraepidermal blister formation at fluences greater than approximately $4 \ J/cm^2$, depending upon skin type.

Lightening of the tattoo begins 7-10 days after exposure, apparently by more than one mechanism. Desquamation (flaking of the epidermis) of some ink occurs, and tattoo ink can sometimes be found in scales after a treatment. However, major tattoo lightening can occur in the absence of apparent desquamation, suggesting other local or systemic means for pigment alteration or removal. It is well known that tattoo ink is transported to regional lymph nodes after tattooing, and therefore likely that lymphatic transport occurs after laser exposure as well. It is also possible that re-packaging or particle size changes make the ink less apparent, by altering absorption and scattering. Some biopsies of almost invisible Q-switched ruby laser-treated tattoos occasionally showed large quantities of tattoo ink.

A dose-response study (Taylor et al., 1990) showed that increasingly good results occurred with a range of fluences from 2-8 J/cm^2, and there was no advantage or disadvantage to multiple exposures within a single treatment session. Multiple treatments are usually required for good results, and there are significant differences between amateur and professional tattoos. In our present experience of approximately 200 patients, 65-75% of amateur tattoos can be removed without scarring, typically in 3-5 treatments of 6-8 J/cm^2 at monthly intervals. For professional tattoos, response is less predicable, and generally worse. Only 25-35% of green, black or blue tattoos can be completely or almost completely removed, although essentially all can be significantly lightened. The usual result is complete clearing in portions of the tattoo after six to eight treatments, with resistant residual areas which can often be excised with a biopsy punch or small ellipse. The rate of scarring in professional tattoos is about 2%. Transient depigmentation or hypopigmentation occurs in essentially all cases, and persists in about 30% of cases for one year or more.

Thus, the Q-switched ruby laser, and probably other short pulsed lasers in the red-near infrared spectrum, offers a single but highly significant advantage over all other present tattoo treatments -- a very low risk of scarring. The disadvantages of Q-switched ruby laser treatment, and hence the challenges for improvement, are: frequent incomplete removal of professional tattoos, very limited ability to remove red or yellow ink, the need for multiple treatments, and a risk of significant prolonged or permanent depigmentation especially in darker skin types. There have been no systematic studies of laser pulse duration, laser wavelength, or possible "adjunctive" techniques. A preliminary comparison of Q-switched vs. normal mode pulses showed that much higher fluences are necessary with normal-mode pulses, which also produced widespread thermal damage. Furthermore, it is unknown whether thermochemical or photochemical alteration of tattoo inks occurs, and the correlation between specific inks and the ease with which they can be removed, if any. The role of "adjuvant" techniques such as epidermal hypopigmenting agents, or stimulation of lymphatic transport, remains to be determined. Tattoo treatment with more penetrating near-infrared wavelengths is also worthy of investigation.

PHOTODYNAMIC THERAPY (PDT) OF SKIN CANCER

It is beyond the scope of this paper to discuss photodynamic photosensitization for tumor treatment, which has become an area of active and diverse research. However, particular studies related to dermatology are worth noting, and may suggest new strategies for other tumor types. The accessibility of skin cancers to both topically-applied or locally-injected photosensitizers, and to optical radiation makes skin cancer a logical application of PDT. In particular, the development of PDT for basal cell carcinomas (epitheliomas) makes sense. These are common tumors which usually advance slowly with a low risk of metastasis, but may present a major surgical challenge. Not infrequently, surgery is disfiguring, and radiation therapy may become the treatment of choice. Squamous cell carcinomas tend to be more aggressive and invasive, but PDT may still become useful given further development; primary melanomas are much less amenable to laser treatments, for reasons discussed above.

McCaughan et al (1989) found a 67% initial response rate after PDT with dihematophorphyrin ether or ester (DHE) given i.v. for the treatment of a variety of skin cancers including basal and squamous carcinomas, melanomas, and metastatic breast lesions. The rate of recurrence for primary skin tumors is unknown. Both the dose and timing of DHE or light exposure can probably be significantly optimized. Clinical experience using other photosensitizers, such as phthalocyanines or rhodamine derivatives is extremely limited at present in dermatology.

At present, laser Doppler velocimetry as a means for estimation of cutaneous blood flow is the only well-established diagnostic medical laser technique in use in skin. There has been significant recent interest, however, in reflectance analysis, skin autofluorescence spectroscopy, confocal scanning laser microscopy of skin, and pulsed photothermal radiometry.

Aging, and the photochemistry and photobiology of long-term changes in dermal connective tissue are important, poorly-understood topics that are central to dermatology. Leffell et al (1988) reported that laser-induced autofluorescence spectra appear to correlate with age and sun-exposure related changes in human skin. Although the results may be in part due to changes in melanization of skin caused by sun exposure, it is likely that accumulation of elastin-associated fluorophore(s) are involved.

Pulsed photothermal radiometry (PPTR) is a technique in which the thermal transient caused by absorption of an optical pulse is monitored by measuring a perturbation in the emitted infrared (blackbody) radiation. PPTR yields a direct measurement of photothermal excitation of skin tissue, and may become useful in some dermatologic laser procedures, with further development. In addition, PPTR can provide a noninvasive measurement of tissue absorption coefficients, even in scattering samples such as skin. Using PPTR, it was shown that scattering can greatly increase optical exposure within tissue, by internal reflectance of back-scattered diffuse light (Anderson et al, 1989b). Because the infrared signal in the 3-12 μm region arises from the outermost layer, the stratum corneum, it is possible to measure absorption in this layer. Stratum corneum provides the major photoprotective and chemoprotective barrier of the skin, and hence PPTR may be a useful noninvasive means for studying cutaneous barrier function.

OVERVIEW

The varied and successful present uses of lasers in dermatology probably represent a small fraction of what is possible. For photothermal, photomechanical, and photochemical interactions and treatments, both fundamental and clinical questions remain. "Smart", limited-thermal-damage laser ablation techniques in skin surgery have yet to be developed and explored. What are the effects of multiple pulses, causing multiple thermal cycles, physically and biologically? How can selective photothermolysis or other laser treatments be devised for removal of disfiguring dermal pigmented lesions, for which there are currently only poor treatments? What are the mechanisms of Q-switched ruby laser treatment of tattoos, and can they be optimized? Can effective tumor cell-selective phototoxic agents be developed for topical or local treatment of skin cancer? Optical spectroscopic and imaging applications in dermatology are at a formative stage, and may well provide alternative ways in which to view the most obvious organ of the body.

REFERENCES

Anderson R.R., and Parrish J.A., 1981, Microvasculature can be selectively damaged using dye lasers: a basic theory and experimental evidence in human skin, Lasers Surg Med., 1:263.

Anderson R.R., and Parrish J.A., 1983, Selective photothermolysis: precise microsurgery by selective absorption of pulsed radiation, <u>Science</u>, 220:524.

Anderson, R.R., and Parrish, J.A., 1981, The optics of human skin, <u>J Invest Dermatol</u>., 77:13.

Anderson R.R., Jaenicke K.F., and Parrish J.A., 1983, Mechanisms of selective vascular changes caused by dye lasers, <u>Lasers Surg Med</u>., 3:211-215.

Anderson, R.R., Margolis R.J., Watenabe S., Flotte, T., Hruza G.J., and Dover, J.S., 1989a, Selective photothermolysis of cutaneous pigmentation by Q-switched Nd:YAG laser pulses at 1064, 532, and 355 nm, <u>J Invest Dermatol</u>., 93:28.

Anderson R.R., Beck H. Bruggemann U., Farinelli W., Jacques S.L., and Parrish J.A., 1989b, Pulsed photothermal radiometry in turbid media: internal reflection of backscattered radiation strongly influences optical dosimetry, <u>Appl Optics</u>., 28:2256.

Anderson R.R., 1987, Carbon dioxide lasers: a broader perspective, <u>Arch Dermatol</u>., 123:566.

Apfelberg D.B., Maser M.R., White D.N., Lash H., Lane B., and Marks M.P., 1990, Combination treatment for massive cavernous hemangioma of the face: YAG laser photocoagulation plus direct steroid injection followed by YAG laser resection with sapphire scalpel tips, aided by superselective embolization, <u>Lasers Surg Med</u>., 10:217.

Barsky S.H., Rosen S., Geer D.E., and Noe J.M., 1980, The nature and evolution of port wine stains: a computer assisted study, <u>J Invest Dermatol</u>., 74:154.

Birngruber R., 1980, Thermal modeling in biological tissues, <u>in</u>: "Lasers in Biology and Medicine," F. Hillenkamp, R. Pratesi, and C.A. Sacchi, eds., Plenum Publishing Co., London.

Clarke R.J., Isner J.M., Donaldson R.F., and Jones G. 2nd, 1987, Gas chromatographic-light microscopic correlative analysis of excimer laser photoablation of cardiovascular tissue: evidence for a thermal mechanism, <u>Circ Res</u>., 60:429.

Dover J.S., Polla L.L., Margolis R.J., Whitaker D., Watanabe S., Murphy G.F., Parrish J.A., and Anderson R.R., 1986, Pulse width dependence of pigment cell damage at 694 nm in guinea pig skin, <u>in</u>: "Lasers in Medicine," Scott R.S., Parrish J.A., and Jaffe N. eds., Proc. SPIE 712:200.

Dover J.S., Smoller B.R., Stern R.S., Rosen S., and Arndt K.A., 1988a, Low-fluence carbon dioxide laser irradiation of lentigines, <u>Arch Dermatol</u>., 124:1219.

Dover J.S., Margolis R.J., Polla L.L. Watanabe S., Hruza G.J., Parrish J.A., and Anderson R.R., 1988b, Pigmented guinea pig skin irradiated with Q-switched ruby laser pulses -- morphologic and histologic findings, <u>Arch Dermatol</u>., 125:43.

Gange R.W., Jaenicke K.F., Anderson R.R., and Parrish J.A., 1984, Effect of preirradiation tissue target temperature upon selective vascular damage induced by 577-nm tunable dye laser pulses, <u>Microvasc Res</u>., 28:125.

Garden J.M., OBanion M.K., Shelnitz L.S., Pinski K.S., Bakus A.D., Reichmann M.E., and Sundberg J.P., 1988, Papillomavirus in the vapor of carbon dioxide laser-treated verrucae, <u>JAMA</u>., 259:1199.

Garden J.M., Tan O.T., Kershmann R., Boll J., Furumoto H., Anderson R.R., and Parrish J.A., 1986, Effect of dye laser pulse duration on selective cutaneous vascular injury, <u>J Invest Dermatol</u>., 87:653.

Garden J.M., Polla L.L., and Tan, O.T., 1988, The treatment of portwine stains by the pulsed dye laser -- analysis of pulse duration and long-term therapy, <u>Arch Dermatol</u>., 124:889.

Goldman L., and Rockwell R.J., 1971, "Laser in Medicine," Gordon and Breach, New York.

Jacques S.L., McAuliffe D.J., Blank I.H., and Parrish J.A., 1987, Controlled removal of human stratum corneum by pulsed laser. J Invest Dermatol., 88:88.

Kochevar I.E., 1989, Cytotoxicity and mutagenicity of excimer laser radiation, Lasers Surg Med., 9:440.

Leffell D.J., Stetz M.L., Milstone L.M., Deckelbaum L.I., 1988, In vivo fluorescence of human skin a potential marker of photoaging, Arch Dermatol., 124:1514.

Levin J.A., Margolis R., Hruza G., Dover J., and Anderson R.R., 1988, Q-switched ruby laser irradiation of lentigines, Lasers Surg Med., 8(2):185 (abstract)

Margolis R.J., Dover J.S., Polla L.L. Watanabe S. Shea C.R., Hruza G.J., Parrish J.A., and Anderson R.R, 1989, Visible action spectrum for melanin-specific selective photothermolysis. Lasers Surg Med., 9:389.

Marshall J., Trokel S., Rotherty S., and H. Schubert, 1985, An ultrastuctural study of corneal incisions induced by an excimer laser at 193 nm. Ophthalmology, 92:749.

McCaughan J.S., Guy J.T., Hicks H., Laufman L., Nims T.A., and Walker J., 1989, Photodynamic therapy for cutaneous and subcutaneous malignant neoplasms. Arch Surg., 124:211.

Mordon S.R., Rotteleur G., Buys B., and Brunetaud J.M., 1989, Comparative study of the "point-by-point technique" and the "scanning technique" for laser treatment of port-wine stain, Lasers Surg Med., 9:398.

Morelli J.G., Tan O.T., Garden J., Margolis R., Seki Y., Boll J., Carney J.M., Anderson R.R., Furumoto H., Anderson R.R., and Parrish J.A., 1986, Tunable dye laser treatment of port-wine stains, Lasers Surg Med., 6:94.

Murphy G.F., Shepard R.S., Paul B.S., Menkes A., Anderson R.R., and Parrish J.A., 1983, Organelle-specific injury to melanin-containing cells in human skin by pulsed laser irradiation, Lab Invest., 49:680.

Olbricht S.M., and Arndt K.A., 1988, Carbon dioxide laser treatment of cutaneous disorders, Mayo Clin Proc., 63:297.

Orenstein A., Nelson J.S., Liaw L.L., and Berns M.W., 1990, Photodynamic therapy of hypervascular dermal lesions. Lasers Surg Med., suppl.2:49 (abstract).

Polla L.L., Margolis R.J., Dover J.S., Whitaker D., Murphy F.G., Jacques S.L., and Anderson R.R., 1987, Melanosomes are a primary target of Q-switched ruby laser irradiation in guinea pig skin, J Invest Dermatol., 89:281.

Reid W.H., McLeod P.J., Ritchie A., and Furguson-Pell M, 1983, Q-switched ruby laser treatment of black tattoos, Br J Plast Surg., 36:455.

Ruiz-Esparza J., Goldman M.P., and Fitzpatrick R.E., 1988, Tattoo removal with minimal scarring: the chemo-laser technique, J Dermatol Surg Oncol., 14:1372.

Scheibner A., and Wheeland R.G., 1989, Argon-pumped tunable dye laser therapy for facial portwine-stain hemangiomas in adults -- a new technique using small spot size and minimal power, J Derm Surg Oncol., 15:277.

Schomacker K.T., Walsh, J.T., Flotte T.F., and Deutsch T.F., 1990, Thermal damage produced by high-irradiance continuous wave CO_2 laser cutting of tissue, Lasers Surg Med., 10:74.

Sherwood K.A., Murray S. Kurban A.K., and Tan O.T., 1989, Effect of wavelength on cutaneous pigment using pulsed irradiation, J Invest Dermatol., 92:717.

Sherwood K.A., Murray S., Kurban A.K., and Tan O.T., 1989, Effect of wavelength on cutaneous pigment using pulsed irradiation, J Invest Dermatol., 92:717, 1989.

Stern J.C., and Lucente F.E., 1989, Carbon dioxide laser excision of earlobe keloids: a prospective study and critical analysis of existing data, <u>Arch Otolaryngol Head Neck Surg</u>., 115:1107.

Tan O.T., Sherwood K., and Gilchrest B.A., 1989a, Treatment of children with port-wine stains using the flashlamp-pumped tunable dye laser, <u>N Engl J Med</u>., 320:416.

Tan O.T., Murray S., and Kruban A.K., 1989b, Action spectrum of vascular-specific injury using pulsed irradiation, <u>J Invest Dermatol</u>., 92:868.

Taylor C.R., Gange R.W., Dover J.S., Flotte T.J., Gonzalez E. Michaud N., and Anderson R.R., 1990, Treatment of tattoos by Q-switched ruby laser, <u>Arch Dermatol</u>., 126:893.

Trokel S.L., Srinivasan R., and Braren B., 1983, Excimer laser surgery of the cornea, <u>Am J Ophthalmol</u>., 96:710.

Walsh J.T. Jr., Flotte T.J., Anderson R.R., and Deutsch T.F., 1988, Pulsed CO_2 laser tissue ablation: effect of tissue type and pulse duration on thermal damage, <u>Lasers Surg Med</u>., 8:108.

Walsh, J.T. Jr., Flotte T.J., and Deutsch T.F., 1989, Er:YAG laser ablation of tissue: effect of pulse duration and tissue type on thermal damage, <u>Lasers Surg Med</u>., 9:314.

Watenabe S., Flotte T.J., McAuliffe D.J., and Jacques S.L., 1988, Putative photoacoustic damage in skin induced by pulsed ArF excimer laser, <u>J Invest Dermatol</u>., 90:761.

Wolbarsht M.L., 1984, Laser Surgery: CO_2 or HF, <u>IEEE J Quant Elect</u>., QE-20:1427.

PHOTOCHEMIOTHERAPY OF TUMOURS: MOLECULAR AND

BIOPHYSICAL BASES

Giulio Jori and Elena Reddi

Department of Biology
University of Padova, Italy

INTRODUCTION

A variety of solid tumours, having different histological and optical properties, have been shown to be responsive to photodynamic therapy (PDT) and this technique is presently under investigation in several clinical centers.[1] Promising results have been obtained especially in the treatment of bladder[2] and lung[3] cancers, as well as in the detection and therapy of early squamous cell carcinomas of the pharynx, oesophagus and tracheo-bronchial tree.[4] In spite of such diversified applications, PDT is still in a developing stage mainly owing to the scarcity of information on its mechanism of action and the interplay of different factors controlling the efficacy and safety of the technique. Thus, the two parameters which should ensure the selectivity of PDT for malignant tissues, i.e. the larger accumulation and longer retention of the systemically injected photosensitizer by tumours as compared with peritumoural tissues, and the degree of light propagation through biological tissue layers, appear to be quite complex problems and are poorly defined.

Photofrin II and haematoporphyrin, the most frequently used tumour-photosensitizers, often give tumour/healthy tissue ratios of drug concentration lower than 3;[5] on the other hand, in vivo mapping of light fluence distribution by interstitially introduced probes or non-invasive techniques requires a more thorough knowledge of the tissue optical properties.[6,7]

Photofrin II is also a rather heterogeneous mixture of monomeric and oligomeric porphyrins with a low molar absorbance in the clinically useful wavelength range, namely 600-900 nm. This has stimulated intensive research aimed at identifying second generation tumour photosensitizers with high purity and improved absorption properties in the red spectral region in order to obtain a greater utilization of the delivered light.

In this paper, we discuss some photobiological and pharmacokinetic approaches which are presently being pursued to overcome the most important limitations of PDT for solid tumours.

Optronic Techniques in Diagnostic and Therapeutic Medicine
Edited by R. Pratesi, Plenum Press, New York, 1991

TARGETS OF PHOTODYNAMIC THERAPY

Molecular and cellular level

Several types of biomolecules and subcellular organelles have been shown to undergo irreversible photodamage upon photosensitization by porphyrins and other macrocyclized light-absorbing dyes.[8] The photoprocesses most frequently observed include the peroxidation of unsaturated lipids and steroids, the photodegradation of guanine residues in DNA and the photooxidation of sulfur-containing and aromatic amino acids in proteins. In some cases, the initial photoproducts originate further thermal reactions propagating the chemical modification to targets not immediately adjacent to the photosensitizer binding site: e.g., intra- and inter-molecularly cross-linked proteins have been isolated from porphyrin-sensitized human erythrocytes.[9] However, in a cell, the nature of the photodamaged sites is often dependent on the physico-chemical features of the photosensitizer, as well as on the experimental conditions adopted for photosensitizer-cell incubation. Thus, water-soluble porphyrins, such as uroporphyrin, do not bind to cells and exert their photosensitizing activity from the bulk aqueous medium;[8] hydrophobic porphyrins partition among apolar subcellular compartments, although the distribution is affected by the incubation time:[10] at short incubations (up to 4 h), the photosensitizer molecules are preferentially bound to loci in the cytoplasmic membrane, while for prolonged incubations (above 12 h) lysosomes and mitochondria are more heavily loaded. The latter distribution is likely to mimic the distribution occurring when PDT is performed in vivo, since irradiation is carried out at about 24-48 h after administration of the photosensitizer.

Actually, PDT-induced damage to experimental tumours involves the inactivation of mitochondrial marker enzymes.[11] In general, binding of the photosensitizer to cells appears to be a stringent requisite for efficient photosensitization to occur.[9,10] On these bases, it is reasonable to find that lipophilic photosensitizers, such as unsubstituted or mono-/di-sulfonated meso-tetraphenyl-porphyrins and phthalocyanines, are more powerful photodynamic agents than the corresponding more polar tri- and tetra-sulfonated derivatives.[12] Analogously, the oligomeric components of Photofrin II have been found to photosensitize cell killing more efficiently than the monomeric components, which possess a smaller level of hydrophobicity.[13] Cell death is thus consequent to impairment of cell respiration, release of lysosomal enzymes, and alterations of membrane permeability leading to efflux of cytoplasmic material.[8,13] Nuclear photodamage has been detected only upon prolonged irradiation and seems to be correlated with cell death; this is in agreement with the lack of mutagenic effects of porphyrin photosensitization and the relatively low degree of photosensitivity exhibited by cells deficient in DNA repair, such as ataxia telangiectasia and xeroderma pigmentosum.

There are no significant differences in porphyrin accumulation and photosensitivity for a variety of normal and transformed cells in vitro. However, endothelial cells display an especially large photolability upon visible light-irradiation in the presence of Photofrin II. This has been ascribed[14] to a less efficient synthesis of stress proteins by these cells as compared with most normal and malignant cells; stress proteins are known to play an important role in protecting cells from cytotoxic processes and/or enhancing the recovery of partially injured cells. These findings suggest that endothelial cell photosensitivity may give a major contribution to the vascular damage occurring during the PDT of tumours.

In vivo level

Both damage of malignant cells and destruction of the vascular system have been invoked to explain the tumoricidal effect of PDT in experimental animals.[1,5]

Ultrastructural and histological analyses of irradiated tumour tissues indicate that the relative importance of the two tissular compartments as PDT targets is modulated by the hydro-/lipo-solubility and transport mechanism of the photosensitizing agent. Thus, photosensitizers having relatively low hydrophobicity (octanol/water partition coefficient lower than ca. 10) are preferentially transported by serum albumin and deposited in the vascular stroma; more hydrophobic photosensitizers become tightly associated with serum lipoproteins and are preferentially released to neoplastic cells of tumour tissues.

In the former case, the photoinduced tumour necrosis is of hemorrhagic nature, with profound alterations of the endothelial wall and stoppage of blood flow:[15] this is evident from micrographs obtained from a tumour specimen taken at short times after the end of PDT (fig. 1). On the other hand, when damage of neoplastic cells prevails over vascular damage (see, for example, fig. 2), oxygen supply to the irradiated tissue efficiently occurs for several hours, thus minimizing the risk of formation of hypooxygenated areas from which recurrences often originate.

As a consequence, several efforts are being made to enhance the probability of direct inactivation of malignant cells. Present evidence suggests that this goal can be achieved by reducing the number of polar peripheral substituents in the macrocycle of the photosensitizer after its incorporation into the phospholipid bilayer of dipalmitoyl-phosphatidylcholine (DPPC) liposomes.[16] The latter vesicles transfer the entrapped dye to lipoproteins to a quantitative extent.

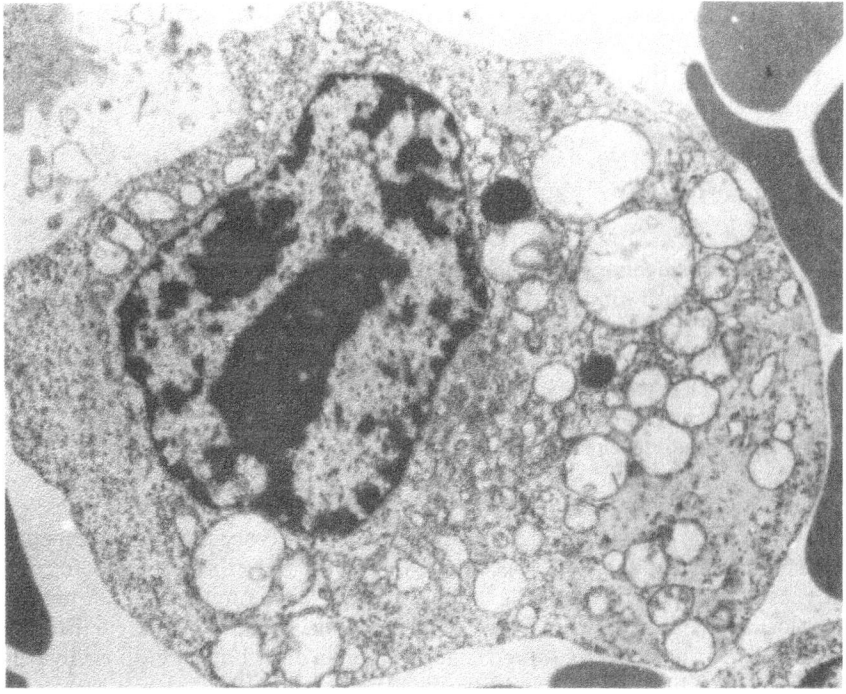

Fig.1. Micrograph taken from a MS-2 fibrosarcoma at 5 h after PDT in the presence of zinc(II)-phthalocyanine (0.2 mg/kg body weight of a Balb/c mouse). The micrograph shows the typical PDT-induced photodamage of malignant cells, including swelling of mitochondria, formation of gaps in the plasma membrane, and extensive endocytoplasmic vesiculation.

229

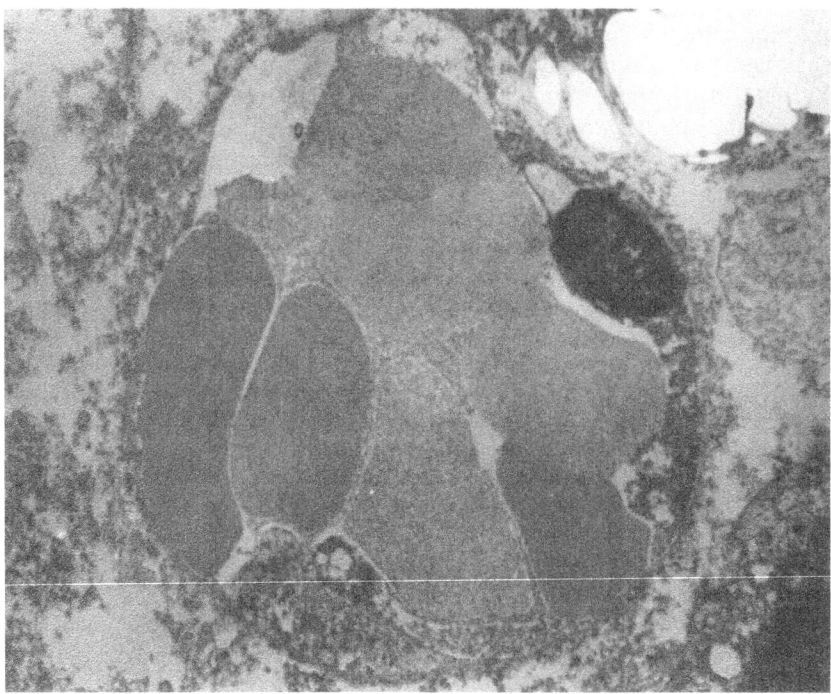

Fig.2. Micrograph taken from a MS-2 fibrosarcoma at 12 h after PDT in the presence of zinc(II)-phthalocyanine (0.2 mg/kg body weight of Balb/c mouse). The micrograph shows typical features of PDT-induced vascular damage, including congestion of the lumen by erythrocytes, loss of the organized structure of the endothelial wall, and alteration of endothelial cells.

The transport mechanism also determines the biodistribution pattern of the photosensitizer:[17] both albumin and high-density lipoproteins (HDL) carry relatively large amounts of the systemically injected photosensitizer to the components of the reticuloendothelial system, such as liver and spleen, and to cutaneous districts from which the dye is cleared at very slow rates. As a consequence, an undesired side effect of PDT is represented by the persistence of skin photosensitivity for several weeks. Moreover, the possible onset of toxicity in heavily dye-loaded tissues should be considered.

On the contrary, LDL and, to a lesser extent, VLDL (very low-density lipoproteins) deliver the complexed photosensitizer to rapidly proliferating tissues, including tumours. Actually, cells having a high mitotic index express a large number of LDL receptors on the cytoplasmic membrane; the LDL-receptor complex is endocytosized via the formation of coated pits.[18] It has been observed that the in vivo distribution of those hydrophobic photosensitizers, which are mainly transported by lipoproteins, is closely correlated with the number of LDL receptors in the various tissues.[19]

In any case, for the safety of PDT, one essential parameter is represented by the ratio of photosensitizer concentration in the tumour and normal tissues. The value of this ratio must be especially high as regards the tumour and the tissue from which the tumour spreads or in which the tumour grows, if PDT is to be applied for the treatment of infiltrating malignancies.

SECOND GENERATION PHOTODYNAMIC SENSITIZERS

The selection and design of more efficient PDT agents for tumours depend on the definition of which molecular characteristics enhance the selectivity of tumour targeting and the photocytotoxic activity. No general answer to this difficult question has been provided as yet owing to the truly large number of variables controlling the result of a PDT treatment. At present, the trend is to synthesize chemically defined photosensitizers with high extinction coefficients in the 650-800 nm interval, where light transmission through animal tissues is maximal.[6]

Three approaches have been developed (see ref. 20 for a detailed discussion) in order to increase the efficiency of red light absorption by porphyrinoid compounds:

i) Extension of the π-electron cloud (fig. 3). This feature can be achieved by insertion of conjugated double bonds into the tetrapyrrolic ring so that the aromatic 18π system is expanded to 22π- or even 34π-electron system: red shifts of the longest wavelength absorption to 1000 nm can be achieved. Alternatively, benzene or naphthalene rings can be condensed with the pyrrole moieties, as it occurs for phthalocyanines and naphthalocyanines, respectively.

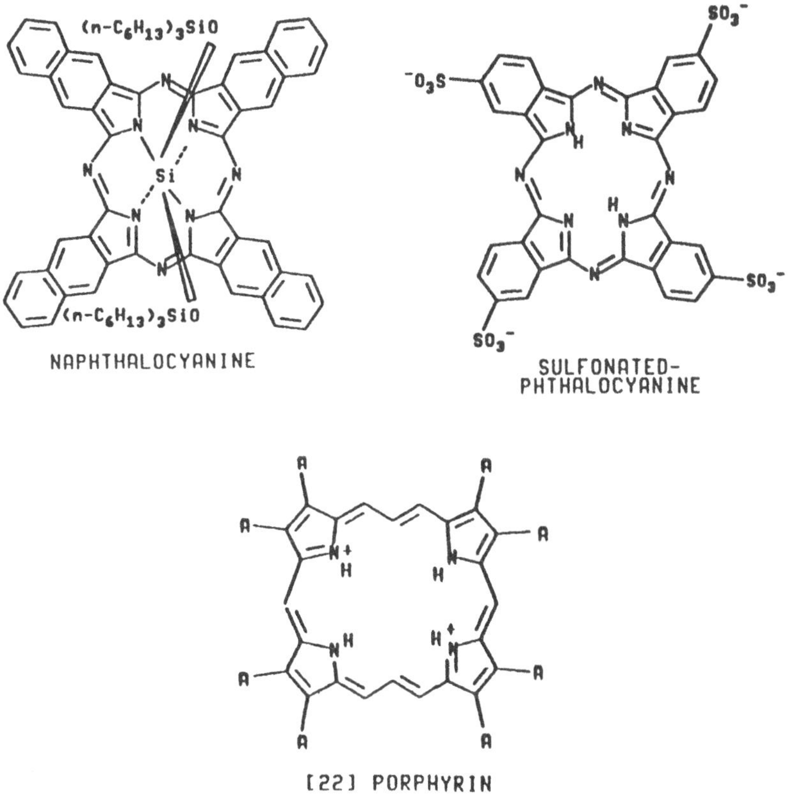

NAPHTHALOCYANINE

SULFONATED-
PHTHALOCYANINE

[22] PORPHYRIN

Fig.3. Examples of porphyrin analogs with expanded conjugation of the electron cloud.

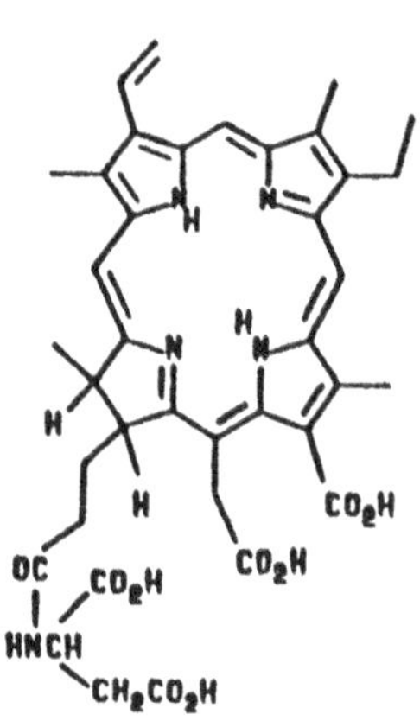

MONO-L-ASPARTYL
CHLORIN e6

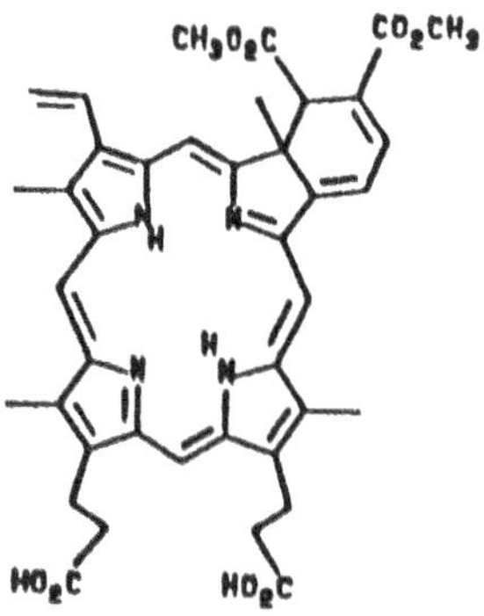

BENZOPORPHYRIN
DERIVATIVE

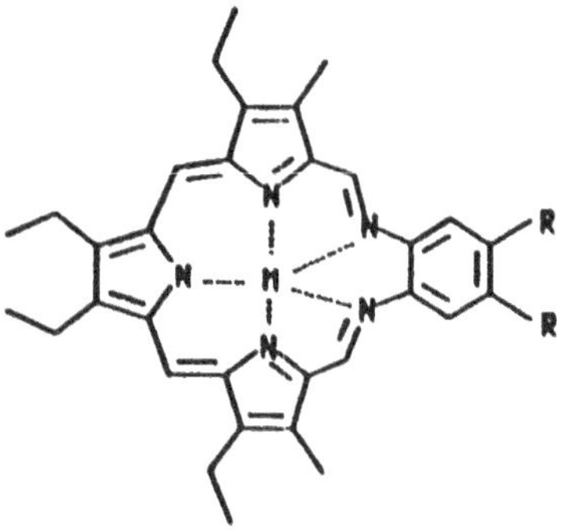

TEXAPHYRIN

Fig.4. Examples of modifications of the porphyrin macrocycle inducing spectral bathochromicity and hyperchromicity.

ii) Modification of the tetrapyrrolic ring macrocycle (fig. 4), by changing the sequence of the inter-ring double bonds (e.g., porphycenes), or replacing the methine bridges by nitrogen atoms (texaphyrins, phthalo- and naphthalo-cyanines), as well as by partial hydrogenation of one pyrrole ring (chlorins, benzoporphyrin derivatives, purpurins).

iii) Addition of selected peripheral substituents, such as phthalocyanines with alkoxy functional groups in the   position of the benzene ring or purpurins having a five- or six-membered exocyclic ring and one meso-substituent with π-electron conjugated to the aromatic electron cloud (fig. 5).

Typical absorption maxima and extinction coefficients for the various classes of photosensitizers are given in Table 1.

Obviously, one should ascertain that the chemically modified far red-absorbing dyes still possess a good phototherapeutic activity. In particular, the lowest-lying excited triplet state should efficiently promote either electron transfer processes   with

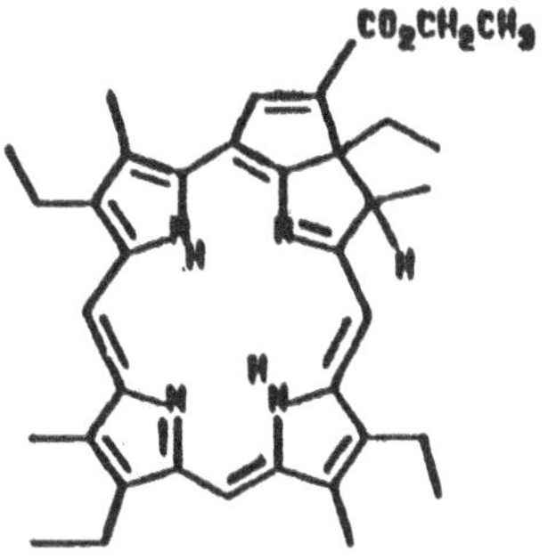

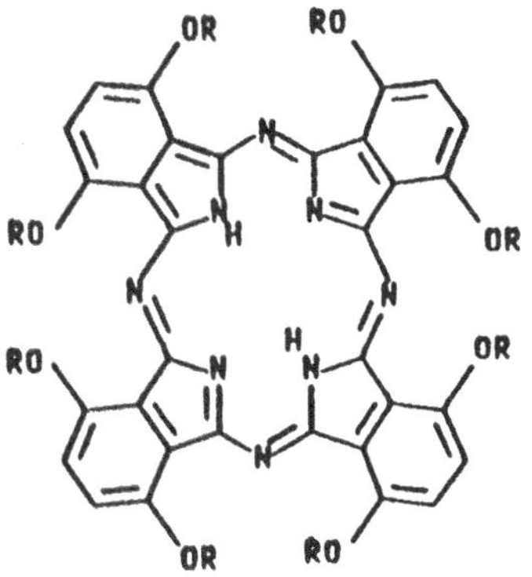

ETIOPURPURIN

ALKOXY-
PHTHALOCYANINE

Fig.5. Examples of peripheral substitution causing a red shift of the absorption spectrum of phthalocyanines and purpurins.

other species present in the surrounding medium (type I photosensitization) and/or energy transfer to ground-state dioxygen which is converted to the highly reactive excited singlet state (singlet oxygen, type II photosensitization). The competition between the two reaction pathways cannot be readily predicted since the potential type I subtrates are undefined and their local concentration is unspecifiable. However, the redox potential of the triplet state can give a reliable indication on the tendency of this transient to act as an oxidant or reductant. On the other hand, the energetics of singlet oxygen production from the triplet photosensitizer can be quantitatively addressed: the energy level of singlet oxygen over its ground state is 22.5 kcal/mole, hence the triplet energy of the photosensitizer must be at least equal to this value.

Table 1. Absorption Properties of Photodynamic Sensitizers in the Red Spectral Region

Photosensitizer	Absorption range (nm)	Extinction coefficient at max (M^{-1} cm^{-1})
Photofrin II	620-640	3,000
Haematoporphyrin	620-640	3,500
34 -porphyrin	960-1000	80,000
Etio-purpurin	640-660	45,000
Benzoporphyrin derivative	660-690	40,000
Bacteriochlorophyll a	770-790	180,000
Zn(II)-phthalocyanine	670-690	150,000
Zn(II)-octabutoxy-phthalocyanine	740-750	180,000
Si(IV)-naphthalocyanine	770-790	240,000

Since the triplet is formed by intersystem crossing from the first excited singlet state, the question arises of how far one can push the singlet manifold 0-0 band of a photosensitizer without impairing its ability to generate singlet oxygen. Toward this aim, one needs to know the singlet-triplet energy gap; typical values range between 10 kcal/mole for porphyrins to 15 kcal/mole for naphthalocyanines.[21]

This feature makes energy transfer from triplet naphthalocyanines (0-0 absorption at 780 nm) to oxygen a slightly endoergonic event, whose efficiency is about one-ninth that to be expected for a diffusion-controlled process. On the other hand, in the case of octabutoxy-phthalocyanines, which also have a singlet-triplet splitting of ca. 15 kcal/mole but absorb at 750 nm, oxygen quenching of the triplet dye with production of singlet oxygen is very efficient.[21]

APPROACHES FOR SELECTIVE TUMOUR TARGETING BY PHOTOSENSITIZERS

The increasingly available information on the mechanisms of photosensitizer accumulation and retention by neoplastic tissues stimulates the development of adequate technology in order to improve the selectivity of tumour targeting. The procedures proposed take advantage of specific properties of tumour cells or tissues as compared to normal cells.[22] As discussed below, three approaches appear to hold favourable prospects.

Photosensitizers with delocalized positive charge

Tumour cell mitochondria exhibit a very large affinity for polycyclic cationic photosensitizers with a delocalized positive charge.[23] Typical examples are kryptocyanines, xanthenes and benzophenoxazines, which are also characterized by intense absorption bands in the red spectral region. The specificity of tumour targeting by those dyes can be very high, although some concern is caused by the possible significant levels of dark toxicity and the generally low photosensitizing activity. However, the latter can be enhanced by combining phototherapy with hyperthermia.[24]

Conjugation of the photosensitizer to monoclonal antibodies

The surface of tumour cells may express some antigens which are not present in normal cells.[25] As a consequence, photosensitizers can be covalently linked to monoclonal antibodies directed against such antigens, thus hopefully yielding a very specific loading of the neoplastic tissue. The experiments performed[26] clearly indicate that the coupling of the photosensitizer to the antibody does not change the photophysical and photochemical properties nor the ability of the antigen to recognize the antibody. Actually, when porphyrin-type compounds were injected to tumour-bearing animals in association with suitable antibodies, quite favourable responses to PDT were obtained[26] even upon administration of photosensitizer doses about one-tenth lower than those normally used. A further development of this technique requires a more precise definition of the amount of photosensitizer which can be transported and delivered to the tumour tissue by the antibody without depressing its specificity.

Conjugation of the photosensitizer to low-density lipoproteins.

As previously mentioned, LDLs are specific carriers of hydrophobic photosensitizers to tumour tissues. Therefore, attempts have been made to enhance and stabilizing the binding of photosensitizers to this lipoprotein class. Such approaches include the incorporation of the administered photosensitizer into liposomal vesicles which are in a quasi-solid state at the physiological body temperature,[27] and the addition of functional groups which increase the lipophilic character of the photosensitizer complex.[28-30] In all cases, the LDL-delivered photosensitizers give remarkably higher tumoural concentrations as compared with free photosensitizers. In particular, LDL delivery reduces the amount of dye accumulated by liver and skin; thus, the risk of cutaneous photosensitivity should be minimized.

The identification or formulation of tumour-specific delivery systems for PDT agents would certainly widen the scope and potential of this technique, since peritumoural tissues would be spared during irradiation. Moreover, this could allow the use of different photosensitizers tailored to the optical and histological features of any given tumour, e.g. to modulate the depth of the photoinduced tissular damage or to control the nature of the subtissular compartments which are affected by PDT.

REFERENCES

1. T.J. Dougherty, Photodynamic therapy (PDT) of malignant tumours, CRC Crit. Rev. Oncol. Hematol. 2:83 (1984).
2. D. Jocham, G. Staehler, R. Baumgartner and E. Ünsold, Die integrale photodynamische therapie beim multifokalen blasenkarzinom, Urologe (A) 24:316 (1985).
3. H. Kato, C. Konaka, J. Ono, N. Kawate and Y. Hayata, Five-year disease-free survival of a lung cancer patient treated only by photodynamic therapy, Chest 80:768 (1986).
4. P. Monnier, M. Savary, C. Fontolliet, G. Wagonieres, A. Chatelain, P. Cornaz, C. Depeursinge and H. Van den Bergh, Photodetection and photodynamic therapy of early squamous cell carcinomas of the pharynx, oesophagus and tracheo-bronchial tree, Lasers Med. Sci. 5:149 (1990).
5. C. Zhou, Mechanisms of tumour necrosis induced by photodynamic therapy, J.Photochem. Photobiol., B: Biol. 3:299 (1989).
6. B.C. Wilson, M.S. Patterson, S.T. Flock, and D.R. Wyman, Tissue optical properties in relation to light propagation models and in vivo dosimetry, in: "Photon Migration in Tissue", B. Chance, ed., Plenum Publ. Corp., New York: 25 (1990).
7. U. Bernini, R. Reccia, P. Russo and A. Scala, Quantitative photoacoustic spectroscopy for cateractous human lenses, J. Photochem. Photobiol., B: Biol. 4:407 (1990).
8. G. Jori and J.D. Spikes, Photobiochemistry of porphyrins, in: "Topics in Photomedicine", K.C. Smith, ed., Plenum Publ. Corp., New York: 183 (1984).
9. T.M.A.R. Dubbelman, A.F.P.M. De Goeij and J. Van Steveninck, Photodynamic effects of protoporphyrin on human erythrocytes; nature of the cross-linking of membrane proteins, Biochim. Biophys. Acta 511:141 (1978).
10. D. Kessel, Sites of photosensitization by derivatives of hematoporphyrin, Photochem. Photobiol. 44:489 (1986).
11. R. Hill, D.B. Smail, R.S. Murant, P.B. Leakey and S.L. Gibson, Hematoporphyrin derivative-induced photosensitivity of mitochondrial succinate dehydrogenase and selected cytosolic enzymes of R3230 AC mammary adenocarcinoma of rats, Cancer Res. 44:1483 (1984).
12. J.E. van Lier and J.D. Spikes, The chemistry, photophysics and photosensitizing properties of phthalocyanines, in: "Photosensitizing compounds: their chemistry, biology and clinical use", G. Bock and S. Harnett, eds., Wiley, Chichester, p.395 (1989).
13. J. Moan, Porphyrin-sensitized photodynamic inactivation of cells, Lasers Med. Sci. 1:5 (1986).
14. C.J. Gomer, A. Ferrario, N. Hayashi, N. Rucker, B.C. Szirth and A.L. Murphree, Molecular, cellular and tissue responses following photodynamic therapy, Lasers Surg. Med. 8:450 (1988).

15. B. Henderson and G. Farrell, Possible implications of vascular damage for tumor cell inactivation in vivo: comparison of different photosensitizers, Proc. SPIE 1065:2 (1989).

16. C. Zhou, C. Milanesi and J. Jori, An ultrastructural comparative evaluation of tumours photosensitized by porphyrins administered in aqueous solution, bound to liposomes or to lipoproteins, Photochem. Photobiol. 48:487 (1988).

17. G. Jori, Pharmacokinetic studies with hematoporphyrin in tumour-bearing mice, in: "Photodynamic Therapy of Tumours and Other Diseases", G. Jori and C. Perria, eds., Libreria Progetto, Padova: 159 (1985).

18. J.L. Goldstein, R.G.W. Anderson and M.S. Brown, Coated pits, coated vesicles and receptor-mediated endocytosis, Nature 279:679 (1979).

19. D. Kessel, Porphyrin-lipoprotein association as a factor in porphyrin localization, Cancer Lett. 33:183 (1986).

20. G. Jori and E. Reddi, Second generation photosensitizers for the photodynamic therapy of tumours, in: "Light in Biology and Medicine", R. Douglas, J. Moan d G. Ronto, eds., Plenum Press, New York, vol. III (1990), in the press.

21. W.E. Ford, P.A. Firey, J.R. Sounik, B. Rihter, M.E. Kenney and M.A.J. Rodgers, Photoproperties of naphthalocyanines, Proc. SPIE 997:105 (1988).

22. G. Jori and E. Reddi, Strategies for tumour targeting by photodynamic sensitizers, in: "Photodynamic Therapy", D. Kessel, ed., CRC Press, Boca Raton, p.117-130 (1990).

23. S.K. Powers, Cationic dyes with mitochondrial specificity for phototherapy of malignant tumours, Proc. SPIE 997:74 (1988).

24. A. Oseroff, D. Ohuoha and G. Ara, Mild hyperthermia synergistically enhances killing of malignant cells by cationic photosensitizers, Proc. SPIE 997:11 (1988).

25. A.R.M. Coates, P. Baird, H. Nicholai and Y. Nitzan, Clinical applications of monoclonal antibodies against mycobacteria, in: "Clinical applications of monoclonal antibodies", R. Hubbard and V. Marks, eds., Plenum Press, New York 167 (1987).

26. D. Mew, C.K. Wat, G.H.N. Towers and J.G. Levy, Photoimmunotherapy: treatment of animal tumors with tumor-specific monoclonal antibodies-hematoporphyrin conjugates, J.Immunol. 130:1473 (1983).

27. F. Ginevra, S. Biffanti, A. Pagnan, R. Biolo, E. Reddi and G. Jori, Delivery of the tumour photosensitizer Zn(II)-phthalocyanine to serum proteins by different liposomes: studies in vitro and in vivo, Cancer Lett. 49:59 (1990).

28. A.M. Richter, S. Cerruti-Sola, E.D. Sternberg, D. Dolphin and J.G. Levy, Biodistribution of tritiated benzoporphyrin derivative, a new potent photosensitizer, in normal and tumour bearing mice, J. Photochem. Photobiol., B: Biol. 5:231 (1990).

29. A. Barel, G. Jori, A. Perin, A. Pagnan and S. Biffanti, Role of high-, low- and very low-density lipoproteins in the transport and tumor-delivery of hematoporphyrin in vivo, Cancer Lett. 32:145 (1986).

30. B. Allison, A. Richter, F. Jiang, S. Jiang, D. Liu and J.G. Levy, Approaches to improved delivery of photosensitizers, Proc. SPIE 1093:153 (1990).

LASERCHEMOTHERAPY OF TUMOURS: CLINICAL ASPECTS

Pasquale Spinelli, Marco Dal Fante, Andrea Mancini

Endoscopy Division, National Cancer Institute
via Venezian 1, 20133 Milan, Italy

Photodynamic therapy (PDT) is a selective, experimental treatment for solid tumors. PDT consists of the activation of a photosensitizing agent by light. The photodynamic reaction induced by light causes damage to the tissue containing the photosensitizer in the presence of oxygen. The idea of treating tumors by photosensitizers is as old as the early 1900s; already in 1903, topic application of eosin and exposition to sunlight was known to produce a response in skin tumors.[1] Later, in 1924, Policard reported reddish fluorescence in animal and human tumors observed under Wood lamp. The presence of fluorescence was attributed to endogenous porphyrins accumulated after infection of the observed tissue by hemolytic bacteria.[2] In 1942, Auler and Banzer[3] reported animal tumor fluorescence after systemic administration of Hematoporphyrin (HP) and in 1960 Lipson and coworkers prepared the Hematoporphyrine derivative (HPD), a mixture of porphyrins obtained by treating HP with acetic and sulphuric acids.[4] They demonstrated that HPD was selectively accumulated by malignant as well as by actively proliferating tissues and produced the first demonstration of endoscopic diagnosis of malignant tissues by detection of fluorescence in the respiratory and in the upper digestive tract.[5] After the development of the laser, diagnosis through fluorescence and particularly PDT have been studied further. Photodynamic reactions have different cellular targets: cross-linking of cellular membrane proteine,[6] inactivation of mitochondrial membrane enzymes[7] and DNA damage[8] have been reported. However, in-vivo observations suggest that necrosis of malignant tumors may be secondary to a damage of the tumor vasculature.[9]

The main parameters involved in PDT are: the photosensitizer, the light for activation, and the selection of patients.

Photosensitizers: an ideal sensitizer should have low toxicity, specific absorption spectrum and tumor selectivity. Biologically photoactive agents can be divided into (a) natural phluorochromes, such as porphyrins, (b) exogenous phluorochromes, such as acridine orange, phluorescin, rhodamine, (c) endogenous phluorochromes, such as flavoproteins and keratine. Most studies deal with the first group of natural phluorochromes and their derivatives, because of their activation with a wavelength (600-690 nm) more deeply penetrating in the biological tissues than the shorter wavelengths needed to activate other phluorochromes. After the previous use of HPD, Di-Hematoporphyrin Ether or Ester (DHE) has now become the most widely employed photosensitizer in clinical studies;[10] DHE is considered the major active constituent of HPD. The drug is injected intravenously and after an interval of 24 to 72 hours it is concentrated in malignant tissues at a variable rate of 3-4 times more than in normal tissue. HPD is administered at dosages of 3 to 5 mg/kg body weight and DHE at 1.5-3 mg/kg body weight (fig. 1).

These drugs present two kinds of limitations which maintain the procedure at an experimental stage: skin photosensitization and low tissue penetration of the light at the wavelength used for the activation of the sensitizer. The photosensitivity to sunlight, due to the drug retained by the skin, may be present for 4 to 6 weeks after the injection. Precautions must be taken to avoid exposure to direct sunlight for 30 days and during this period patients are advised to stay indoors, cover exposed parts and protect eyes from sun rays and strong fluorescent or incandescent lighting.

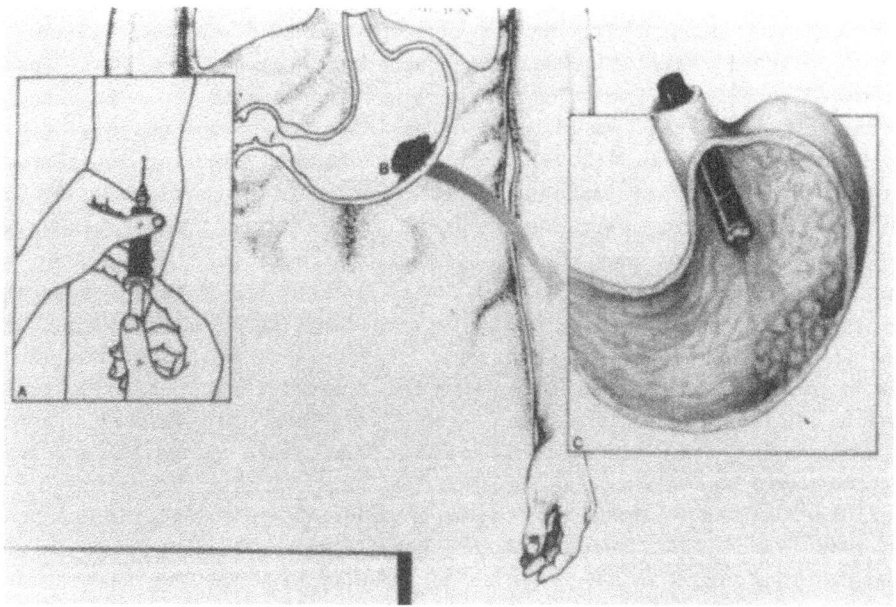

Fig.1. After i.v. injection (a), the photosensitizer is retained by malignant tissue (b). Endoscopic laser irradiation (c) causes the selective necrosis of the tumor.

The future tendency will be to improve the selectivity of photosensitizers and consequently use smaller amounts of drugs, reducing cutaneous sensitization. An improved selectivity may be obtained by means of inclusion of the drugs into liposomes,[11] or their linkage with monoclonal antibodies.[12,13] Furthermore, the possible use of new drugs, now in the experimental phase, having a high absorption coefficient in the near infrared would improve light penetration into biological tissue, inducing necrosis of larger volumes of tumor.[13]

Among the new drugs, some are compounds resulting from modification of porphyrins: modifying the structure of DHE by converting one or more of the porphyrin rings to chlorin (DHEC), or linking HP to chlorin, or modifying the ester and acid functions (Benzoporphyrine).[14] Of great interest is also the use of phthalocyanines, which are porphyrin-like compounds with a main absorption band in the red and which have been experimentally demonstrated as being very efficient as photosensitizers. The action spectrum for chloroaluminium phthalocyanine (ClAlPC) is a narrow band centered around 680 nm. ClAlPC appears to be about 50 times more efficient than HPD and the red-shifting of its action spectrum allows a better light penetration into the irradiated tumors.[13]

Light for activation: there are different possibilities of obtaining a sufficient amount of light to be used for PDT. The activation of porphyrins is usually obtained with a 630 nm wavelength. This wavelength can be produced by filtered lamps, to be used for surface application,[15] but, when an intracavitary tumor must be treated by endoscopic systems and light must be trasmitted on fiberoptics it needs special properties such as intensity, coherence and monochromaticity,[16] that are characteristic of lasers.

Generally speaking, in PDT applications laser sources allow the photobiological responses produced by laser-tissue interaction to be quantitatively and qualitatively different from those caused by conventional light sources.[17] The advantages of lasers can be summarized as follows: a) intensity: important to produce effects requiring large energy doses; varying peak and average power separately, one can induce thermal or photodynamic effects. High peak powers can produce localized thermal damage and high average powers are more likely to produce thermal damage over larger tissue volumes. Low energy density, with no or negligible thermal effect, is responsible for photodynamic action; b) coherence: important for direction and focusing on small areas such as the proximal end of a flexible optic fiber; c) monochromaticity: allows chromophore selection within tissues and selective photobiological responses.

The most used lasers for PDT are: 1) Ar-laser: (488-514 nm), limited penetration; 2) Dye lasers, namely Rhodamine-B laser (630 nm, tunable), the most extensively used for PDT; 3) Gold vapour laser (628 nm).[18] A comparison of gold-vapour and dye-lasers for PDT has be made and the gold-vapour laser appears to be simpler and easier to install and run, although it requires a larger diameter fiber for light delivery. This less flexible fiber, when inserted into the operative channel, can reduce the manoeuvrability of endoscopes. The wavelength of the dye laser is tunable, whereas that of the gold vapour is fixed; it can be turned into a copper vapour laser at 510 and 578 nm and used to pump a dye laser.

New laser devices are being studied at the present time; of particular interest are the tunable dye lasers, allowing different wavelengths to be pruduced. Future directions will be the development of diode lasers, which are small, simple and easy to use.[19,20]

The major difficulty in light irradiation during endoscopic treatments is to convert the unidirectional laser beam into an isotropic illumination of the lesion. In endoscopic applications, irradiation of tumors can be carried out by inserting the fiber into the tissue or keeping the fiber distant from the tissue itself. The evaluation of the energy delivered by the fiber is different in the two situations: if a sharp cut fiber is inserted into a tissue the energy is expressed simply in Joules; if a circularly radiating fiber is inserted into a tissue the energy is expressed in Joules/cm of inserted fiber; if the fiber is kept at a distance, the energy is expressed in Joules/cm^2. Between the extremity of the fiber and the tissue, different light diffusing devices can be used, like diffusing solutions, sapphire tips or microlenses.[21,22] Diffusing solutions can be contained in diffusing balloons attached to the end of a fiber, or to the end of an endoscope. These devices have the aim of obtaining a homogeneous light distribution on the tumor surface.

Selection of patients: From July 1982 to May 1990, at the Endoscopy Division of the National Cancer Institute in Milan, 32 patients (24 males, 8 females) were submitted to endoscopic PDT. A total of 43 treatments were given: 10 patients were submitted to two sessions of PDT, and 1 patient to three treatments. All patients were treated after a surgical resection was excluded on the basis of a multidisciplinary discussion and after an informed consent about the modality of treatment was given by the patients.

In 14 patients lesions were located in the tracheo-bronchial tree, and were histologically classified as squamous cell carcinoma in 13 cases, and as atypical squamous metaplasia in the remaining case. In 8 of these patients the indication for PDT was the palliative reopening of the obstructed airways due to advanced tumors (n=7), or to recurrence after surgery (n=1). In 5 patients tumors were at an early stage and PDT was performed with curative purpose. In the last patient, affected by a precancerous lesion, PDT was considered the treatment of choice.

Gastrointestinal (GI) malignant tumors were treated in 17 patients: 2 patients had advanced squamous cell carcinomas of the esophagus, and 7 patients advanced adenocarcinomas of the rectum. In this group of patients the purpose of PDT was recanalization of the lumen. The remaining 8 patients were treated for early stage squamous cell carcinomas of the esophagus (n=3), or for early stage adenocarcinomas of the stomach (n=5).

One patient suffering from a multicentric T1 urinary bladder carcinoma received PDT in order to avoid a total cistectomy.

Details of treated patients are listed in Tables 1 and 2.

TABLE 1. Advanced Stage Carcinomas: Patients Treated by PDT for Palliation of Obstruction

Case No.	Age	Sex	Tumor location*	Histology**	E.E.D.§	Combined therapy°	Follow-up	Current status
1.	52	M	RMB & carina	Sq. ca.	640 J/cm^2	–	4 months	alive
2.	65	M	RMB & carina	"	90 "	Laser	5 "	"
3.	58	M	RMB	"	235 "	RT	10 "	"
4.	56	M	LMB	"	120 "	Laser	5 "	"
5.	60	M	trachea	"	210 "	"	10 "	"
6.	56	M	LMB & RMB	"	120 "	CT	1 "	"
7.	63	M	trachea	"	200 J/cm	Laser, RT	20 "	dead
8.	47	F	LULB	"	400 "	–	8 days	"
9.	47	M	esophagus	"	230 J/cm^2	–	5 months	alive
10.	70	F	esophagus	"	170 "	Laser	35 "	alive
11.	33	F	rectum	Ad. ca.	95 "	–	2 "	dead
12.	73	M	rectum	"	720 "	–	6 "	"
13.	93	F	rectum	"	185 "	–	6 "	"
14.	76	F	rectum	"	150 "	–	10 "	"
15.	80	F	rectum	"	80 "	–	20 "	alive
16.	82	M	rectum	"	150 "	Laser, RT	52 "	"
17.	52	M	rectum	"	55 "	–	2 "	dead

(*) RMB, right main bronchus; LMB, left main bronchus, LULB left upper lobe bronchus

(**) Sq. ca., squamous cell carcinoma; Ad. ca., adenocarcinoma

(§) E.E.D., estimated energy dose

(°) RT, radiotherapy; Laser, Nd:YAG laser photocoagulation; CT, chemotherapy

TABLE 2. Early Stage Carcinomas: Patients Treated by PDT with Curative Purposes

Case No.	Age	Sex	Tumor Location°	Endoscopic Appearance§	Histology*	No. of PDT sessions	Results**	E.E.D.***	Combined therapy°°	Follow-up
1.	54	M	LULB	flat	Sq. ca.	1	CR	330 J/cm^2	-	10 months
2.	55	M	RLLB & RULB	flat	"	1	CR	600 "	-	36 "
3.	57	M	LULB	not visible	"	1	CR	90 "	-	51 "
4.	54	M	LULB	flat	"	1	CR	120 "	RT	13 "
5.	54	M	LULB	flat	"	1	PR	100 J/cm^2	Surgery	2 "
6.	44	M	RULB	flat	Sq. met.	1	CR	90 J/cm^2	-	58 "
7.	55	M	esophagus	IIa	Sq. ca.	2	CR	290 "	Laser, RT	48 "
8.	77	F	esophagus	IIc	"	3	CR	300 "	Laser	46 "
9.	81	M	esophagus	IIb	"	2	CR	200 "	-	15 "
10.	67	M	stomach	I	Ad. ca.	2	CR	200 "	-	14 "
11.	69	M	stomach	IIc	"	1	CR	110 "	-	36 "
12.	77	M	stomach	III	"	1	PR	120 "	Laser	15 "
13.	72	M	stomach	IIa	"	1	CR	100 "	-	22 "
14.	78	F	stomach	III	"	1	CR	80 "	-	2 "
15.	58	M	bladder	papillary	Trans. ca.	1	NR	40 "	-	1 "

(°)   LULB, left upper lobe bronchus; RLLB, right lower lobe br.; RULB, right upper lobe br.

(§)   see ref. 36 for esophageal cancer; see ref. 27 for gastric cancer

(*)   Sq. ca., squamous cell carcinoma; Sq. met., squamous metaplasia; Ad. ca., adenocarcinoma;
      Trans. ca., transitional cell carcinoma

(**) criteria of evaluation explained in the text

(***) E.E.D., estimated energy dose

(°°) RT, radiotherapy; Laser, Nd:YAG laser photocoagulation

In all cases of advanced tumors treated by PDT, a reduction of the tumor mass more than 50% of the initial volume was achieved. In these patients it was possible to obtain the reopening of the bronchial or the gastrointestinal lumen with an improvement of symptomatology. In early stage cancers results of the treatment were divided into: CR, complete response or total disappearance of the lesion at endoscopy and at histological examination; PR, partial response or total disappearance at endoscopy with positive biopsy; NR, no response.

Early tracheo-bronchial cancers were treated in 5 patients: 4 CR, and 1 PR were observed. Median follow-up was 20 months (range 10–51 months). The estimated energy dose (E.E.D.) delivered on the tumor surface was 330 and 600 J/cm^2 in the first 2 patients of our series, respectively; in the last 3 patients lower energy doses, ranging between 90 and 120 J/cm^2 were used. In 1 patient PDT was performed to eradicate a precancerous lesion; atypical squamous metaplasia was lighted with an energy of 90 J/cm^2. A CR was obtained, and in the treated area the growth of the normal cylindrical epithelium was observed. No recurrence occurred during the follow-up.

Early GI tumors were treated in 8 patients during 13 PDT sessions: 7 CR, and 1 PR were noted. Median follow-up was 18.5 months (range 2–48 months). An E.E.D. of 40 J/cm^2 to 170 J/cm^2 was delivered in all cases. The patient who responded partially received a dose of 120 J/cm^2. In 2 patients showing CR, a recurrence occurred 12 and 8 months later, respectively. In the first case Nd:YAG laser photocoagulation was carried out; in the second patient laser treatment was associated with radiotherapy.

The urinary bladder tumor was lighted with an energy dose of 40 J/cm^2. No response was observed.

No major complications were noted after treatments. Patients treated for esophageal carcinoma experienced pain in the treated area, lasting 3–4 days. A minor hemorrhage was observed in few cases after treatments of advanced cancer, expecially in the lower GI tract.

Hayata, at Tokyo Medical College, started with clinical endoscopic applications of PDT in 1980 and has, up to now, accumulated the widest experience in the world in the various fields.[23] In two international enquiries in 1984 and in 1986, we collected data from respectively 467 and 918 patients. Enquiries suggest that the number of centers working in the area of PDT is increasing. Geographically, the centers have expanded all over the world in recent years. Up to 1984, 467 patients had been treated in 8 centers, between 1984 and 1986 the total number had risen to 912 and the number of centers to 20. The laser sources used show that 4 groups are using the new gold vapour lasers and that activation by Nd:YAG laser photoradiation has been abandoned.

Practically unchanged are: a) the photosensitizers used, b) the time interval between drug injection and irradiation, c) the modality of irradiation. Regarding the anatomical areas irradiated, the number of bladder treatments is increasing. Regarding the stage, early tumors exceed advanced ones; the power and the energy of treatments tend to decrease, probably because of a relative optimization of treatment parameters. The results of the 1986 enquiry show a complete response (CR) in 61% of early stage and in 7% of advanced tumors treated, partial response (PR) in 33% early and 80% advanced and no response (NR) in 6% early and 13% advanced.[24] Up to now, more than 3,000 patients have been treated, but mainly in uncontrolled trials. Results suggest that tumor histology plays a small role in determining response to PDT and that some lesions such as slowly growing, poorly vascularized and high pigmentated tumors do not respond well to this therapy.[19]

Current indications for endoscopic PDT in the gastrointestinal tract include treatment of early stage cancers in high risk patients.[25] In the esophagus, lesions are usually treated with cylindrical diffusing fiber, but tumors located in the upper, as well as in the lower esophageal sphincters require particular devices. At these levels, the introduction of a Savary dilator or of an endoluminal light delivery system[26] allows the sphincter to be kept open; in this way it is possible to obtain a homogeneous distribution of the light.

Early gastric cancers are treated by PDT when they are type IIb or IIc, or type III according to the classification of the Japanese Society for Gastroenterologic Endoscopy.[27] Types I or IIa are better treated by Nd:YAG laser photocoagulation. In the stomach, the lesions are usually irradiated by bare fiber or sapphire micro-probe.[21]

At present, the indications for PDT in the rectum are very limited. However, recently published results[28,29] have demonstrated that, after e.v. injection, adenomas present an uptake of HPD similar to that of carcinomas. This observation led us to consider the applicability of endoscopic PDT mainly in the case of flat colorectal adenomas. In fact, endoscopic resection or laser photocoagulation of these lesions is followed by a high rate of recurrence due to the presence of adenomatous tissue not macroscopically visible during the treatment.[30]

Bronchoscopic PDT is performed as a conservative treatment of superficial cancers in high risk patients. Japanese authors[31] use PDT before surgery to reduce proximal extension of the tumor. In this way, some inoperable cases become resectable and more limited resections can be performed. Light is delivered to the surface of the tumor by cylindrical diffusing fiber that can be inserted into the lesion, or can be used for a surface illumination.

Cleanup bronchoscopy is necessary 72 hours after laser irradiation in order to remove necrotic tissue that could obstruct the airways.
In the urinary bladder, PDT is used in the treatment of carcinomas in situ; a whole bladder irradiation is performed because the tumor is usually multicentric. In order to obtain a homogeneous distribution of light dose, the use of a bulb tip fiber, or of a light scattering medium has been proposed.[32,33] In the former case, the bladder is filled with a normal saline solution and the bulb of the fiber is positioned at the center of the viscerum. In the latter, the bladder is filled by a lipidic physiological suspension (Intralipid 1:1000) and the position of

the fiber tip is at a quarter of the distance between the center and the neck of the bladder.

Side effects and complications after PDT are infrequent. Among the former, pain and oedema at the irradiated site have been reported, mainly in the esophagus and stomach where they can be responsible for a temporary dysphagia or obstruction, respectively. NSAID are useful in these cases.

Major complications are hemorrhage and perforation due to extensive wall necrosis.

Severe reduction of urinary capacity, due to bladder shrinkage, is the most important complication after whole bladder irradiation; it seems to be related to the light dose and to the pressure used to distend the bladder. Light doses of 20 Joules/cm^2, pressure of less than 30 cm column of water and the use of normal saline, or lipidic physiological suspension, can reduce the incidence of irreversible shrinkage and prolonged inactive bladder symptoms.[19]

CONCLUSION

Indications for PDT are changing from the first attempts. PDT seems to be more reliable in treating small cancer lesions, superficially extended over large areas, multicentric and with undefined borders. PDT can be used as a curative and as a palliative treatment. It can treat cancer at various stages, from precancerous lesions, to cancer both at early[34] and at advanced stages.[35] PDT seems to be the ideal therapy for cancer, but many problems become evident on detailed study.

The selection of patients is very important for PDT, because for palliation of advanced cancers deeply infiltrating the walls of hollow viscera, thermal laser action is usually preferred to obtain a controlled and quicker necrosis. In such a way, the incidence of the most serious complications such as perforation and hemorrhage can be limited.

Another comment concerns the location of the cancer: the best location in relation to the possibility of treatment is in large hollow viscera that can be filled with a refractive medium, as for example the urinary bladder. A quite homogeneous distribution of the light power can be obtained in this way. On the contrary, in a narrow channel with a stenotic tortuous segment, as in the esophagus or colon-rectum, light distribution is a problem.

The future tendency is to develop treatments in the field of early cancer and precancerous lesions. At present, treatment protocols involving combination of traditional therapies are submitted to international evaluation.

REFERENCES

1. H. Tappenier, A. Jesionek, Therapeutische versuche mit fluoreszierenden stoffe, Muench. Med. Wochenschr. 1:2042 (1903).

2. A. Policard, Etudes sur les aspects offerts par des tumeurs experimentales éxaminées à la lumière de Woods, Cr. Soc. Biol. 91:1423 (1924).

3. H. Auler and G. Banzer, Untersuchungen uber die rolle der porphine bei geschwulstkranken menschen und tieren, Z. Krebsforsch. 53:65 (1942).

4. R. L. Lipson and E. J. Baldes, The photodynamic properties of a particular hematoporphyrin derivative, Arch. Dermatol. 82:508 (1960).

5. R. L. Lipson, E. J. Baldes and A. M. Olsen, Hematoporphyrin derivative: a new aid for endoscopic detection of malignant disease, J. Thorac. Cardiovasc. Surg. 42:623 (1961).

6. J. Moan, Porphyrin-sensitized photodynamic inactivation of cells: a review, Lasers Med. Sci. 1:5 (1986).

7. R. Hilf, D. B. Smail, R. S. Murant, et al., Hematoporphyrin derivative-induced photosensitivity of mitochondrial succinate dehydrogenase and selected cytosolic enzymes of R3230 AC mammary adenocarcinomas of rats, Cancer Res. 44:1483 (1984).

8. C. J. Gomer, DNA damage and repair in CHO cells following hematoporphyrin photoradiation, Cancer Lett. 11:161 (1980).

9. T. J. Dougherty, Photodynamic therapy (PDT) of malignant tumors, CRC Crit. Rev. Oncol/Haemat. 2:83 (1985).

10. T. J. Dougherty, W. R. Potter, and K. R. Weishaupt, The structure of the active component of hematoporphyrin derivative, in: "Porphyrin localization and treatment of tumors," D. R. Doiron and C.J. Gomer, eds., Alan R. Liss, Inc., New York (1984).

11. G. Jori, Pharmacokinetics studies with hematoporphyrin in tumor-bearing mice, in: "Photodynamic Therapy of Tumors and other Diseases," G. Jori and C. Perria, eds., Progetto Publ., Padova (1985).

12. D. Mew, C. K. Wat, G. H. N. Towers, and J. G. Levy, Photoimmunotherapy: treatment of animal tumors with tumor specific monoclonal antibody-hematoporphyrin conjugates, J. Immunol. 130:1473 (1983).

13. R. Kol, E. Ben-Hur, E. Riklis, R. Marko, and I. Rosenthal, Photosensitized inhibition of mitogenic stimulation of human lymphocytes by aluminium phthalocyanine tetrasulphonate, Lasers Med. Sci. 1:187 (1986).

14. A. M. Richter, E. Sternberg, E. Waterfield, D. Dolphin, and J. G. Levy, Characterization of benzoporphyrin derivative, a new photosensitizer, Proc. SPIE 997:132 (1988).

15. T. J. Dougherty, J. E. Kaufman, A. Goldfarb, K. R. Weishaupt, D. G. Boyle, and A. Mittleman, Photoradiation therapy for the treatment of malignant tumors Cancer Res. 38:2628 (1978).

16. P. Spinelli and M. Dal Fante, Photodynamic therapy in G.I. tract, Acta Endosc. 15:69 (1985).

17. J. A. Parrish, "Photomedicine potential for lasers. An overview from lasers in photomedicine and photobiology," R. Pratesi and C.A. Sacchi, eds., Springer-Verlag, Berlin (1980).

18. A. L. Mc Kenzie and J. A. S. Carruth, A comparison of gold-vapour and dye lasers for PDT, Lasers Med. Sci. 1:117 (1986).

19. T. J. Dougherty, Photodynamic therapy - New approaches, Sem. Surg. Oncol. 5:6 (1989).

20. R. Brancato, L. Giovannoni, R. Pratesi, and U. Vanni, New lasers for ophthalmology: retinal photocoagulation with pulsed diode lasers, Proc. SPIE 701:365 (1986).

21. P. Spinelli and M. Dal Fante, Contact laser endoscopic surgery with sapphire microprobes, Proc. SPIE 701:331 (1986).

22. V. Russo, Optical fiber delivery systems for laser medical applications, in: "Photodynamic Therapy of Tumors and other Diseases," G. Jori and C. Perria, eds., Progetto Publ., Padova (1985).

23. Y. Hayata, H. Kato, C. Konaka, J. Ono, and N. Takizawa, Hematoporphyrin derivative and laser photoradiation in the treatment of lung cancer, Chest 81:269 (1981).

24. P. Spinelli and M. Dal Fante, PDT - State of the art, in: "Laser Optoelectronics in Medicine," W. Waidelich and P. Kiefhaber, eds., Springer-Verlag, Berlin (1987).

25. P. Spinelli, S. Andreola, R. Marchesini, E. Melloni, V. Mirabile, P. Pizzetti, and F. Zunino, Endoscopic HpD-laser photoradiation therapy (PRT) of cancer, in: "Porphyrins in tumor phototherapy," A. Andreoni and R. Cubeddu, eds., Plenum Publishing Corporation, (1984).

26. J. T. Allardice, A. C. Rowland, N. S. Williams, and C. P. Swain, A new light delivery system for the treatment of obstructing gastrointestinal cancers by photodynamic therapy, Gastrointest. Endosc. 35:548 (1989).

27. Japanese Study for Gastroenterological Endoscopy, in: "T. Murakami, Pathomorphological diagnosis, definition and gross classification of early gastric cancer, T. Murakami, ed., University of Tokyo Press, Tokyo (1971).

28. M. Dal Fante, G. Bottiroli, and P. Spinelli, Behaviour of haematoporphyrin derivative in adenomas and adenocarcinomas of the colon: a microfluorometric study, Lasers Med. Sci. 3:165 (1988).

29. R. M. Cothren, R. Richards-Kortum, M. V. Sivak, Jr., et al., Gastrointestinal tissue diagnosis by laser-induced fluorescence spectroscopy at endoscopy, Gastrointest. Endosc. in press.

30. E. M. H. Mathus-Vliegen and G. N. J. Tytgat, Nd:YAG laser photocoagulation in colorectal adenoma. Evaluation of its safety, usefulness, and efficacy, Gastroenterol. 90:1865 (1986).

31. H. Kato, C. Konaka, J. Ono, et al., Preoperative laser photodynamic therapy in combination with operation in lung cancer, J. Thorac. Cardiovasc. Surg. 90:420 (1985).

32. R. C. Benson, Jr., Laser photodynamic therapy for bladder cancer. Mayo. Clin. Proc. 61:859 (1986).

33. D. Jocham, G. Staehler, C. Chaussy, et al., Integral dye-laser irradiation of photosensitized bladder tumors with the aid of a light-scattering medium, Prog. Clin. Biol. Res. 170:249 (1984).

34. H. Tajiri, N. Daizukono, S. N. Joffe, and Y. Oguro, Photoradiation therapy in early gastrointestinal cancer, Gastrointest. Endosc. 33:88 (1987).

35. J. S. Mc Caughan, T. E. Williams, and B. H. Bethel, Palliation of esophageal malignancy with photodynamic therapy, Am. Thorac. Surg. 40:113 (1985).

36. M. Nishizawa, T. Okada, T. Hosoi, et al., Detecting early esophageal cancers, with special reference to the intraepithelial stage, <u>Endoscopy</u> 16:92 (1984).

LASER ANGIOPLASTY : STATE OF THE ART

Herbert J. Geschwind

Cardiac Catheterization Laboratory
University Hospital Henri Mondor
INSERUM U2, Créteil, France

Since the development of balloon angioplasty, interventional cardiovascular medicine has been increasingly used for recanalization of obstructed coronary and peripheral arteries. Balloon angioplasty relieves arterial stenosis by inflating a balloon into the obstruction.[1] It acts by disrupting the atheromatous plaque, fissuring the intima and stretching the media.[2] This controlled injury heals by retraction of the plaque, fibrosis of the media and neo-intimal formation.[3]

First limited to proximal, single vessel disease, the indications of the technique have been expanded to multiple vessel disease, acute myocardial infarction and total occlusions.

The method is also applied to arteries of the lower limb such as iliac, femoral, popliteal, and below the knee arteries. The procedure offers a less morbid, less expensive and equally effective alternative to bypass surgery. However, the primary success rate is only 70 % for obstructed coronary arteries and 50 % for total occlusions. Moreover, the rate of restenosis is 30 % at 6 month follow-up.[4] In addition, the smaller the arteries the less effective and more hazardous the procedure.

By contrast, laser irradiation, transmitted through flexible optical fibers inserted into obstructed vessels, destroys by vaporization obstructing lesions, cratering the plaque by a thermal mechanism.[5-7] Thus, it was thought that laser angioplasty could become an alternative or complementary procedure to balloon angioplasty.

A recurring problem with optical fiber delivery systems is imprecision in guiding and positioning. Vessel perforation, even at low energy levels, results when an improperly positioned optical fiber directs sufficient laser energy into the vessel wall. Manipulation of optical fibers inside arteries can also be hazardous. The needle-like end can inadvertently perforate the vessel wall at curves or branch points.[8-10] It also appears that to be effective, laser created recanalizations of obstructed arteries have to be wide enough to avoid any significant residual stenosis, thus decreasing the risk of reocclusion.[11]

Thus, the aims of a laser catheter delivery system are to : 1) make holes large enough to relieve obstruction by > 70 %, 2) avoid major damage to the arterial wall such as perforation, 3) protect fiber tip against melting and back burning, 4) leave an arterial surface that is neither thrombogenic nor prone to late restenosis.

Optronic Techniques in Diagnostic and Therapeutic Medicine
Edited by R. Pratesi, Plenum Press, New York, 1991

The means required to meet these goals is a laser catheter that allows for : 1) coaxial laser emission in the artery toward the target obstruction, 2) creation of wide holes through the occlusion, 3) cooling perfusion during laser emissions when continuous wave lasers are used or delivery of pulsed lasers that avoid thermal damage to the adjacent tissue, 4) penetration of the catheter tip into the atheroma for a pilot hole, 5) widening of the laser created tunnels using balloon angioplasty.

To determine if Argon or neodymium yttrium aluminum garnet (Nd:YAG) is a safer, more efficient laser angioplasty source, we created craters in atherosclerotic human cadaver arteries with bare fibers and we compared the results with those obtained with ball tip fibers and sapphire control probes.

Three optical fiber systems were used : i) a bare optical fiber, 0.20 mm core diameter with an outer diameter of 0.25 mm, ii) a lensed ball type catheter[12] consisting of a silica fiber 0.2 mm in diameter on the end of which a 1 mm diameter lens was made by heating the silica. The lensed fiber was inserted into a 5 French balloon catheter with the ball fiber tip maintained 3 mm beyond the catheter tip. The diameter of the ball was greater than that of the taped distal end of the catheter. iii) A contact laser probe[13] made of a selected physiologically neutral synthetic sapphire crystal with great mechanical strength, low thermal conductivity and a high melting temperature of 2050°C. The contact probe was screwed into a metal connector attached to an 8 French woven-dacron catheter. A similar 0.2 mm core diameter optical fiber was inserted into the catheter and maintained in close contact with the sapphire. The sapphire contact probe with an outer diameter of 2.2 mm had a round shape which resulted in high power density and broad energy delivery. During laser emissions, saline perfusate was circulated at a rate of 10 ml/min through the catheter to prevent excessive heating of the fiber tip.

LASER SOURCES

The proximal end of all fibers were first connected to a continuous wave Nd:YAG laser 1064 nm, then, to an argon ion laser 488-514 nm.

In order to compare the effects of the three laser delivery systems on tissue, the same energy level was used for both sources of laser. Approximately 50 joules were delivered for each emission consisting of 25 watts with an emission time of 2 seconds for Nd:YAG and 17 watts for 3 seconds for the argon laser.

The volume of tissue removed was significantly greater for argon than for Nd:YAG when bare fibers and lensed fibers were used. The volume of tissue removed was similar when sapphire probes were used. The rim of carbonization was remarkably similar regardless of the laser source and delivery mode. There were no significant differences observed between the two laser sources regardless of the delivery catheter, although the extent of thermal injury including the zone of vacuolization was slightly higher for argon than for Nd:YAG.

The efficiency was low for bare fibers, higher for ball fibers and very high for sapphire probes. The efficiency of argon was superior to that of Nd:YAG except for sapphire probes. The highest efficiency was obtained with the Nd:YAG and sapphire probe and the lowest efficiency was obtained with the Nd:YAG and bare fiber.

Argon was more efficient than Nd:YAG for bare fibers and lensed tips. The lesser efficiency of Nd:YAG was confirmed using lensed fibers, but not confirmed using the sapphire probe. The volume of tissue removed using the sapphire probe was similar for both lasers but since more energy was used with argon, due to better transmission of the identical energy input, the efficiency appeared to be lower. However, this device appeared to be the most efficient, regardless of laser used, which was probably due to a highly convergent beam with low tissue back scattering effect. It appears that efficiency is related to the surface of the catheter tip, although further studies are required to substantiate this hypothesis. In this respect, optically modified laser catheter tips have the advantage of higher efficiency regardless of the laser wavelength used since they can be attached to very thin optical fibers that are easier and safer to use than thick fibers because of higher flexibility.

The efficiency of the lasers was more dependent on the delivery system than on the source. Side effects were similar and resulted from the delivery system.

Thus, the choice of the delivery system for laser angioplasty is critical and should be made for the laser catheter rather than the laser source.

The extent of thermal damage was limited with each laser source. The damage did not extend as the diameter of craters was increased by modified fibers. Thus, there was no direct relationship between the size of the holes and the extent of side effects. The limited thermal injury was probably due to the high energy density that was obtained with the modified tips. The position of the tips in contact with tissue was likely to reduce the beam divergence and dispersion which, it was anticipated, would result in limited side effects.[4] We used relatively short durations of emission which could be achieved because of both high input power and high energy density delivered at the modified tips. This may account, in part, for the reduced thermal damage.

GUIDANCE OF LASER ANGIOPLASTY

The experimental studies and human treatments have been carried out using conventional fluoroscopy with repeat injections of contrast medium. Complementary guidance may be achieved through direct observation. Angioscopes allow direct visualization of the anatomical structures during laser emissions.[15-16]. Although the diameter is now 1 mm, packaging of these fiber optic and lens systems creates technical problems due to insufficient flexibility, distortion of image and small field. Good visualization requires saline perfusion and maximal illumination. In intra-operative procedures, perforation due to laser emissions was not detected by the angioscope. In only 5/11 patients were the images good enough for adequate structure recognition.[17] However, recent devices are promising since they are very thin, flexible and allow tiny arterial structures to be visualized. They were also able to provide precise data on the pathophysiology of such events as unstable angina and myocardial infarction.[18] Histological follow-up after laser angioplasty in two patients revealed thermal injury to the inner quarter of the arterial vessel wall and no evidence of thrombus formation.[19]

To improve the visualization of the spatial relationships of the catheter and the anatomical structures during laser angioplasty, we studied high frequency ultrasonography in obstructed human cadaver arteries embedded in agar and perfused with blood, in atherosclerotic rabbit aortas and in patients either percutaneously or perioperatively.[20] Varying positions of the laser catheter were recognized. A cloud of echoes was seen during laser emissions due to target tissue vaporization. The dimensions of tunnels created into the obstructions correlated well with anatomic size. Perforations of the arterial wall were recognized as a gap in the wall and predicted when the fiber tip was not coaxial and close to the vessel wall. In patients, the long axis view compared well with fluoroscopy for catheter location and arterial wall was defined without contrast medium. Ultrasound guidance results in better scanning in various positions and angles, wall defect visualization and continuous and prolonged visualization of plaque resection.

Endovascular ultrasound devices are currently being used for improved guidance and vessel wall analysis. They are able to accurately identify the thickness of echogenic intima, less echogenic media. Both the specificity and sensitivity of the method appear to be high when compared with histologic data. These systems will be able to determine the amount of the atheromatous tissue to be removed with laser angioplasty.

Specific fluorescence intensities due to wall thickness and extent of atherosclerosis were detected with low power argon-ion laser irradiation at 476 nm. The normal artery has three characteristic fluorescence peaks at 520, 550 and 595 nm with well defined valleys between the peaks, whereas atherosclerotic artery exhibits peaks at the same three wavelengths with less well defined valleys. In addition, the intensity ratio of the 600 and 550 nm peaks is greatly diminished in atherosclerotic spectra. Safety of ablation of atherosclerotic lesions may be improved by identifying arterial wall structure both in thickness and presence of atheroma. In recent trials in patients with occluded peripheral arteries we were able to identify atheroma[21] with spectral analysis so that a pulsed dye laser could be utilized to ablate the targets.[22]

CONSEQUENCES OF LASER IRRADIATION

The consequences of laser irradiation on vessel wall have been assessed.

1. Analysis of the healing process showed craters after laser radiation to be filled with coagulum of blood and cellular debris within 4 days, with healing occurring without inflammatory response. Reendothelialization occurred between 7-14 days and completed by 1-2 months. No accelerated atherosclerosis was observed. The healing process with argon and excimer is similar but initial damage and scarring to surrounding tissue was shown to be more extensive with argon than with excimer irradiation. Controversial results were presented on the thrombogenicity of channels obtained with pulsed and continuous wave lasers.

2. The size of debris after continuous wave laser emission resulting from vaporization of plaque was small. The laser generated photo-products of atherosclerotic segments were hydrogen, water vapor, light hydrocarbon, fragmentation products suggesting thermal degradation.[23,24] Therefore, the risk of thermal embolization after laser irradiation seems to be low and photo products are likely to be quickly absorbed. On the other hand, the size and number of debris resulting from pulsed laser emission were found to be greater than with cw emission, due to the ability of the excited shock waves to disrupt the tissue and to break down also calcified tissue.[25]

PULSED DYE LASER ANGIOPLASTY GUIDED BY SPECTROSCOPY

Pulsed laser emissions, in contrast to continuous-wave lasers, have been shown to penetrate calcified plaques and reduce unwanted thermal damage in normal tissue adjacent to the obstruction.[26,27] Spectroscopic detection of tissue fluorescence allows accurate discrimination of normal from atherosclerotic tissue in vitro and in vivo studies.[28,29] This method serves as an additional guidance system to complement fluoroscopy in improving the specificity and sensitivity of laser treatment in arterial occlusions. Therefore a prospective clinical trial was designed to test the safety and efficacy of the laser device in patients with total occlusion of iliac superficial femoral or popliteal arteries.

The major components of the laser system included a diagnostic laser, a computer detection subsystem for detection and targeting of plaque, a treatment laser, and a fiber delivery system.

The diagnostic laser consisted of a helium-cadmium laser operating at a wavelength of 325 nm and a pulse duration of 20 to 100 msec/pulse. The light emitted from the distal end of the optical fiber in contact with tissue induced fluorescence and was transmitted back through the same fiber to a spectroscopic detection system.

The treatment laser consisted of a pulsed dye laser operating at 480 nm, 2 μsec, 50 mJ/pulse, 5 Hz repetition rate.

The laser catheter was a disposable system component consisting of a 200 μm optical silica fiber wrapped with a metal coil and marked with a radiopaque tip.

The procedure was performed by means of the standard percutaneous Seldinger approach of the ipsilateral or the contralateral common femoral artery with a 6 French or an 8 French arterial introducer sheath. Attempts to recanalize the obstructed artery were performed with a guidewire that was gently advanced toward the obstruction through the arterial sheath without any catheter assistance so that the guidewire could be moved freely within the arterial lumen.

When mechanical attempts failed to penetrate the occlusion, the guidewire was removed. The optical fiber was then inserted into a modified Van Andel catheter, and both were advanced through the sheath and positioned near the occlusion. The distal end of the fiber was placed against the occlusion. Since little if any thermal effects was expected to occur with pulsed emission, no saline flush was used during the procedure. The optical fiber was advanced 3 to 5 mm beyond the tip of the Van Andel catheter under spectroscopic and fluoroscopic monitoring.

When plaque was detected the system was activated, allowing the probe-and-treat sequence to cycle at 5 Hz.

The laser fiber was then gently advanced with continuous spectroscopic guidance. Care was taken to advance the fiber at a reasonable rate. The fiber was never allowed to advance mechanically. The Van Andel catheter was slightly bent at the distal end before insertion into the artery and was rotated into the artery so that the fiber could be directed to various sites when either a normal tissue signal or a media signal was detected. When healthy tissue was detected and the therapeutic laser inhibited, the laser fiber was manipulated until plaque spectra were obtained. Media were detected when the fiber was directed to an eccentric position through the bent Van Andel catheter or the deflated balloon catheter after attempts to find the true lumen of the artery in the center of the vessel had failed. The actual rate of advancement during the probe and fire process was 0.5 mm/sec so that 1 minute was required to penetrate a 3 cm totally occluded artery. However, the probe and fire sequences were not emitted continuously. Time was taken to reposition the fiber, advance the catheter, and detect tissue. The Van Andel catheter was advanced through the recanalized occlusion over the laser fiber. The laser fiber was removed and replaced by a 0.035 inch guidewire.

RESULTS

Primary recanalization of the occluded arterial segment was achieved in 82 % of patients (n=66). Balloon dilatation was achieved in all but two patients with satisfactory angiographic results. In one patient the balloon catheter could not be advanced through the recanalized channel. Mean ankle brachial indexes improved from 0.49 before to 0.95 after the procedure.

Complications resulting from treatment or diagnostic laser consisted of mechanical perforation (n=6), early reocclusion (n=7), dissection (n=6). There was no burning or pain during laser emissions. There were no laser-induced spasms, no clinically detectable emboli.

Failures were due to inability to penetrate the true lumen, to advance the balloon catheter over the fiber or the guidewire, or to penetrate calcified plaques. The patency rate at a mean 18 month follow-up was 64 % including 90 % and 55 % for iliac and superficial femoral arteries, respectively.

Induced tissue fluorescence has been used to differentiate between atheroma and normal tissue. A wavelength of 325 mm has been shown in vitro studies to detect atheroma and thrombus with a high degree of sensitivity and specificity. Results of preliminary in vivo experiments have confirmed the ability to detect plaque by means of a computerized detection system that evaluated the laser-induced fluorescence spectra. In this system tissue fluorescence was induced five times per second, allowing pulse analysis as a potential rapid and reproducible method of targeting tissue ablation.

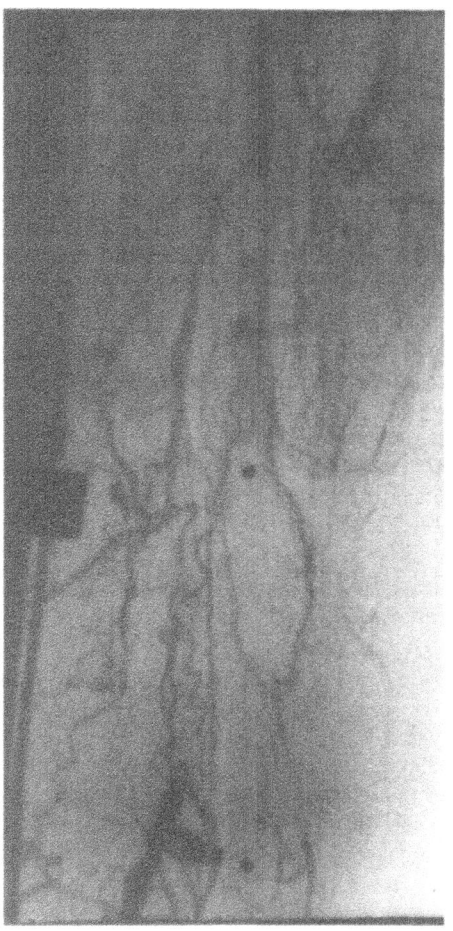 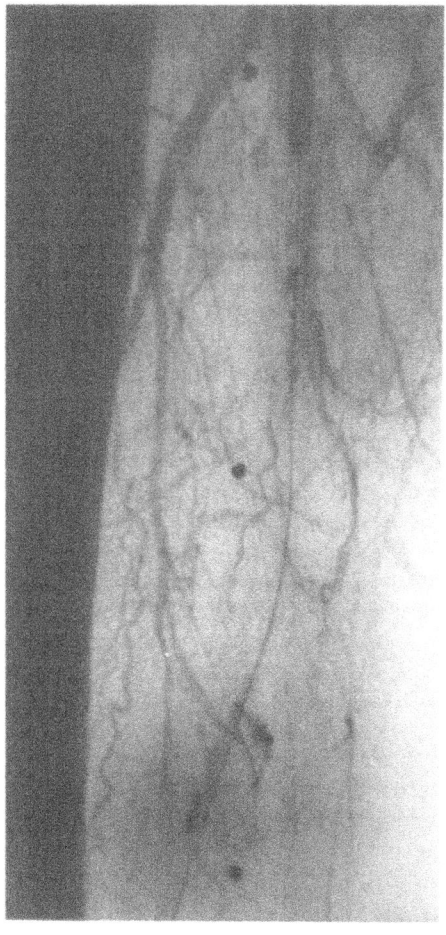

Fig.1   Angiogram of a totally occluded right superficial femoral artery.

Fig.2   A laser catheter has been passed through the occlusion.

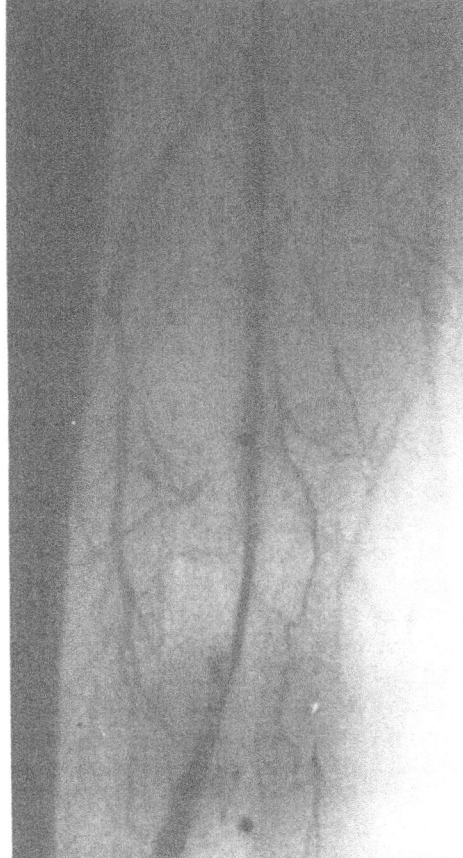 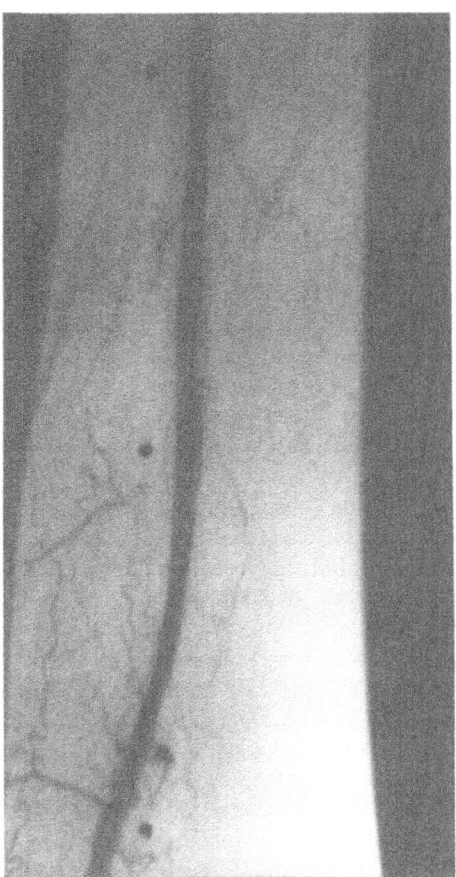

Fig.3   A multifiber laser catheter "over
the wire" created a channel through the
obstruction.

Fig.4   Final result after balloon
angioplasty. Effective recanalization
of the occluded artery.

The use of an ablation wavelength of 480 nm enhanced the safety and efficacy of
the procedures because of the high absorption coefficient of the atheroma relative to
healthy tissue.[30]

Laser recanalization of totally obstructed arteries resulted in a small channel that
always required balloon angioplasty for the establishment of adequate distal flow.
Improvement in design with regard to fiber size, flexibility, and steerability will be
required if this system is to be applied to smaller or more tortuous arteries such as
infrapopliteal or coronary vessels. Procedures that use laser alone must be able to
achieve much greater tissue ablation and therefore luminal size than is presently
possible with the device used in this series of patients.

LASER CORONARY ANGIOPLASTY

For this reason, multifiber catheters advanced over conventional guide wires that
allow for wide channels, are under investigation. Recent reports presented results
obtained with the multifiber catheters advanced through the arterial obstructions over
a guidewire with a 308 nm excimer source operating at 60 mJ/mm^2, 20-40 Hz, 60-240
nsec.[31] This is a mechanical guidance which is likely to increase the safety of the
procedure, since little if any vessel wall perforation is likely to occur. The technique is

currently being applied in coronary artery stenoses.[32] Patients selected for such a procedure include proximal lesions, diffuse lesions, long rigid lesions, restenoses in straight portion of the artery. More than 1000 patients have been submitted to such a treatment with a primary success rate of 83 %, a low rate of complications including spasm, acute closure, dissection. Forty one percent of the patients could be left on laser stand-alone treatment, without any complementary balloon angioplasty. The restenosis rate is still unknown even though preliminary data suggest a rate of 20-30 %.

In our department a pilot study on laser recanalization of obstructed coronary arteries is underway using a pulsed mid-infrared Holmium:YAG laser operating at 600 mJ/pulse, 3 Hz, 250 μsec coupled into a new multifiber catheter consisting of 40 fibers concentrically arranged around a 0.018" central lumen. Promising results were obtained without any complication.

CONCLUSIONS

From these studies it can be concluded that single bare fibers are hazardous to use because of increased risk of vessel wall perforation unless a smart guiding system is associated. Protecting systems of the fiber tip are still under investigation. The effectiveness of laser delivery is still insufficient due to narrow tunnels so that complementary balloon angioplasty has to be used. However, multifiber catheters over a guide wire are being extensively used in peripheral and coronary lesions with a high primary success rate and few complications. The best source of laser has not yet been determined although it is thought that the adequate wave length of specific absorption by atheroma should be investigated and there is a trend to use either ultraviolet (excimer) or infrared (Holmium:YAG) lasers. Although thermal damage should be avoided to obtain controlled laser angioplasty short term emissions were shown to result in an adequate healing process without extensive thrombus formation. Visualization and identification of atherosclerotic tissue have to be enhanced to specifically destroy diseased tissue. Attempts to recanalize obstructed peripheral arteries in humans have been achieved with bare single or multiple fibers. Thermal angioplasty with argon or Nd:YAG heated metal probes or sapphire tips has also been performed. New devices using spectral analysis and pulsed lasers are being developed. Regardless of the device used, recanalization in most cases was insufficient so that complementary balloon angioplasty had to be done to obtain adequate patency of the artery. However, a significant number of patients with coronary obstructions are being treated by laser angioplasty either completed by dilatation or left as a stand-alone technique.

REFERENCES

1. A.R. Gruntzig, A. Senning, W. Siegenthaler, Non-operative dilatation of coronary artery stenosis : percutaneous transluminal coronary angioplasty, N.Engl.J.Med. 301:61 (1979).
2. P.C. Bloch, Mechanism of transluminal angioplasty, Am.J.Cardiol. 53:69C (1984).
3. A. Cragg, W.R. Castaneda-Zuniga, K. Amplatz, Pathophysiology of transluminal angioplasty, Sem.Interv.Radiol. 1:241 (1984).
4. D.R. Holmes, R.E. Vliestra , H.C. Smith et al., Restenosis after percutaneous transluminal coronary angioplasty (PTCA). A report from the National Heart, Lung and Blood Institute Registry, Am.J.Cardiol. 53:77C (1986).
5. G. Lee, R. Ikeda, J. Kozina, D. Mason, Laser dissolution of coronary atherosclerotic obstruction, Am.Heart J., 102:1074 (1981).
6. G.S. Abela, S. Normann, D. Cohen, R. Feldamn, E. Geiser, R. Conti, Effects of carbon dioxide, Nd:YAG and argon laser radiation on coronary atheromatous plaques, Am.J.Cardiol. 50:1199 (1982).

7. D.S.J. Choy, S. Stertzer, H. Rotterdam, N. Sharrock., M. Bruno, Laser coronary angioplasty : experience with 9 cadaver hearts, Am.J.Cardiol. 50:1209 (1982).

8. F. Crea, A. Fenech, W. Smith, R. Conti, G.S. Abela, Laser recanalization of acutely thrombosed coronary arteries in live dogs : early results, J.Am.Coll.Cardiol. 6:1052 (1985).

9. F. Crea, G.S.S. Abela, A. Fenech, W. Smith, C.J. Pepine, R. Conti, Transluminal laser irradiation of coronary arteries in live dogs : an angiographic and morphologic study of acute effect, Am.J.Cardiol. 57:171 (1986).

10. J.M. Isner, R.F. Donaldson, J.T. Funta, et al., Factors contributing to perforations resulting from laser coronary angioplasty : observation in an intact human post-mortem preparation of intra-operative laser coronary angioplasty. Circulation, 72 (suppl. II):191 (1985).

11. Geschwind H.J., Boussignac G., Vieilledent C., Teisseire B., Removal of atheromatous plaques with Nd:YAG laser, J.Am.Coll.Cardiol. 5(2):545 (1985).

12. H.J. Geschwind, A.M. Aita, S. Ramee, et al., Lensed fibers for laser assisted balloon angioplasty, J.Am.Coll.Cardiol. 9(2):177A (1987).

13. H.J. Geschwind, J.D. Blair, D. Mongkolsmai, Development and experimental application of contact probe catheter for laser angioplasty, J.Am.Coll.Cardiol. 9:101 (1987).

14. S.N. Joffe, Contact neodymium:YAG laser surgery in gastro-enterology : a preliminary report, Lasers Surg.Med. 6:155 (1986).

15. G. Lee, R.M. Ikeda, D. Stobbe, C. Ogata, A. Embi, M.C. Chan, R.L. Reis, D.T. Mason, Intraoperative use of dual fiber optic catheter for simultaneous in vivo visualization and laser vaporization of peripheral atherosclerotic obstructive disease, Cathet.Cardiovasc.Diagn. 10:11 (1984).

16. G.S. Abela, S.J. Normann, D.L. Cohen, et al., Laser recanalization of occluded atherosclerotic artery in vivo and in vitro, Circulation 71:403 (1985).

17. G.S. Abela, J.M. Jeeger, E. Barbieri, D. Franzini, A. Fenech, C.J. Pepine, R. Conti, Laser angioplasty with angioscopic guidance in humans, J.Am.Coll.Cardiol. 8:184 (1986).

18. C.T. Sherman, F. Litvack, W. Grundfest, et al., Coronary angioscopy in patients with instable angina pectoris, N.Engl.J.Med. 315:913 (1986).

19. H.J. Geschwind, M. Fabre, B.R. Chaitman, et al., Histopathology after Nd:YAG laser percutaneous transluminal angioplasty of peripheral arteries, J.Am.Coll.Cardiol 8:1089 (1986).

20. H.J. Geschwind, G. Williams, A. Labovitz, M.J. Kern, J. Stern, M. Vandermael, H.L. Kennedy, High frequency ultrasound guidance of laser angioplasty, Circulation 74(suppl. II):468 (1986).

21. M.B. Leon, L.J. Prevosti, P.D. Smith, J.A. Swain, C.L. Mcintosh, H.J. Geschwind, W. Mok, D. Murphy-Chutorian, R.F. Bonnet, In vivo laser induced fluorescence plaque detection : preliminary results in patients, Circulation 76(suppl. IV): 408 (1987).

22. H.J. Geschwind, J.L. Dubois-Rande, B.R. Bonner, et al., Percutaneous pulsed laser angioplasty with atheroma detection in humans, J.Am.Coll.Cardiol. (in press) 1988.

23. J.M. Isner, R.H. Clark, R.F. Donaldson, A.S. Sharon, Identification of photo products liberated by in vitro argon irradiation of atherosclerotic plaque, calcified cardiac valves and myocardium, Am.J.Cardiol. 55:1192 (1985).

24. C. Vieilledent, H.J. Geschwind, G. Boussignac, B. Gaujour, B. Teisseire, Debris after laser arterial recanalization, Laser Med.Surg. 6:31 (1986).

25. L.G. Prevosti, J.A. Cook, M.B. Leon, R.F. Bonner, Comparison of particulate debris size from excimer and argon laser ablation, Circulation 76(suppl. IV):410 (1987).

26. P. Selzer, D. Murphy-Chutorian, R. Ginsburg, L. Wexler, Optimizing strategies for laser angioplasty, Invest.Radiol. 20:860 (1985).

27. L.I. Deckelbaum, J.M. Isner, R.F. Donaldson, R.H. Clarke, S. Laliberte, A.S. Aharon, J.S. Bernstein, Reduction of pathologic tissue injury using pulsed energy delivery, Am.J.Cardiol. 56:662 (1985).

28. L.I. Deckelbaum, M.L. Stez, K.M. O'Brien, F.W. Cutruzzola, A.F. Gmitro, Fluorescence spectroscopy guidance of laser ablation of atherosclerotic plaque. (Abs.), J.Am.Coll.Cardiol. 11:107 (1988).

29. M.B. Leon, L.G. Prevosti, P.D. Smith, J.A. Swain, C.L. McIntosch, H.J. Geschwind, W. Mok, D. Murphy-Chutorian, R.F. Bonner, In vivo laser-induced fluorescence plaque detection : preliminary results in patients. (Abs.), Circulation 76(suppl. IV):408 (1987).

30. M.R. Prince, G.M. La Muraglia, P. Teng, T.F. Deutsch, Preferential ablation of calcified arterial plaque with laser-induced plasmas, IEEE.J.Quantum Electron. QUE-23:1783 (1987).

31. F. Litvack, J. Margolis, N. Eigler, Rothbaum, T. Linnemeier, S. King, J. Douglas, W. Untereker, L. Hestrin, S. Cook, D. Tsoi, T. Goldenberg, J. Segalowitz, W. Grundfest, J. Forrester, Percutaneous excimer laser coronary angioplasty : results of the first 110 procedures, J.Am.Coll.Cardiol. 15(2):25A (1990).

32. T.A. Sanborn, R.A. Hershmann, R.M. Siegel, R.A. Schatz, M.J.M. Schnee, D.R. Leachman, J.A. Bittl, Percutaneous coronary excimer laser-assisted balloon angioplasty: initial multicenter experience, J.Am.Coll.Cardiol. 15(2):25A (1990).

UV LASER ANGIOPLASTY: CLINICAL ASPECTS

Alberto Fracasso and Vincenzo Gallucci

Department of Cardiovascular Surgery
University of Padua
Padua, Italy

INTRODUCTION

With the introduction in the 1970s of balloon angioplasty in the peripheral and later in the coronary arteries, the concept of nonsurgical revascularization became a reality. To deal with the limitations of this technique several new mechanical and optical technologies have been recently developed. The lasers ablative capacity very soon appeared particularly adapted to reopening partially or totally occluded vessels, and different sources of light were tested.

Many experimental and clinical data have proved the clear superiority of pulsed excimer lasers over the traditional dye laser, in treating obstructive atherosclerotic disease. The excimer laser is a pulsed gas laser that uses a mixture of rare gas and halogen as the active medium, to generate pulses of short wavelength, high energy, ultraviolet light.

The main quality of the UV lasers consists in their ability to cut with extreme precision even densely calcified tissues. This capacity for "clean" ablation results from the absence of thermal damage to the tissue immediately surrounding the target area and the relatively low penetration power of the UV wavelength. The exact mechanism of the ablative process is not yet completely understood: it depends more or less on three effects produced by the UV pulsed radiation on the biological tissue: 1) direct bond breaking of the peptide bonds of organic molecules (photoablation); 2) tissue fragmentation by shock waves generated by high peak power laser pulses (photodisruption); 3) thermal mechanism with optimization of heat dissipation.

Besides the tissue-laser interaction, in the assembly of an integrated system it is necessary, when the exact wavelength is to be chosen, to consider the laser-catheter interaction. The clinical application of this class of lasers was temporarily held up by the difficulty in the delivery system of the very short wavelength and very high peak energy to the target tissue. Thanks to the intense research of many groups of physicists, it has recently been possible to assemble flexible silica fibers, able to carry the high peak power (megawatt) characteristic of pulsed lasers with sufficient resistance to optical breakdown. The safety limit is calculated by the difference between the ablative and the breakdown thresholds. It is possible to rise this limit with longer pulse duration and consequently with lower peak power.

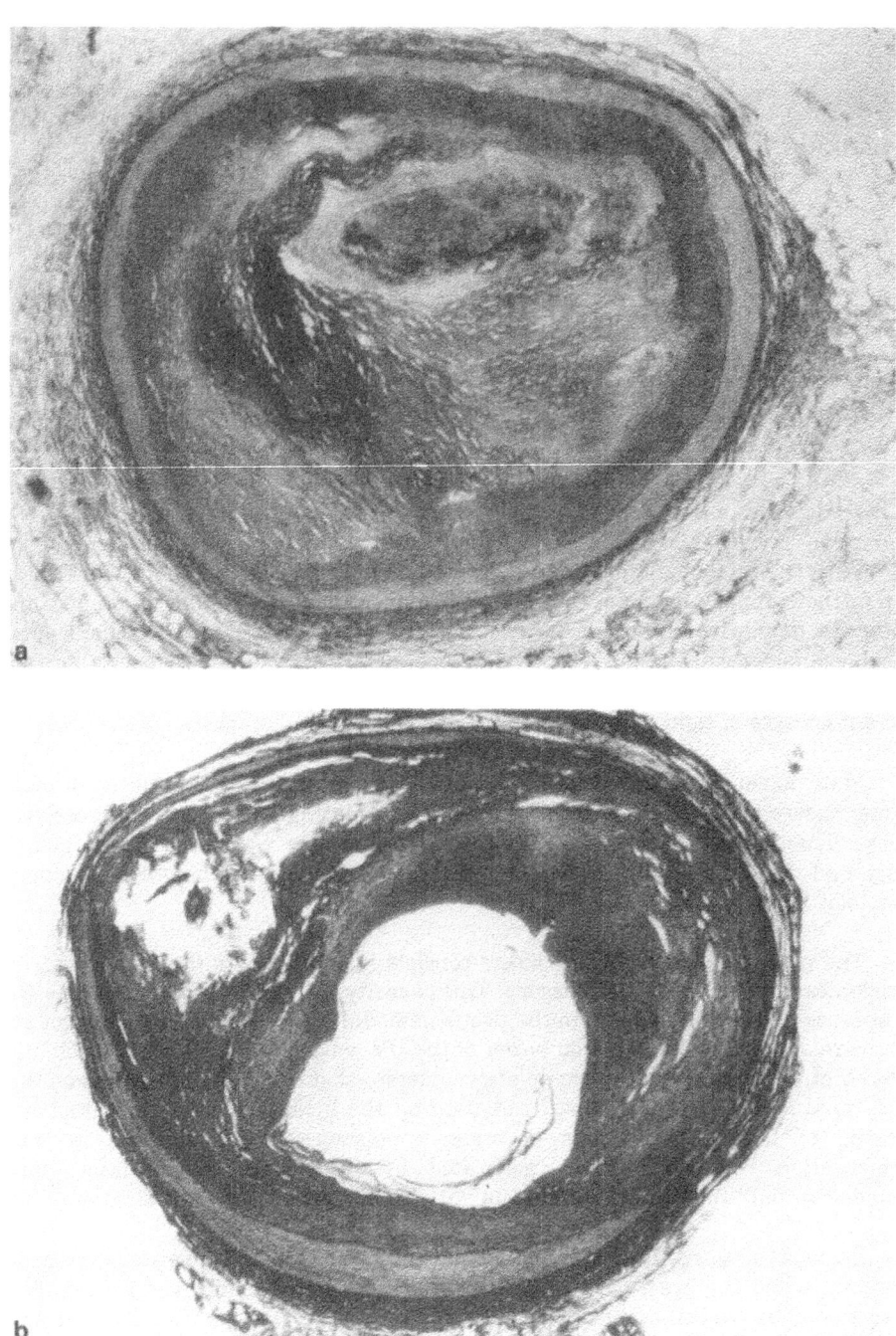

Fig.1. Cross-section of totally occluded coronary artery before (a) and after (b) UV laser recanalization. Van Gieson stain; magn. x 6.3

With these characteristics in mind, it was possible to assemble an integrated system, realized by the I.E.Q.- CNR and El-En in Florence. It is a pulsed XeCl excimer laser coupled with a 3m long catheter, with the following technical characteristics: emission wavelength: 308 nm; repetition rate: 1-100 Hz; pulse duration: 70 ns; energy per pulse: 100 mJ; energy density: 3 J/cm^2.

It is electrically compatible with operating devices, with a quick connection system to change and collimate the different catheters; the gas filling procedure is completely automatic, with only one control switch. In our experiment and clinical trials it was coupled with a monofiber silica fiberoptic 600 microns in diameter, or with a group of three 600 micron fibers, with a total diameter of 1.8 mm.

After the experimental demonstration of the physical adequacy of the system, it was necessary, before its clinical testing, to complete a long experimental protocol to demonstrate:

a) the effective capacity to ablate atherosclerotic plaques and to advance the catheter through obstructed segments of human fixed coronary arteries, or through human fresh coronary arteries from transplanted hearts or human freshly explanted femoral arteries (Fig.1 a,b);

b) the tissue healing process after laser irradiation of arterial segments of rabbits killed 2-3 weeks after treatment.

These trials demonstrated the possibility to open with single fiber or multifiber catheters large new lumina, whose diameter was close to that of the original vessel. To eliminate the tissue fragments in the neolumen, produced by multifiber catheters, it was necessary to retract the fiber end from the catheter tip (Fig.2). At the same time it was observed that the low tissue damage observed in vitro was confirmed by the rapid healing process of the treated rabbit arterial segments after 14-21 days.

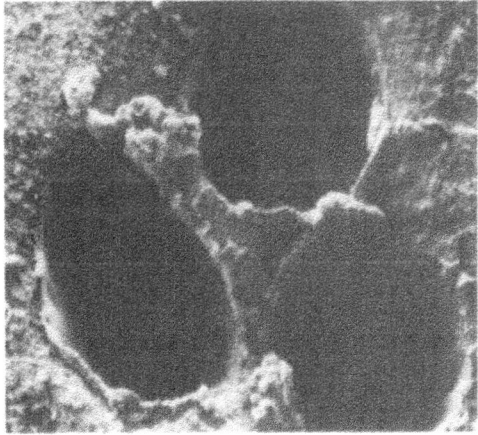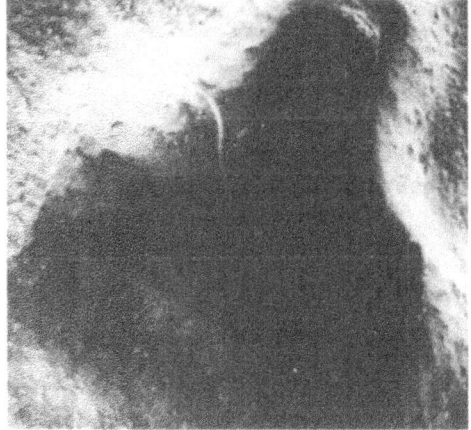

Fig.2.   Scanning electron photomicrographs of aortic wall segments after UV laser irradiation with 3-fiber catheter: left: excavation shape with catheter tip in direct contact with target tissue; right: after fiber retraction; note the absence of tissue fragments in the middle of the neolumen.

Therefore on June 1989 we began the clinical application of excimer lasers on coronary and peripheral vessels, with the purpose of improving the completeness of revascularization, reducing the number of grafts, and to satisfactorily recanalize arterial segments not amenable to classical surgical techniques.

MATERIALS AND METHODS

Between June and December 1989 at the Institute of Cardiovascular Surgery of the University of Padua, seven patients with critical coronary stenoses were treated with excimer laser angioplasty during an aorto-coronary by-pass operation.

The patients, six male and one female, ranging in age from 57 to 73 years (mean 63) suffered from class III or IV angina all with three coronary vessel disease. The laser angioplasty was attempted on 11 critical stenoses (left descending anterior 5, right coronary artery 1, left circumflex 1, first diagonal branch 1, intermediate branch 1, second obtuse marginal branch of left coronary artery 1); four lesions were concentric stenoses and seven eccentric, six completely calcified. The same procedure was used in all cases: in total cardiopulmonary by-pass, moderate hypothermia, during a single aortic cross clamping period, a short arteriotomy 3-4 mm long was performed, at the site of future distal anastomosis. A single fiber catheter (600 microns) in 4 patients, and a multifiber catheter (1.8 mm) in 7 others, was gently advanced through it, until it reached the occluding plaque. This was localized with the help of coronary arteriography, digital palpation and direct observation, and was located in 9 cases proximal to the arteriotomy site.

Pulsed laser energy was then applied during the slow advancement of the catheter at a repetition rate of 25-50 Hz for 5 to 20 seconds, until the lesion was apparently crossed. A graduated probe was then introduced to control the result of the angioplasty, and finally a saphenous vein aorto-coronary graft was applied to the vessel. The whole procedure, from arteriotomy to anastomosis, lasted usually 4 to 5 minutes.

The intraoperative results are summarized in Table 1. An angiographic control was performed on all patients 7-10 days postoperatively, with selective injection in the native coronary vessel and in the grafts. In three cases the procedure was considered successful (residual stenoses less than 40%): in a right coronary artery the stenosis was reduced from 75% to 20%, in a first diagonal branch from 75 to 10, and in a marginal branch from 80% to nearly 0%.

Angiography revealed dissection of the left descending artery at the site of laser application in two patients, with good patency of the vessel; complete thrombotic occlusion proximal to the site of anastomosis was evident in one circumflex and in two left descending arteries; all these patients presented vessel perforation, immediately repaired during the laser procedure. All the grafts were patent and without stenoses. The post-operative course was uneventful, but two patients had an asymptomatic myocardial infarction, eight a new Q-wave on the ECG.

Table 1. Intraoperative results of UV laser coronary angioplasty in 11 stenoses

Probe patency	8 (73%)
Improvement	3 (27%)
Perforation	5 (45%)

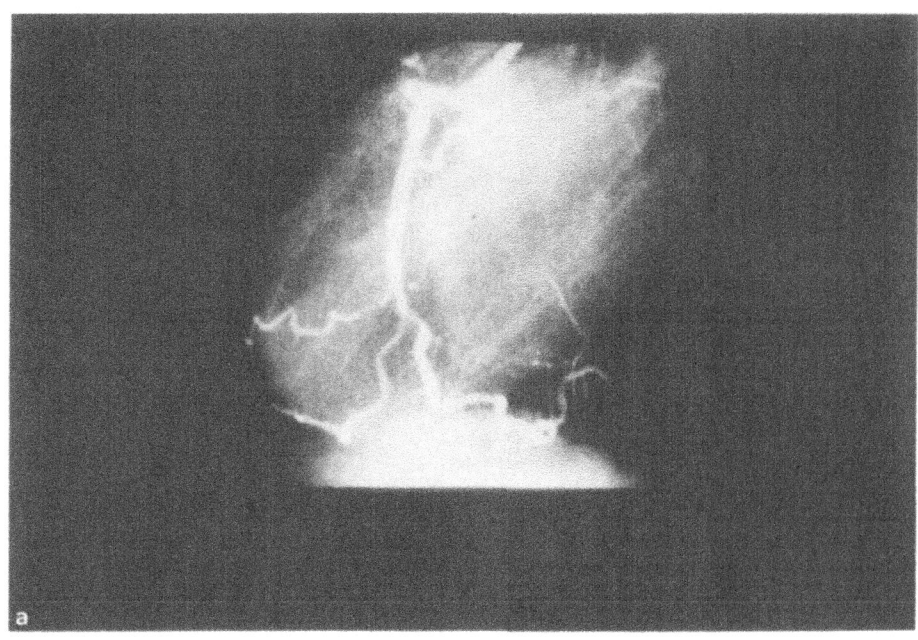

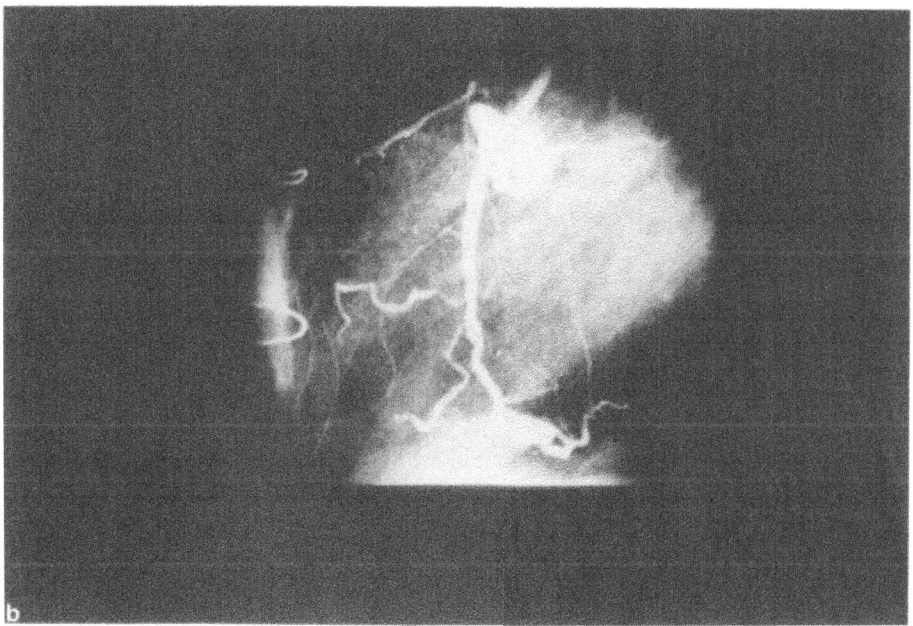

Fig.3.   (a) Prelaser selective injection of right coronary artery discloses a 75% stenosis
in the middle portion of the vessel. (b) Postoperative control with only a 20%
residual stenosis.

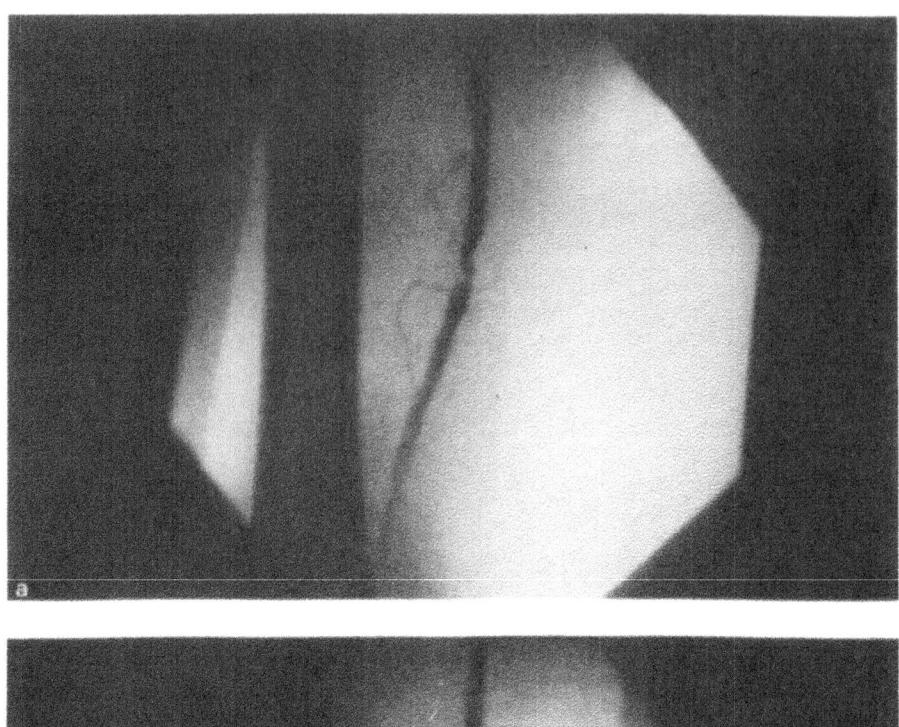
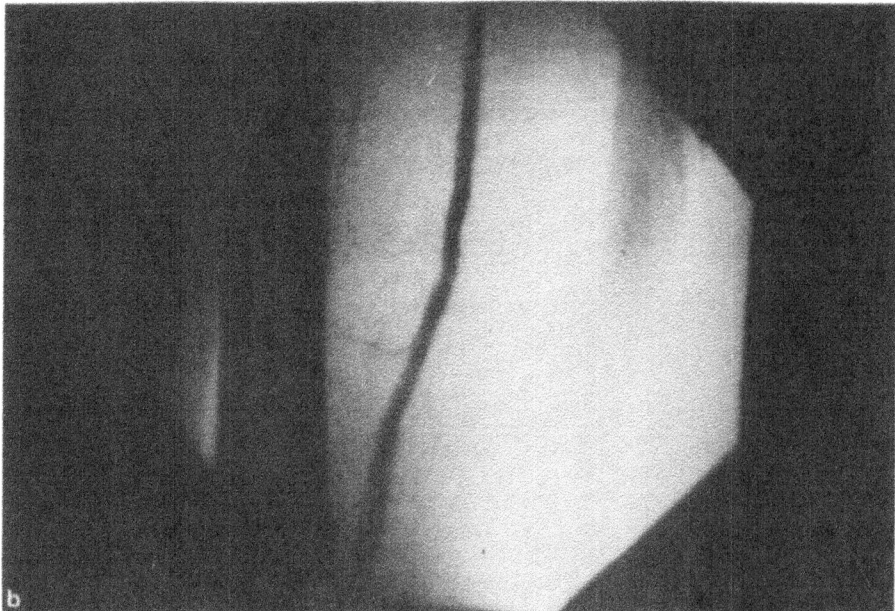

Fig.4.   (a) Baseline angiogram demonstrates critical stenosis of the superficial femoral artery. (b) Angiogram after UV laser angioplasty, without balloon dilatation, showing excellent lumen dimension.

Eleven patients affected by advanced ischemic pathology of the lower limbs were treated with a similar laser angioplasty procedure . Five iliac arteries and 16 femoro-popliteal segments, totally occluded, were lased with percutaneous or surgical approach and local anaesthesia. The catheter advancement was performed under intermittent fluoroscopic control; in a few cases angioscopy or high frequency echography was utilized.

The iliac obstructions were, with one exception, extending to the entire vessel; in three cases the superficial femoral artery was obstructed from its origin to the distal third; 13 femoral segments were occluded from the proximal third to the popliteal region. The iliac stenoses were treated in the operating room, in order to be able to proceed to immediate surgical intervention in the case of perforation.

Table 2. UV laser peripheral angioplasty: location of lesion.

Iliac artery	5	Femoral artery	16
long segment	4	proximal	3
short segment	1	distal	13

Table 3. UV laser peripheral angioplasty: results and complications

Improvement	18
Perforation	2
Inability to advance	1

In four cases of iliac pathology, after the laser procedure a conventional balloon angioplasty was performed to obtain a wider neolumen.

All the patients were angiographically controlled after 30 days. The results are the following: 1 localized popliteal perforation, 1 perforation of the femoral artery at the Hunter canal, 13 recanalizations with good neolumen, 1 failure to advance the catheter through a completely calcified obstruction.

COMMENT

From many clinical data it is now well demonstrated that laser irradiation is capable of achieving improved luminal patency in even the most severely narrowed and diffusely diseased arteries. Also, long segments of completely occluded peripheral vessels are recanalized with a high percentage of success utilizing laser energy with or without the adjunction of balloon angioplasty.

In the experimental studies comparing laser angioplasty with conventional balloon angioplasty, the laser procedure resulted in significantly fewer restenoses and a significantly larger luminal diameter than was the case with balloon dilatation.

The first clinical application of laser energy in the peripheral and coronary arterial tree was reported in 1983 by Ginsburg and Choy, with an argon laser. For some years light energy was produced by argon or Nd-YAG thermal lasers, with continuous emission and a different delivery system: the bare fibers were modified with terminal lenses or a metal cap, to reduce the incidence of perforation and to obtain a larger neolumen. But follow-up studies of the first patients treated with these techniques demonstrated a high percentage of early reocclusion, due principally to the thermal damage on the vessel wall, characterized by a proximal layer of carbonization with a peripheral zone of necrosis and vacuolization. The consequence of these alterations was an easier apposition of platelets, with secondary thrombotic occlusion.

The observation, from the first studies of Garrison and Srinivasan, that high power ultraviolet radiation is able to produce a "clean"ablation of tissue and no damage to surrounding cells, favored the experimental and clinical application of pulsed UV excimer lasers. The first clinical results obtained with this type of laser source were published in 1989 by Adler, Litvack and Grundfest, who treated 26 peripheral lesions and by Karsh and Litvack, who performed the first laser assisted coronary angioplasty, with a high percentage of success.

Our good results obtained in a small series of patients have demonstrated the efficacy of our excimer laser equipment coupled with a single or multiple fiber catheter, and the technical feasibility of nearly complete vaporization of atherosclerotic plaques in the coronary and peripheral tree without the adjunction of conventional balloon angioplasty.

However, the relatively large number of perforations, especially in the treatment of coronary artery obstructions, has prompted us to examine the characteristics of this complication, which is secondary not to the damage to the surrounding tissues, as in the case with the thermal laser, but to a direct mechanical or lasing action of the catheter not constantly centered in the lumen. Contributing factors to perforation are: eccentric plaques, tortuosity of the vessel, branch origin, plaque composition, but, more than anything else, catheter stiffness. Obviously the consequence of a perforation on the coronary wall can be dramatic, due to the ratio of vessel to catheter diameter and to the myocardial consequences secondary to acute coronary closure, even though the damage can be repaired.

From our preliminary experience it appears evident that the guiding system of the catheter along a tortuous vessel is far from perfect, even with the direct control of the surgeon in the operating room on a still heart.

Many devices are experimentally and clinically being tested to obtain an easier and more reliable manipulation of the optical fiber, also in view of a percutaneous utilization, with the heart beating. The most promising techniques are: angioscopy, high frequency two-dimensional echography, and laser-excited fluorescence spectroscopy. With the last technique, in which our group of physicists is actively engaged, it is possible, with different silica fibers grouped in the same catheter, to discriminate between atherosclerotic plaque and normal tissue. Other important progress in the delivery system is being obtained with new very flexible catheters with up to 19 thin optical fibers around a guide wire, to improve the coaxial alignment within the vessel lumen.

In conclusion, laser angioplasty, both in the peripheral and especially in the coronary artery occlusions, is gaining widespread acceptance as a complement to, and sometimes a replacement of, balloon angioplasty. In the specific case of multiple coronary stenoses, it could enable the surgeon to extend the revascularization procedures to previously almost forbidden vessels. Unfortunately the incidence of coronary perforations due mainly to calcific eccentric stenosing plaques remains a difficult problem to overcome, despite careful catheter handling and direct vision advancing. Immediate repair allows an uneventful course of the operation.

The technique is very promising, initial results quite interesting, but a generalized application seems to us certainly impossible at the present time.

REFERENCES

Abela G. S., 1988, Laser recanalization: a basic and clinical perspective, J. Thoracic Cardiovasc. Surg., 36:137.

Adler L., Litvack F., Grundfest W., 1988, Excimer laser-balloon angioplasty treatment of peripheral vascular occlusions and stenoses, Radiology, 169:140.

Choy D. S., Stertzer S., Rotterdam H. Z., Sharrock N., Kaminov I. P., 1982, transluminal laser catheter angioplasty, Am. J. Cardiol., 50:1206.

Cragg A. H., Gardiner G. A., Smith T. P., 1989, Vascular application of laser, Radiology, 172:925.

Fuller T. A., 1984, The characteristics in operation of surgical lasers, Surg. Clin. N. Am., 64:843.

Garrison B. J., Srinivasan R., 1984, Microscopic model for the ablative photodecomposition of polymers by far ultraviolet radiation (193 nm), Appl. Phys. Lett., 44:849.

Grundfest W. S., Litvack F., 1987, The current status of angioscopy and laser angioplasty, J. Vasc. Surg., 5:667.

Lee G., Ikeda R. M., 1985, Limitations, risks and complications of laser recanalization: a cautious approach warranted, Am. J. Cardiol., 56:181.

Samborn T. A., 1988, Laser angioplasty. What has been learned from experimental studies and clinical trials?, Circulation, 777:78.

Welch A. J., Torres J. H., Cheong W. F., 1989, Laser physics and laser-tissue interaction, Tex. Heart Inst. J., 16:141.

PART 5

FUTURE DIRECTIONS

INITIAL APPLICATIONS AND POTENTIAL OF MINIATURE
LASERS IN MEDICINE

R. Pratesi

Istituto di Elettronica Quantistica del CNR and
Dipartimento di Fisica dell'Universita'
Firenze - Italia

INTRODUCTION

The availability of near-infrared (NIR) and visible (VIS) coherent light-emitting diodes (CLEDs) and CLED-pumped solid-state lasers (Microlasers) of suitable output power will permit revolutionary changes and important progress in the diagnostic and therapeutic use of light. As foreseen[1-3], the application of CLEDs to Medicine is already expanding rapidly.

In this Section initial applications and potential of miniature lasers in the medical field will be briefly discussed. A short comment on the advantages and limitations of these lasers is also given.

MERITS AND LIMITATIONS OF MINIATURE LASERS

Diode lasers

CLEDs offer several important advantages over the traditional laser sources.

First of all, the extremely high electrical-to-optical conversion efficiency allows the extraction of high optical power from extremely small lasing volumes. The drastic reduction in the physical size of the laser sources leads to consequent miniaturization of the laser head and power supply. Higher optical efficiency also means the dissipation of less thermal power, and hence smaller sized cooling systems. Very compact laser sources are now available, which will permit the construction of new generations of optical systems for biomedical applications.

Cost reduction associated with mass production of CLEDs, typical for solid-state electronic components, represents a potential factor for increasing the diffusion of optical techniques in Medicine.

The structure of the laser head (i.e., semiconductor laser chip, cavity mirrors realized directly onto the end facets of the crystal; simple, efficient and compact temperature control by thermoelectric coolers; electronically controlled pump injection current) permits improved laser stability, reliability, and ruggedness. Close-coupling of CLEDs (or CLED bars) to optical fibers (or fiber bundles) directly assembled by the manufacturer means that endoscopic applications of these lasers are already possible.

The small physical size, electrical and cooling requirements which are trivial compared to most existing medical lasers, the lack of gas or dye reservoirs and the reduced maintenance, all combine to make portable surgery and therapy possible, together with the consequent extension of laser therapy to remote and/or third world countries.

The main limitation of present generation CLEDs is represented by the poor spatial quality of the laser beam. The peculiar geometry of the lasing structure gives rise to strongly divergent and astigmatic beams, with asymmetric power distribution. This makes collimation and focusing of CLED light difficult, with reduction of the usable power and the increase of system size. Low-cost microlenses with diffraction limited performance have been very recently developed for collimating light from high-divergence CLEDs.[4]

High-power CLEDs are still limited to the narrow 780-850 nm wavelength region. A growing effort is being devoted to extend laser wavelengths into the green-blue region and toward 3 μm, but commercial devices suitable for many medical applications will not become available for some time to come. Nonlinear frequency conversion of CLEDs and CLED-pumped optical parametric oscillators (OPOs) will cover in part the existing gap in the CLED emission spectrum.

Microlasers

The medical solid-state lasers will benefit from the all-solid-state structure of Microlasers, which permits better stability and reliability, reduced size and maintenance. To date Microlasers offer superior beam quality (TEM_{00} mode) and very narrow bandwidths with respect to CLEDs. Moreover, unlike semiconductor lasers, Microlasers permit the storage of large amounts of energy in the upper laser level, and thus short pulse operation with high energy content.

Quick developments have been achieved with Nd:YAG Microlasers, and several commercial models are already suitable for medical applications. High average power Nd:YAG Microlasers (~ 100 W) will become available in the near future: this will permit the complete replacement of flash-pumped Nd:YAG lasers in the medical field by the new models. Frequency doubling of these lasers will also offer enough average power to replace Ar lasers in many medical applications.

Only Nd:YAG Microlasers are sufficiently developed at present to offer a reliable alternative to the flashlamp systems. Considerable research is being undertaken to produce new and more perfect laser crystals for operation in the NIR (1-3 μm) at suitable output power. As for CLEDs, NIR Microlasers suitable for pumping with existing CLEDs will not become available for several years. Nonlinear frequency techniques, including OPO, will compensate in part for the lack of direct laser lines from Microlasers.

APPLICATIONS TO THERAPY

Ophthalmology

Photocoagulation of the retina is currently carried out by using the blue-green or red lines of ion lasers: transmission of these wavelengths through the eye is high, and absorption by melanin and hemoglobin is large enough to permit coagulation with short exposure times. For treatments in the macular region, wavelengths on the tail of the absorption spectrum of xanthophyll (peak at ~ 460 nm) are preferred in order to avoid undesirable heating of the delicate structure containing this pigment. The red line of krypton ion laser at 647 nm is useful in this respect.

The photomedical treatment of some pathologies of the anterior segment of the eye (posterior capsolotomy, iridotomy, synechiotomy, vitreolysis) employs the Nd:YAG laser operated in "giant-pulse" (Q-switched) mode. The transmission of its 1.06 μm radiation by the cornea and lens is still sufficiently high to ensure a safe procedure. The absorption by melanin and the large penetration depth in tissue at 1.06 μm are utilized to obtain chorioretinal photocoagulations with contact transcleral procedure, using the pulsed or cw mode operating Nd:YAG laser connected to an optical fiber.

Recently, new potential applications of laser techniques in ophthalmology have been reported. In particular, photoablation of corneal tissues has been demonstrated by using UV excimer lasers. The extremely small penetration depth in the cornea of a 3 μm radiation has also been utilized to demonstrate the ablative capability of IR lasers, such as HF and Er:YAG lasers emitting at 2.9 and 2.94 μm, respectively.

Of all medical applications, ophthalmology needs the lowest intensity (cw) for therapeutic treatments, and has therefore been the first to benefit from the availability of CLEDs.[1] Puliafito[5] and Brancato[3,6] reported the first successful retinal coagulations with cw CLEDs at ~ 800 nm with endoscopic and transpupillary approaches, respectively. Retinal coagulation with pulsed CLEDs at 950 nm have also been demonstrated.[7] The introduction of CLED sources in ophthalmology has been welcomed as the beginning of a new era for retinal photocoagulation.[8] Following these initial studies, detailed animal and clinical investigations have been performed by several groups in the UK, USA, Italy, and Japan.[9-17] Several endoscopic and slit-lamp transcleral photocoagulators are now commercially available.

The initial clinical experience and the limitations of present CLED photocoagulators are discussed elsewhere in this Volume. We only recall here that the poor beam quality of present generation, high-power CLEDs and the limited power (~ 2 W cw) emitted by the narrow active area CLEDs or CLED arrays available on the market, do not permit the efficient delivery of CLED light to the target. Hence, multiple-CLED systems have been introduced to offer a suitably extended range of usable power for photocoagulation.

For the above reasons, single-spatial mode, frequency doubled (cw) Nd:YAG Microlasers are likely to become the near future candidates for replacing Ar lasers in eye photocoagulation. Microlasers emitting several watts at 532 nm are already present on the market.

Future applications of miniature lasers in ophthalmology will regard other consolidated therapies and presently investigated experimental procedures, such as the photodisruptive treatments of transparent ocular structures, the PDT of ocular tumors, photoablative treatments of the cornea, etc..

High-peak power from mode-locked CLEDs has been reported. Q-switched and/or mode-locked CLED-pumped Nd:YAG lasers will represent an improvement with respect to the flash-pumped Nd:YAG lasers in terms of reduction of size and increased efficiency and reliability. Q-switched Nd:YAG/YLF Microlasers with energy and peak power outputs suitable for photodisruptive applications are already commercially available. Their use for medical investigation is underway.

Red emitting CLEDs are particularly promising in conjunction with the new generation of red-absorbing dyes (such as phthalocyanines and cryptocyanines) for photosensitized tumor therapy.

Pulsed NIR lasers emitting near the peak of the liquid water absorption at 2.9 μm are of interest as possible sources for laser surgery of the cornea. The reported success in controlling corneal ablation with Er:YAG lasers[18] indicates a potential application for future CLEDs emitting at ~ 3 μm and/or Er:YAG Microlasers.

Tumor therapy

i) Endoscopic photocoagulation. Nd:YAG Microlasers with cw/average power approaching 20 W are being developed by several companies and will become available commercially by the end of 1991. These models may be of clinical interest for applications, in particular endoscopic photocoagulation, where the standard 100 W Nd:YAG lasers are used at reduced power levels. With the current effort in the development of CLEDs and Microlasers, and on the basis of the rate of progress observed during the last two years, we expect that Nd:YAG Microlasers will replace the standard models within 1 to 2 years.

High-power CLED bars coupled to fiber bundles are already being developed for remote pumping of solid-state lasers. With CLED bars 1 cm long about 100 W cw has been reported so far at laboratory level. Coupling to fiber bundles with 1 to 2 mm diameter is feasible. Thus, fiber-coupled-bars should become available soon with output power in the 50-100 W range. These lasers are potentially interesting for endoscopic photocoagulation. However, the large penetration depth of the 800-900 nm radiation may represent a limit for the treatments already consolidated with Nd:YAG lasers. It may, on the other hand, in conjunction with appropriate NIR absorbing chromophores, open up new applications for selective photocoagulation.

ii) Long-wavelength Photodynamic Therapy. Photodynamic cancer therapy (PDT) is still receiving increasing attention in order to improve our understanding of the basic mechanisms, the useful range of photosensitizers, and the technology of the clinical systems.

Today, most PDT is performed using dye lasers, which are not ideal clinical instruments, and a derivate of haematoporphyrin (HpD) which has a not yet fully known composition. Moreover, porphyrins absorb poorly in the "therapeutic window", where deep penetration of red and NIR light occurs. The ongoing development of tunable solid state and diode lasers will be important for the further progress of PDT, especially in the use of second generation photosensitizers (such as phthalocyanines and cryptocyanines). Phthalocyanines are characterized by strong light absorption in the red region of the spectrum and good photosensitizing activity. These dyes are accumulated and retained by malignant tissues after systemic administration in experimental animals, and induce tumour necrosis following red light irradiation. Naphthalocyanines absorb strongly in the 740-780 nm spectral range where high power CLEDs are available at output power of several watts cw. Preliminary research is underway with a 780 nm CLED and Si(IV)-naphthalocyanine.[19]

Thus, if the present research effort on these dyes leads to useful clinical results, laser PDT will no longer suffer from the severe limitation imposed on its diffusion by the high cost of suitable laser sources.

iii) Laser-induced hyperthermia (LIHT). Tumor therapy by laser-induced hyperthermia (LIHT) has been investigated for several years with encouraging results. In most cases Nd:YAG lasers have been used, in particular with interstitial procedure.

As in most medical applications, selectivity is a crucial parameter for an optimized use of laser radiation. In PDT of tumors good selectivity is ensured by the larger accumulation and retention of exogenous photosensitizers in the tumor tissue with respect to the surrounding healthy tissue (see preceding chapters in this Volume). In LIHT an analogous selectivity could be given by suitable chromophores to be topically or systemically administered to the patients. The spectral region corresponding to the "phototherapeutic window", i.e. 600-1000 nm, is preferred due to the deeper penetration of light into the tissues. Red and NIR CLEDs are therefore ideal candidates for these applications, in conjunction with non-photosensitizing dyes such as some metallo derivatives of porphyrins and porphyrinoid compounds, azo dyes and triphenylmethano derivatives, with significant absorption at wavelengths > 600 nm.[20]

Photosurgery

i) <u>Contact laser surgery</u>. The contact technique for tissue coagulation and vaporization was introduced by Joffe and coworkers.[21] A synthetic sapphire probe is attached to the distal end of a quartz fiber connected to a Nd:YAG laser, and allows contact laser surgery. The mechanism of contact surgery is based on a combined action of thermal and light coagulation produced respectively by the high temperature of the laser-heated tip and by the laser light transmitted through the sapphire tip. Ceramic, opaque, materials have also been used recently.[22]

Probes with different geometric shapes deliver a variety of beam patterns that allow coagulation, vaporization, and cutting of tissue at much lower power requirements than does the conventional noncontact method. Clinical results demonstrate that endoscopic and open surgery applications can be performed at laser powers ranging from 10 to 20 W.

The contact technique can take advantage of present developments of high-power CLEDs and Microlasers, as already reported in Ref.22.

ii) <u>Ablative surgery</u>. NIR lasers (1.3-3.0 μm) are presently being evaluated comparatively for photoablative surgery in many applications in view of a potential replacement of UV excimer lasers.[23,24]

Substantial energy and power doses are required for tissue ablation, and are beyond the capability of present generation Microlasers. However, efficient lasing has been reported for several laser crystals with Tm, Ho, and Er doping with CLED pumping (or with Ti:sapphire laser simulation of CLED pumping), thus suggesting a medium-term future availability of Microlasers for ablative microsurgery.

iii) <u>Microsurgery</u> : <u>Tissue welding</u>. Laser tissue "welding" techniques have been considered as potentially useful in microsurgery to replace standard procedures for microvascular anastomosis. Several types of lasers, Ar, Nd:YAG, CO_2, have been used so far. The low cw power level required for this technique (0.5-2 W, cw) makes CLED sources suitable for tissue soldering experiments.

Recently, the first application of an 808 nm CLED has been reported in the case of welding of aortotomies/abdominal aorta cuts in rabbits.[25,26] Since laser emission from AlAsGa CLED occurs in the "optical window", where tissue absorption is very low, direct welding of the vascular wall was not possible even at the maximum power density available (~ 10 W/cm^2). Indio-cyanine green (ICG) dye, which maximally absorbs at ~ 790 nm, was therefore topically applied to the target tissue to enhance laser power uptake leading to positive results with power densities of ~ 5 W/cm^2.

Enhancement of red and NIR laser absorption of CLED radiation by exogenous pigments represents an interesting technique for obtaining selective effects,[27] with minimal damage to adjacent tissues, and has already been used for coagulation of the retinal vessels.[5]

The contact method has also been experimented in laser assisted microvascular anastomoses by CLED heating a ceramic tip with 200 μm diameter output end.[22] Only 60 mW of laser power at 830 nm was necessary to perform end-to-end anastomoses with short time (3-5 s) exposures.

Dermatology

The availability of high-power CLEDs emitting at 800 nm should be of interest for selective treatment of several vascular disorders in combination with enhanced blood absorption by exogenous dyes. The use of ICG was already considered for this aim, but disregarded due to the fast elimination of ICG by the organism.[28,29] Other dyes and/or appropriate clinical procedures could be found to make this technique possible.

NIR Microlasers (Nd:YAG at 1.06, 1.32 μm; Ho:YAG at 2.1 μm; Er:YAG at 2.94 μm) have potential applications in dermatology. Q-switched Nd:YAG Microlasers should be evaluated for selective removal of tattoos as a further extension of pulsed laser technique on the basis of the positive results reported with ruby lasers.[30]

Low-power laser therapy ("biostimulation")

The use of laser radiation at very low power density for inducing alterations in cell and tissue functions and beneficial effects in humans is still a controversial and intriguing problem.[28]

When controlled studies have proved the clinical efficacy of low-power irradiation in some selected pathologies and determined its action spectrum, low power CLEds (or even LEDs, if available with sufficient power outputs in the useful spectral range, as for red and NIR LEDs) will be ideal sources for this application.

OPTICAL FIBERS FOR LASER SURGERY

Two important criteria which determine the effectiveness of a laser source in medical applications are the absorption coefficient of the tissue specimen at the operating wavelength and the availability of a suitable fiber optic delivery system.

With regard to transmission, flexibility, durability, and nontoxicity, the best optical fibers for medical applications are silica based fibers. These fibers have demonstrated excellent performance in the wavelength region between 0.3 and 2.1 μm. Much effort is therefore being dedicated to the development of solid-state lasers with emission wavelengths in the vicinity of 2 μm. At the same time, new optical fibers capable of extending the use of NIR lasers to 3 μm are being developed. Promising results have been reported for single-crystal sapphire fibers.[32]

Sapphire (α-Al_2O_3) has long been recognized as a good material for optical components due to its transparency range (240-4000 nm), high melting temperature, and favorable mechanical and chemical properties.

Sapphire single-crystal fibers with diameter of 110 μm and lengths over 2 m have been grown. The measured absorption loss at 2936 nm is 0.88 dB/m with a damage threshold higher than 1.2 kJ/cm^2 for 110 μs long pulses. Ablation of post-mortem arterial tissue has been demonstrated with incident fluences at the fiber surface well below the damage threshold.[32]

The sapphire fibers are well suited for medical applications due to their nontoxicity, chemical inertness, and mechanical strength.

PHOTOTHERAPY WITH LIGHT-EMITTING DIODES (LEDs)

LEDs are very efficient, extremely compact sources of narrow-band visible and IR radiation. High-efficiency visible LEDs are being increasingly developed due to requirements for a variety of optical applications. Usual peak radiation wavelengths are in the green, yellow, and red. The development of high intensity LEDs at other wavelengths is only a technological and economic problem. LEDs are potential sources for light therapy of various pathologies.

Coherence or incoherence

The medical applications of lasers do not use in a direct way any of the coherence properties of laser light. Spatial coherence is important to obtain highly collimated beams, which in turn allow the concentration of almost all the power emitted by the source into very small focal spots, and the efficient coupling to optical fibers. Therapeutic effects connected to the high degree of spatial and temporal correlation of laser light have not been reported to date. Moreover, the possibility of taking advantage of potential biological effects depending in a direct way on the temporal and spatial coherence of laser light is reduced by the severe scattering properties of living tissues, which degrade quickly the coherence of laser light as it propagates through them.

Therefore, when illumination of large external areas of the body is needed, incoherent sources may be conveniently used in place of lasers, if enough useful power is available.

At present, there are two important therapeutic applications of incoherent light: 1) photochemotherapy of psoriasis, the so-called PUVA-therapy (see the relative papers on this topic in this Volume), and 2) phototherapy of hyperbilirubinemia of the neonate (see the relative paper on this topic in this Volume). Fluorescent lamps are prevalently used as light sources in both cases.

Phototherapy of jaundice

Fluorescent lamps currently used for phototherapy have good electrical to optical efficiency and are cheap. Their replacement by more advanced (and hence more expensive) light sources requires important improvements in the clinical protocol, and, in particular, the wish to avoid the presence of potentially hazardous, not therapeutically useful, spectral components. In fluorescent lamps the high intensity Hg-lines are always present together with the continuous fluorescent emission of the phosphor. The glass envelope of the lamp cuts off the residual UV-C and UV-B lines, while the plastic sheet of the illuminator and/or incubator block all wavelengths below 380 nm. However, the intense 405 and 436 nm Hg-lines cannot be easily filtered out without strongly reducing the light intensity within the useful phototherapy spectral region. Moreover, aging of the lamps may produce the consumption of the phosphor, with increased spill-over of these lines.

Broad-band tungsten-halogen lamps are also employed in phototherapy. They require filters to select the useful spectral band, so as to avoid loading of the infant with heating and toxic radiation.

The future design of an efficient phototherapy unit will very likely be based on the use of solid-state lamps.[33]

A blue LED with peak wavelength at 480 nm is already available at very low radiation power (Siemens SFH-710). Its spectral characteristics appear suitable for efficient phototherapy. In fact, a combined effect of the spectral dependence of bilirubin absorption, light transmission through skin tissues, and quantum yields for photoisomer production, indicate that more efficient excretion of bilirubin by the organism should occur with light irradiation in the 480-510 nm spectral interval.[34]

Recently, a phototherapy blanket, which incorporates light diffusing plastic fibers, has been introduced to improve the delivery of therapeutic radiation to the infant skin.[35] Again, the all-solid-state approach is appealing since it provides a more efficient, practical, and safer therapeutic procedure.

Phototherapy of tumors

High-efficiency red LEDs emitting in excess of 10 mW at 660 nm are available at low cost. Still greater power is available in the far red-NIR region. Multi-LED or integrated LED-array panels could provide the suitable irradiation pattern and irradiance for photodynamic and/or hyperthermic therapy of superficial tumours.

Coupling of single LED chip to individual optical fibers represents a realistic possibility for multiple fiber implantation into tumor masses; fiber bundles may also permit the delivery of enough power to internal tumors with standard endoscopes.

COMMERCIAL SITUATION

The situation at a commercial level regarding CLEDs and Microlasers is evolving rapidly. The number of manufacturers has increased substantially, and higher performance models are being introduced onto the market.

In this Section a very concise indication of the characteristics of Microlasers presently available is given with specific regard both to those models which are of potential interest for medical applications and to those which are already being used. Tables 1-3 show a selection of Microlasers and CLEDs which offer the highest performances in this regard.

Although end-pumped designs are still the most diffused geometry, side-pumped configurations are now available for higher average power output.

CW Nd:YAG Microlasers are available at several output powers for both TEM_{00} single transverse mode and multimode (MM) operation. The maximum output power is 3 W MM and 2 W TEM_{00} at 1064 nm. Aircooling is used. This figure is expected to move to 15-20 W for the TEM_{00} model (liquid-cooled) in the near future.

Single longitudinal mode Nd:YAG Microlasers have maximum power output of 100 mW at 1064 nm and 50 mW at 1320 nm, respectively. These lasers have linewidth of the order of 5 to 25 kHz and short term frequency jitter of less than 25 to 500 kHz. Microchip lasers have also become available commercially with output powers of 500 mW at 1064 nm and 250 mW at 1319 nm, respectively. The linewidth is extremely narrow (< 3 kHz) and the frequency stability very high. These microchip lasers can be pulsed (Q-switched or modelocked) and frequency doubled with specific options given by the manufacturer.

TABLE 1. Nd:YAG CW Micro Laser

Output wavelength (nm)	1064	1064	1064	1064	1319	1064	1320	1064/1319	1064/1319
Output power (mW)	3000	2000	150	300	150	350	175	500/250	50/25
Beam diameter (mm)	1	1	0.2	----	----	0.5	0.5	----	----
Beam divergence (mrad)	1	1	7	----	----	4	5	----	----
Spatial mode	MM	TEM_{00}	TEM_{00}	TEM_{00}	TEM_{00}	TEM_{00}	TEM_{00}	TEM_{00}	TEM_{00}
Pumping geometry	T	T	L	L	L	L	L	L	L
Cooling system	Air/Water	Air/Water	----	----	----	----	----	----	----
Linewidth (kHz)	----	----	<10 (50 ms)	5	5	----	----	<3	<3
Amplitude stability (peak to peak)	----	----	± 2%	----	----	± 2%	± 2%	----	----
Frequency scanning range (between modehoops) (GHz)	----	----	9	8	8	----	----	----	----
Frequency scanning rate	----	----	>10 MHz/s	1 MHz/V	1 MHz/V	----	----	----	----
Total frequency scanning (GHz)	----	----	----	40	40	----	----	20-200	----
Frequency stability	----	----	----	----	----	----	----	<200 kHz	$1:10^9$
Model	LPD3000M	LDP2000	PRO1000-500	NPRO 122-1064-300F	NPRO 122-1319-150F	1064-350P	1320-175P	Microchip-03	Microchip-02
Manufacturer	Laser Diode Products	Laser Diode Products	Electro-Optics	Lightwave Electronics	Lightwave Electronics	Amoco Laser Co.	Amoco Laser Co.	Micracor	Micracor

MM : Multi-mode ; T : transverse pumping ; L : longitudinal pumping

279

TABLE 2. Pulsed Nd:YAG/YLF Microlaser

Laser material	Nd:YAG	Nd:YAG	Nd:YLF	Nd:YLF	Nd:YAG	Nd:YAG
Output wavelength (nm)	1064 h	1064	1047	1047	532	532
Output energy/pulse	20 mJ	10 mJ	----	50 mJ	1 mJ	3 mJ
Pulse width	250 μs	40 ns	<20 ps	<7 ns	200 μs	35 ns
Average output power	----	----	<200 mW	----	----	----
Repetition rate	1-10 Hz	1-10 Hz	250 MHz	0-50 kHz	1-10 Hz	1-10 Hz
Pulse option	gain switched	EO Q-switch	AO Modelocker	AO Q-switch	gain switched	EO Q-switch
Beam diameter/dimensions	2 x 2 mm	2 x 2 mm	<410 μm	170 μm	2 x 2 mm	2 x 2 mm
Beam divergence (mrad)	3	3	1.6	7.8	3	3
Spatial mode	TEM$_{00}$	TEM$_{00}$	TEM$_{00}$ (dl)	TEM$_{00}$ (dl)	TEM$_{00}$	TEM$_{00}$
Stability (rms) Amplitude	---	----	1%	1%	---	---
Energy	---	----	---	2%	---	---
Model	LDP 150	*LDP 150	130-1047-200	110-04	LDP 150 D	*LDP 150D
Manufacturer	Laser Diode Products	Laser Diode Products	Lightwave Electronics Inc.	Lightwave Electronics Inc.	Laser Diode Products Inc.	Laser Diode Products Inc.

TABLE 3. CW and pulsed GaAlAs diode lasers

Output wavelength (nm)	820-860	820-860	770-840	820-860	790-830	----	----
Output power (mW)	100	1000	1000	3000	3000	10000	----
Total conversion efficiency (%)	30	26	----	30	----	25	>30
Emitting dimension, w x h (μm)	3 x 1	200 x 1	200 x 1	500 x 1	600 x 1	10000 x 1	200 x 1
Beam divergence, $\theta_\perp$, $\theta_{\parallel}$ (deg)	30,10	40,10	28,13	40,10	28,15	40,10	25,10
Spatial mode	TEM_{00} (dl)	----	----	----	----	----	----
Linewidth	<10 MHz	----	----	----	----	----	----
Quasi-cw, peak output power (W)	----	----	----	----	----	----	2
Energy per pulse (J)	----	----	----	----	----	----	1
Pulse width (maximum) (ms)	----	----	----	----	----	----	500
Duty cycle (maximum) (%)	----	----	----	----	----	----	20
Model	SDL-5410	SDL-2460	SLD-304	SDL-2480	SLD-305	SDL-3490S	SDL-2270
Manufacturer	Spectra Diode	Spectra Diode	Sony	Spectra Diode	Sony	Spectra Diode	Spectra Diode

Pulsed Microlasers are available with output energy up to 20 mJ/pulse in long pulse operation (250 μs) at low repetition rate (1-10 Hz). Q-switched models can supply output energy up to 10 mJ/pulse with 40 ns pulse duration at 1-10 Hz repetition rate.

Manufacturers are moving to Q-switched models with 1 to 2 mJ at 1 kHz and 6 to 7 W average power at repetition rate of 30-50 kHz. Short (5 ns) giant pulses with 100 μJ/pulse are available. Q-switched lasers remain among the highest-priced lasers, yet are already competitive with equally energetic flashlamp pumped lasers.

Commercial modelocked models of Nd:YLF lasers generate 100 mW average power with 10 and 70 ps pulse duration (without compression) at 250 and 100 MHz repetition rate, respectively.

Narrow linewidth operation is achieved with ring lasers : one model generates 300 mW at 1064 nm (extended to 1 W in 1991) and 150 mW at 1319 nm with bandwidth of 5 kHz and total frequency scanning of 40 GHz. Microchip lasers with bandwidths of less than 3 kHz and total frequency scanning range of 20-200 Ghz are offered. Future developments of Nd:YAG Microlasers with nonplanar ring resonator (NPR) geometry are expected to lead to 15 W cw operation with subhertz bandwidth.

Frequency doubled Nd:YAG Microlasers are available with 140 mW power, with scheduled extension to 1-2 W TEM_{00} in 1991.

APPLICATIONS TO DIAGNOSIS

Optoelectronics is finding growing interest in medical diagnostics and biological research. The advantages offered by diode lasers and diode-laser-pumped solid-state lasers will introduce these miniaturized sources into an increasing variety of systems for many laboratory and clinical applications, permitting more compact, reliable, and cheaper devices.

A few examples of this straightforward evolution in the field of medical diagnostics is presented in the following Sections.

Reflectance spectroscopy

The progress of this technique for noninvasive analysis of tissue pigments, oxygen content, etc., is described elsewhere in this Volume. Multiple modulated CLEDs at discrete wavelengths can be efficiently used to perform time-dependent (frequency domain) diffuse reflectance spectroscopy *in vivo*. Since both the light sources and detection system are relatively simple and inexpensive, clinical applications appear feasible and affordable.

Doppler velocimetry

Light scattering techniques are receiving increasing attention in view of the variety of diagnostic possibilities in many fields of Medicine and Biology. The state-of-the-art is briefly discussed elsewhere in this Volume.

Thus far, accurate and remote measurement Doppler systems have made use of gas laser technology (He-Ne and CO_2 lasers). Commercial systems for monitoring and imaging blood flow by coherent light scattering employ low power He-Ne lasers. Only recently have CLEDs been used for Doppler measurements. Most of this work was restricted to the use of 800-900 CLEDs.

Mocker et al.[36] reported the achievement of an eye-safe Doppler velocimeter at 1.54 μm using InGaAsP CLEDs reduced in linewidths by external cavity stabilization. Linewidth of the CLEDs was ~ 40 kHz allowing for a sub-cm/s velocity resolution. Further miniaturization will be possible by using integrated passive cavity lasers[37] or fiber extended cavities.[38]

Transillumination imaging

This technique is evaluated as an interesting procedure for non-invasive diagnosis of breast cancer. Future development of transillumination imaging will implement a time of flight gating technique for improving spatial resolution of the imaging system. This type of system would require sufficiently powerful sources and sensitive detectors at appropriate wavelengths. Laser diodes are a low cost light source that can be operated to produce pulses of appropriate peak power. However, present high-power CLEDs may not emit optimal wavelengths for diagnostic imaging, and further development is needed to cover a wider spectral range.

NIR angiography

As reported elsewhere in this Volume, NIR fluorescence angiography is a potential technique for visualizing vasculature beds which are accessible by direct and/or endoscopic vision.

Diode lasers with emission wavelengths matching the absorption peak of ICG at ~ 790 nm are allowing improved performance of present fundus camera for viewing choroidal circulation, and of several scopes for evaluating internal organ circulation.

Laser-induced-fluorescence (LIF) diagnosis

Tissue diagnostics based on fluorescence laser spectroscopy is emerging as an interesting complement to other modalities. Signals from tumor-seeking agents as well as the natural tissue fluorescence (autofluorescence) can be utilized for demarcating diseased tissue from surrounding unaffected areas. LIF may also have clinical use as a feedback system for laser lithotripsy procedures.

UV or violet Hg lines and Kr laser line are used to excite the fluorescence of endogenous and/or exogenous chromophores. Frequency doubled CLEDs and frequency tripled Nd:YAG Microlasers (39 mW at 429 nm[39], 180 mW at 349 nm[40], respectively, at laboratory level) will be introduced onto the market in the near future, and will play an important role in the development of more advanced diagnostic systems.

Optical trapping of cells

As an example of the very many applications of lasers in Biology, we cite the recent use of CLEDs for optical trapping techniques.

Noncontact optical trapping of single biological cells is currently receiving considerable interest. This new technique is expected to play an important role in experimental cell biology and immunology as well as in other areas of biotechnology and bioscience.[41]

The first nondestructive, reliable, and versatile single-beam optical trapping of biological cells by means of 0.83 and 1.33 μm high power CLEDs has been reported by Shunichi et al.[42] NIR high power CLEDs offer the advantage of low absorption in NIR wavelengths for preventing optically induced damage.

ACKNOWLEDGMENTS

Work supported by the Special CNR-Project "Tecnologie Elettroottiche".

REFERENCES

1.   R. Pratesi, Diode lasers in photomedicine, IEEE J.Quant.Electr. 20:1433 (1984).
2.   R. Pratesi, Semiconductor (diode) lasers: basic principles and potential applications
     in the biomedical field Photobiochem.Photobiophys., Suppl., 57-75 (1987).
3.   R. Brancato and R. Pratesi, Applications of diode lasers in ophthalmology, Lasers
     Ophthalmol. 1:119 (1987).
4.   Laser Focus World 27:13 (1991)
5.   C.A. Puliafito, T.F. Deutsch, J. Boll, and K. To, Semiconductor laser endophoto-
     coagulation of the retina, Arch. Ophthalmol. 105:424-427 (1987).
6.   R. Brancato, R. Pratesi, G. Leoni, G. Trabucchi, L. Giovannoni, and U. Vanni,
     Retinal photocoagulation with diode laser operating from a slit lamp microscope,
     Lasers Light Ohphtalmol. 2:73-78 (1988).
7.   R. Brancato, R. Pratesi, L. Giovannoni, and U. Vanni, New lasers for ophthalmolo-
     ology: retinal photocoagulation with pulsed diode lasers, SPIE vol. 701, 365-366
     (1987).
8.   A. Raven, R.M. Lee, and C.R. Keeler, Diode lasers: a new era in retinal photocoagu-
     lation.I. Basic principles and instrumentation, Second International Congress
     on Laser Technology in Ophthalmology, Lugano, May 24-27 (1989).
9.   R. Brancato, R. Pratesi, G. Leoni, G. Trabucchi, and U. Vanni. Semiconductor diode
     laser photocoagulation of human malignant melanoma, Am. J. Ophthamol. 107:
     295-296 (1989).
10. J.D.A. McHugh, J. Marshall, M. Capon, S. Rothery, A. Raven, and R.P. Naylor,
     Transpupillary retinal photocoagulation in the eyes of rabbit and human using
     a diode laser, Lasers Light Ohphtalmol. 2:125-143 (1988).
11. J.D.A. McHugh, J. Marshall, T.J. Ffytche, A.M. Hamilton, A. Raven, and C.R.
     Keeler, Initial clinical experience using a diode laser in the treatment of retinal
     vascular disease, Eye 3:516-5127 (1989).
12. J.D.A. McHugh, J. Marshall, T. Ffytche, A. Hamilton, and A. Raven, Clinical and
     histopathological results of transpupillary diode laser retinal photocoagulation,
     Invest. Ophthalmol. Vis. Sci. (suppl.) 30: 371 (1989).
13. J. Duker, J. Federman, H. Schubert, and C. Talbot, Semiconductor diode laser
     endophotocoagulation of the retina, Ophthalmic Surgery 20: 717-719 (1989).
14. S. Okamoto, H. Takahashi, Y. Fukado, and T. Ozawa, Laser diode application for
     transscleral photocoagulation, Lasers and light in Ophthalmol. 3: 29-37 (1990).
15. J.D.A. McHugh, J. Marshall, T.J. Ffytche, A.M. Hamilton, and A. Raven, Macular
     photocoagulation of human retina with a diode laser: a comparative
     histopathological study, Lasers Light Ohphtalmol. 3:11-28 (1990).
16. J.S. Schuman, J.J. Jacobson, C.A. Puliafito, R.J. Noecker, and W.T. Reidy,
     Experimental use of semiconductor diode laser in contact transscleral
     cyclophotocoagulation in rabbits, Archives of Ophthalmology 108:1152-1157
     (1990).
17. J.S. Schuman, C.A. Puliafito, and J.J. Jacobson, Semiconductor diode laser
     peripheral iridotomy, Archives of Ophthalmology 108:1207-1208 (1990).
18. Q. Ren, K.T. Schoemaker, T.F. Deutsch, and R. Birngruber, Comparative study of
     erbium laser cutting of corneal tissue, Digest of Technical Papers, CLEO 90,
     paper CTUH90 (May 1990), p.166.
19. G. Jori, private communication (1990).
20. G. Jori, and J.D. Spikes, Photothermal sensitizers: possible use in tumor therapy
     J.Photochem.Photobiol.(B),Biology 6:93-101 (1990).

21. N. Daikuzono, and S.N. Joffe, Artificial sapphire probe for contact photocoagulation and tissue vaporization with the Nd:YAG, Med. Instrum. 19:173-178 (1985).

22. N. Unno, S. Sakaguchi, and K. Koyano, Microvascular anastomosis using a new diode laser system with a contact probe, Lasers Surg. Med. 9:160-168 (1989).

23. S. Mordo, F.X. Roux, S. Mondragon, C. Fallet-Bianco, F. Sahafi, and J.M. Brunetaud, A comparative study of coagulation effects on the cortex of the rat using Nd:YAG (1.32 μm), Nd:YAG (1.06 μm) and CO_2 lasers, Lasers Med. Scie. 5:293-296 (1990).

24. A. Chalton, M.R. Dickinson, T.A. King, and A.J. Freemont, Erbium:YAG and Holmium:YAG laser ablation of bone, Lasers Med. Sci. 5:365-373 (1990).

25. M.C. Oz, R.S. Chuck, J.P. Johnson, S. Parangi, L.S. Bass, R. Nowygrod, M.R. Treat Indocyanine green dye-enhanced welding with a diode laser, Surg. Forum 40:316-318 (1989).

26. M.C. Oz, J.P. Johnson, S. Parangi, R.S. Chuck, C.C. Marboe, L.S. Bass, R. Nowygrod, and M.R. Treat, Tissue soldering by use of indocyanine green dye-enhanced fibrinogen with the near infrared diode laser, J. Vascular Surgery 11:718-725 (1990).

27. R.S. Chuck, M.C. Oz, T.M. Delohery, J.P. Johnson, L.S. Bass, R. Nowygrod, and M.R. Treat, Dye-enhanced laser tissue welding. Lasers Surg. Med. 9:471-477 (1989).

28. R. Pratesi, unpublished (1987).

29. R.R. Anderson, personal communication (1987).

30. C.R. Taylor, R.W. Gange, J.S. Dover, T.J. Flotte, E. Gonzales, N. Michaud, and R.R. Anderson, Treatment of tattoos by Q-switched ruby laser, Arch. Dermat. 126:893-899 (1990).

31. J.R. Basford, Low-level laser therapy: controversies and new research findings Lasers Surg. Med. 9:1-5 (1989).

32. D.H. Jundt, M.M. Fejer, and R.L. Byer, Characterization of single-crystal sapphire fibers for optical power delivery systems, Appl.Phys.Lett. 55:2170-2172 (1989).

33. R. Pratesi, and M. Scalvini, A solid-state-lamp approach to phototherapy, Med. Biol. Environ. 11:467-468 (1983).

34. R. Pratesi, G. Agati, and F. Fusi, Phototherapy for neonatal hyperbilirubinemia, Photodermatology 6:244-257 (1989).

35. T. R. C. Sisson, personal communication, 1987.

36. H.W. Mocker, and P.E. Bjork, High accuracy laser Doppler velocimeter using stable long-wavelength semiconductor lasers, Appl.Opt. 28:4914-4919 (1989).

37. K. Matsuda, T. Fujita, J. Ohya, M. Ishino, H. Sato, H. Serizawa, and J. Shibata, Single longitudinal mode operations of long, integrated passive cavity GaAsP lasers, Appl.Phys.Lett. 46:1028-1030 (1985).

38. K.Y. Liou, R.T. Ku, T.M. Shen, and P.J. Anthony, Oscillation frequency tuning characteristics of fiber-extended cavity distributed-feedback lasers, Appl.Phys.Lett. 50:380-382 (1987).

39. W. J. Kozlovsky, W. Lenth, Stable, diffraction-limited 429-nm output from a frequency doubled GaAlAs diode laser, Digest of Technical Papers, CLEO 90, paper CTHN2 (May 1990), p.448-450.

40. T. M. Baer and D. F. Head, High peak power Q-switched Nd:YLF laser using a tightly folded resonator, Digest of Technical Papers, CLEO 90, paper CMF2 (May 1990), p.24.

41. A. Ashkin, K. Schutze, J.M. Dziedzic, U. Euteneuer, and M. Schliwa, Force generation of organelle transport measured in vivo by an infrared laser trap, Nature 348:346-348 (1990).

42. S. Shunichi, O. Masashi, and U. Tohoku, Optical trapping and manipulation of biological cells using compact semiconductor diode lasers in the near IR region, Digest of Technical Papers, CLEO 90, paper CTUH98 (May 1990), p.174.

FUTURE TRENDS IN LASER MEDICINE

John A. Parrish

Department of Dermatology
Harvard Medical School
Massachusetts General Hospital
BOSTON, MA 02114, U.S.A.

THERAPEUTIC LASER-TISSUE INTERACTIONS

Lasers could come to occupy a unique and very important position in the armament of medicine. They are the brightest known sources of light (man made or natural) on earth and as such can have extremely powerful effect on biological systems. Lasers are to other sources of light as music is to noise. The uniqueness of lasers derives from a number of special properties, viz., high intensity, coherence (which permits focusing to a spot size of one micrometer or less), collimation, extreme mono-chromaticity, and the ability to be delivered in very brief pulses. These collective properties can cause unique alterations of biological materials.

The therapeutic use of lasers began shortly after their invention, and now lasers are considered the tool of choice for certain surgical procedures. For example, laser photocoagulation has found wide application in ophthalmology, where the optical properties of the eye permit precise focusing of the beam onto the target of interest. By this technique, conditions such as senile macular degeneration, angioproliferative retinopathy, and retinal detachment can be safely and effectively treated. The laser is also of value in general surgery, where its ability to achieve coagulation while cutting can effect rapid hemostasis within the operative field, saving tine and limiting blood loss. The speed and precision with which the laser can produce its effects also make it a valuable tool for removing diseased tissues. Minimal divergence, which allows for an operative site (spot size) often less than 50 μm, gives the surgeon access to tissue he could not otherwise reach. In dermatology, laser irradiation of disfiguring cosmetic lesions is of proven value. Lasers presently have a more limited role in orthopedics, neurosurgery, and other medical and surgical specialties.

We now realize that present treatments can be improved, and that applications far exceed their present uses. Three scientific and technical fields, all important to laser therapy, have converged to increase our capabilities in recent years. The field of laser technology has advanced largely from needs of the materials processing industry and from massive investments by the defense and communications industry. Research and development over the past decade has resulted in several new lasers capable of tunable wavelengths, extremely brief pulse-durations, and unmatched power. The second field is endoscopy. The development of fiberoptics has permitted flexible endoscopy of the respiratory, urinary, and gastrointestinal tracts and other regions. This remarkable advance has enabled the endoscopist to visualize tissues directly and thereby target them for laser treatment. Finally, progress has been made in our fundamental understanding of laser-tissue interactions and the optics of turbid tissues. It now appears that these fields could cross-fertilize rapidly and productively.

We are now seeing a "reversal of traffic" in laser applications to medicine. In the past, industrial demands led to development of laser systems which physicians had to adapt to their use as best they could. Now, increased knowledge of laser-tissue interactions favors the development of new laser technology specifically designed for particular medical applications. The flow of information is now bidirectional, with salutary effects for both medicine and the laser sciences.

Most therapeutic uses of lasers are explained by thermal or photomechanical mechanisms rather than photochemical reactions. An exception is the photodynamic therapy of cancer with appropriate exogenous photosensitizers. Our understanding of photochemical tumor destruction is growing, but still limited. New delivery systems, such as microspheres, may provide a more general system for delivery of photosensitizers to tumors. Photosensitizers conjugated to monoclonal antibodies or tumor-localizing dyes lead to enhanced selectivity and higher photosensitizer concentrations. For this application, non- coherent sources may also be used but lasers may be more practical light delivery sources.

Because thermally-driven reactions require that the energy first be randomized within a large ensemble of molecules occupying various vibrational energy levels, heating by a variety of means (light, electricity, flame, friction) has similar and often very nonspecific effects. The heat is now well-confined spatially but spreads within tissues by conduction, and sometimes convection, causing nonspecific injury with blurred margins and transition zones.

Therapeutic laser-tissue interactions result from a variety of thermal effects. Many enzymes are heat-labile. Above 60° to 70°, structural proteins, including collagens, are denatured. Above 70° to 80°, nucleic acids are denatured and membranes become permeable. Coagulation necrosis is useful for causing hemostasis due to the denaturation of plasma proteins and the closing of vessels. Above 100°, vaporization of tissue water occurs followed by carbonization of the dry mass. Rapid vaporization in confined spaces is useful for physically separating or ablating tissues. Rapid localized heating causes large thermal transients and shock waves which may propagate, causing mechanical damage.

Because of the unique properties of lasers we can begin to think of thermal effects in biological materials as having a selectivity that most thermal chemistry does not permit. By altering laser delivery variables, we can also manipulate the mechanisms of injury. Indeed, because there may be combinations of thermal and photochemical mechanisms and, as in some cases, of "photodisruptive surgery", we may be unsure which mechanism is predominant. In fact, many of the nonlinear effects at high fluences blur our simple, intuitive notion of the difference, on a molecular level, between thermal chemistry and photochemistry. Thermal effects of absorbed light from lasers afford new options no longer confined to the radiation of high quantum energy needed for electronic energy transitions. We may employ long wavelength visible and infrared radiation, which penetrates much more deeply into tissues.

Although the mode of damage is important, it is the spatial confinement which mainly dictates which cells, organelles, or tissues will be affected. Selective photothermolysis is a simple scheme for confining thermally mediated radiation damage to chosen pigmented targets at the ultrastructural, cellular, or tissue levels. The confinement of damage can be as precise as with microbeam techniques, but millions of targeted structures are damaged simultaneously in vivo without precise aiming. This may be particularly useful in turbid tissue which, unlike the eye, limits the precision with which isolated structures, including overlying or immediately neighboring cells, are spared, potentially reducing widespread destruction and nonspecific fibrosis. This technique relies on selective absorption of a brief radiation pulse to generate and confine heat to certain pigmented targets. Targets must have greater optical absorption at a particular wavelength than the surrounding tissues for this technique to be effective.

For example, two biological targets have proven useful for selective photothermolysis. Red blood cells are dense packages of hemoglobin, and melanosomes (pigment granules within skin cells) are dense 1 µm organelles filled with melanin. After appropriately brief laser pulses of suitable wavelength, these packages of chromophores can act as internal heat-generating sites. Red blood cells can be targeted by using brief pulses of energy and the damage can be confined to blood vessels or even to individual cells, the degree of spatial confinement of the thermal injury is dependent on pulse duration: the shorter the pulse duration, the more localized is the injury. Using this approach, one can also confine damage to the organelle level sparing other subcellular structures. Melanosomes can be exploded or disrupted in vivo without immediate effects to the remainder of the cell. This represents in vivo spatial confinement on the order of one micrometer in millions of targets simultaneously - a far cry from that of most thermal reactions.

To realize the potential of selectivity with lasers, we must learn more about the fundamental mechanisms of laser-tissue interactions. Mechanisms must be described, quantified, isolated, and then manipulated so that we can cut, seal, drill, cook, explode, ablate, stimulate, move or separate biological tissues. This must be done in such a way that the biological response works for us in effecting a therapeutic, not merely a destructive, result.

As much as 10% of red and near infrared radiation reaching the scalp enters the brain. Thus, we increase the range of potential targets to include sites not accessible by fiberoptics. The relatively deep penetration of long-wavelength visible and near-infrared radiation presents us with a "therapeutic window", a spectrum in which delivery of appropriate exogenous chromophores to targets of interest could lead to extremely high selectivity because of the absence of endogenous chromophores. Also, the optical properties of tissue might be altered selectively with a "priming" laser, providing new targets for a second laser. The host response (e.g., inflammatory cascade, repair processes) could be manipulated to improve beneficial effects and diminish unwanted side effects.

Certain laser-tissue interactions can be expected to lead to unique results in biological systems. The capability of using macromolecules, organelles, or sites on cell membranes as local internal heaters presents a lesion to the host not seen before in evolution. Because the injury is not seen with other sources, microscopically spatially confined selective injury was not a possibility during the 200 years when classical pathology, microscopic anatomy, and medicine were formulated.

There is much work to be done in describing and characterizing this form of injury. An even greater challenge is to understand the biologic response to lasers sufficiently well enough to obtain therapeutic responses by manipulating the properties of the laser beam. Examples of reasonable short term goals include selective destruction of micrometastases in cancer therapy, "noninvasive" hemostasis, correction of eye refractive power through reshaping the cornea, in vivo cell surgery, and improved tissue-specific hemostatic surgery.

Pulsed laser sources produce a considerably greater range of laser-tissue interactions than continuous wave (cw) sources. Pulsed lasers can initiate nonlinear processes that alter the interaction mechanisms at work. The efficiency of ablation or photochemical reactions can be greater than with cw sources. Stepwise control of the tissue effect is much easier with pulsed rather than cw lasers. For spatial confinement of thermal and mechanical effects, pulsed lasers are required. Although some pulsed laser sources have been used in medicine for decades, understanding of pulsed laser interactions with tissue is incomplete. It is most likely that most new diagnostic and therapeutic uses of lasers will require pulsed lasers.

Studies of pulsed infrared laser ablation have verified the basic ideas of spatial confinement of thermal tissue damage by using short pulses. Studies of thermal damage vs pulsewidth and wavelength (optical penetration depth) have provided the basis for using lasers for debridement. Studies of thermal damage are needed to understand the mechanisms that control the intensity, selectivity, and spatial confinement of coagulative effects. Repetitive pulse selective effects allow confinement of thermal damage to cellular, subcellular, and potentially molecular targets, while avoiding unwanted photoacoustic effects. Pulsed excitation of photosensitizers may contribute to the development of photodynamic therapy by initiating new processes or increasing the efficiencies of photoprocesses induced by cw radiation.

The study of laser-induced optical breakdown has already provided understanding of the physical processes involved in ophthalmic photodisruption and laser lithotripsy. Increased understanding of the role of laser-induced acoustic effects is important for the application of UV excimer laser ablation to corneal sculpting.

Pulsed lasers can, when focused, produce nonlinear effects, such as optical breakdown, plasma formation, and two-photon absorption, not encountered using cw lasers. These nonlinear effects can be deleterious, as when they cause damage to optical components and fibers. However, they can lead to medically useful results, such as plasma-mediated disruption in ophthalmic photodisruption and stone fragmentation in laser lithotripsy. In addition, a number of different interaction mechanisms can operate simultaneously. Lasers can produce heating which may lead to vaporization, ablation, and coagulation. Photoacoustic effects (generation of acoustic and shock waves) can be initiated by heating, optical breakdown, or by ablation. Photochemical processes form the basis of tumor destruction by photodynamic therapy (PDT), but may be unwanted ancillary effects of ablation with UV lasers. During the past four years substantial progress has been made in understanding the way these different mechanisms operate in tissue, leading in turn to more controlled utilization of lasers in medicine. The concept of selectivity has been studied and we have learned much about limiting the laser interaction to a specific spatial region, to specific natural chromophores such as melanin, carotenoids, or oxyhemoglobin, or to specific exogenously-dyed subcellular structures, for example mitochondria. Out of this work have come several major laser-based medical treatments.

Tissue removal forms the basis of most laser surgery. Studies of pulsed laser ablation have shown how it differs from cutting with cw lasers, typically Nd:YAG lasers and CO_2 and Er:YAG lasers has shown that thermal damage zones could be reduced to as little as 5-50 µm, depending on wavelength, pulsewidth, and tissue type. For comparison, cutting with cw CO_2 lasers produces damage zones for several hundred µm. Pulsed excitation minimizes thermal diffusion of the deposited energy out of the laser heated zone.

During the last five years several new pulsed solid state infrared lasers have become commercially available. These lasers emit at wavelengths strongly absorbed by water, the primary constituent of most tissue. They include the Er:YAG laser, of interest because its 2.9 μm output wavelength is strongly absorbed by water, the Ho:YAG laser emitting at 2.1 μs, and the Co:MgF$_2$ laser, a tunable laser operating in the 1.75-2.5 μm wavelength region. These lasers have been used to study pulsed infrared ablation over a spectral range in which the optical absorption of water varies by over two decades. The studies have shown that thermal confinement of tissue damage to zones some ten of μm thick is possible.

Optical pump-probe and flash photography experiments show that pulsed laser ablation is an explosive event. This suggests that mechanical reaction forces could result in unexpected tissue damage. Indeed, studies of UV excimer laser ablation of skin have shown organelle damage several hundred microns beyond the penetration depth of the light. This damage has been attributed to photoacoustic effects.

Plasma-mediated effects are also being used in medicine. Pump-probe spectroscopic, and flash-photographic studies of laser fragmentation of kidney stones show that optical breakdown is necessary for stone fragmentation and that shock waves confined by water are involved in fragmentation. Laser-induced plasma-mediated processes have had a major impact on ophthalmic laser surgery of the anterior eye.

The photochemical aspects of laser irradiation can be important when UV excimer lasers are used for tissue ablation. Corneal reshaping using ablation by pulsed 193-nm excimer laser radiation is currently being clinically evaluated. The 193 nm photons are absorbed by protein in the tissue as well as by cellular DNA. The photochemistry (monophotonic or biphotonic) of proteins excited at 193 nm should be studied because it may be important for understanding the 193 nm induced ablation of tissue. 193 nm radiation may produce damage in cells primarily at the level of the plasma membrane because absorption by cytoplasm attenuates the radiation reaching the nucleus. Cellular mechanisms for protection against pulsed-laser thermal injury are poorly understood. Heat-shock response, when induced in human fibroblasts, protect against pulsed CO$_2$ laser lethal injury. In contrast to conventional hyperthermia, the brief thermal cycles induced by pulsed laser exposure do not produce a heat shock response.

SELECTED SPECIFIC EXAMPLES OF NOVEL LASER APPLICATION

The Wellman Laboratories of Photomedicine within the Department of Dermatology, Harvard Medical School at Massachusetts General Hospital includes a multispecialty group of investigators committed to laser medicine research. We have attempted to choose projects that can make a significant improvement in medical care and that can be carried from basic research to applied research to clinical trials within one to three years. These programs serve as examples of possible future trends in laser medicine and surgery.

Coronary Laser Thrombolysis

The treatment of acute myocardial infarction has made significant progress with the evolution of fibrinolytic drugs. The number of patients who fail to reperfuse the thrombosed coronary artery appears to be 30%-40% of those treated with fibrinolytic drugs. Ideally, one might use a laser-catheter to acutely disrupt and remove thrombus without disrupting arterial walls. There are several reports of argon-ion laser thrombolysis in animal models, using quartz fiber optic catheters. These catheters are stiff and difficult to make, and the argon laser is not ideal.

We previously investigated the possibility of selective thrombus ablation using a pulsed dye laser, which showed that strong preferential ablation of thrombus was possible. This led to work on thrombosed grafts, but a suitable catheter was necessary for coronary application. Using a unique, liquid-core design, we have recently been able to approach the coronaries with a simple, flexible catheter that flushes debris and blood from the optical field.

Coronary arteries in 26 dogs were traumatized with forceps proximal to a 90% external band stenosis to promote platelet adhesion during a subsequent twenty minute perfusion period. Thrombin and whole blood was then injected proximally creating a mixed platelet and fibrin thrombosis which matured for 1 to 6 hours to form an occlusive coronary artery thrombis. A pulsed dye laser delivering 480 nm radiation in one microsecond pulses at 9 J/cm^2 was used for laser thrombolysis in 22 dogs. Thrombi treated with the laser pulses were completely removed in all dogs without perforation, and reperfusion established within 100 seconds using repetition rates between 2 and 5 Hz. In 4 control animals there was no perfusion during a similar observation time. Thrombi treated with the catheter prior to laser application in 13 animals had no mechanically induced reperfusion. Sixteen of seventeen coronary arteries followed for 1 to 2 hours remained widely patent.

Final tests of the fluid filled catheter in chronic dog experiments are underway in collaboration with Baxter, Inc. The start of human clinical trials is anticipated in the near future.

Photodilatation of Acute Arterial Spasm

Strokes caused by cerebral arterial spasm following intracranial bleed are a common cause of death and morbidity. Presently there is no acceptable intervention or therapy which has been successful in reversing this form of spasm and only supportive therapies are available.

Recently we investigated two separate mechanisms by which arterial spasm can be relaxed by light exposure. One of these is a classical, although poorly understood, photochemical reaction, which is reversible and occurs with low-intensity visible light. The other is a prolonged (>5 hours) relaxation caused by photoacoustically-generated waves from microsecond laser pulses. This latter mechanism is capable, in animal modeLs, of reversing tight spasm induced by trauma or drugs, and is not blocked by neurotransmitter or vasoactive agents. It shows significant promise for clinical applications.

In recent experiments, a one microsecond pulsed dye laser delivers 480 nm radiation to femoral arteries in 22 New Zealand white rabbits. At pulse energies of 5 mJ, all vessels could be dilated with a 65% average increase in diameter. Pulsed energies above 10 mJ resulted in focal spasm, and higher energies caused perforation of the vessels with a mean perforation energy of about 14 mJ.

The physical and cellular processes underlying photovasodilatation are being systematically investigated. The process is not optimized at present, but clearly works well in animal experiments. Preliminary studies of cerebral vessel spasm release by laser photovasodilatation are being pursued.

Laser Diagnosis and Treatment of Burn Injuries

Laser debridement of burns has been tried by multiple investigators over the past 20 years but none were successful. The main limitation was the inability of the laser ablated bed to support a skin graft, slow tissue removal rates, and technical difficulty. In terms of burn diagnostics, laser Doppler blood flow measurements have been shown effective in predicting burn depth, but are time consuming and hence not readily adaptable for clinical use.

Using pulsed laser systems we have identified a set of laser parameters that permit rapid, hemostatic removal of burn eschar but allows normal graft take. We have also developed a fluorescence technique for discriminating partial from full thickness burns. We are presently developing and testing a video burn diagnostics system suitable for clinical evaluation. In addition, we are continuing to study the burn ablation process which is critical for designing a clinical burn debridement system. We are also exploring the effect of growth factors on laser-debrided wounds and the use of cultured autologous keratinocyte grafts.

A laser system to identify, then rapidly and accurately remove eschar while leaving a bed suitable for grafting would be a major advance in burn treatment. Furthermore, this system could find use in the treatment of other lesions requiring grafting such as chronic skin ulcers and skin cancer surgery.

Laser Lithotripsy of Renal and Biliary Calculi

Laser lithotripsy is an old idea but one that was not practical until the availability of high intensity lasers. The Wellman Laboratories pioneered the technique in 1984. Laser lithotripsy is now an FDA approved procedure for the treatment of biliary and urinary calculi.

Our present research is directed towards developing a safer and more effective laser system for lithotripsy. We are also experimenting with methods for performing lithotripsy without the need for direct visualization.

Laser lithotripsy has become a major advance in the treatment of distal ureteral stones. It is effective treatment for patients with large common duct stones.

Treatment of Vascular Malformations by Selective Photothermolysis.

Conventional argon laser and other therapy for portwine stains and telangiectasias produces scarring. In 1981 Drs. Anderson and Parrish developed the concept of selective photothermolysis, and demonstrated that histologically-selective thermal necrosis of vessels is achieved with yellow laser pulses in the microsecond domain. This led to FDA approval in 1985 of a non-scarring, effective treatment, which is particularly useful for portwine stains in pediatric and adult patients. Because of histologic selectivity, the incidence of scarring is much lower than with other treatment. Treatment of hemangiomas, which are proliferating hamartomas that almost always spontaneously regress, has become a controversial issue. Given a non-scarring treatment and high parental anxiety, the ethics of performing laser treatment vs. letting nature take its course can be complex. Phlebectasias on the female leg are common, do not respond well to this or other therapy, and remain a therapeutic challenge.

Our present aims at Wellman Laboratories include refining laser treatment parameters for portwine lesions in children and adults, developing a means for individualizing treatment parameters, developing a laser robotic device to replace the expensive unreliable dye laser with a cheaper, solid state, and more versatile but equally selective "smart" device, developing an effective treatment for phlebectasia of the leg, and examining pulse trains as a more controlled means for treatment.

Laser Treatment of Vascular Lesions of the Colon

Several lasers have been used to treat of vascular lesions of the colon but have been limited by an unacceptably high complication rate. We have developed a method whereby the colonic vasculature is thrombosed, without damaging surrounding tissue. The same laser parameters developed for selective photothermolysis of portwine lesions in skin are used. This permits selective injury of the colonic microvasculature without perforation. A clinical trial is in progress. This technique has the potential to provide a new treatment of vascular lesions of the gastrointestinal tract.

Tattoo Removal

About 10% of the US population has at least one tattoo; the majority of these are unwanted. Until recently, there has been no non-scarring alternative to surgical removal of tattoos. CO_2 laser treatment is grossly scarring. Starting in the 1960's, ruby laser pulses were used for experimental treatment of tattoos. In Glasgow, a group starting in 1981 reported several cases of non-scarring tattoo removal with Q-switched ruby laser pulses, but did not extend their work until recently. Last year, we studied about 60 tattoos in a pilot dose-response study and found that approximately 75% of amateur tattoos and 25% of professional tattoos could be removed with good to excellent cosmetic results. The incidence of scarring was only 2%. On the sole basis of our small study, the FDA has recently approved this treatment, and Q, switched lasers for tattoo removal will be available. MGH is presently in a phase-II clinical trial of up to 500 patients.

Ophthalmic Applications: Sclerostomy for Glaucoma

Filtration surgery remains the procedure of choice to treat glaucoma which is not controlled by medical therapy or laser trabeculoplasty. Conventional filtration surgery involves punching a hole into the sclera and requires hospitalization since the complications and morbidity can be significant. Closure of the newly-created filtration channel by subconjunctival fibroblast proliferation is a major problem, which may be reduced by decreasing surgical trauma. By performing filtration surgery non-invasively via an ab-interno approach with a laser we hope to first reduce complications and improve efficacy by reducing surgical trauma.

The goal of this project is to develop a clinically acceptable technique to perform filtration surgery non-invasively with μs laser pulses. Iontophoretic staining with Methylene blue of the scleral site to be ablated is necessary to provide enough 670 nm light deposition.

Ab-intero scleral drilling was carried out using a 200 μm spot size either 1.2-μs and 40 rabbit eyes at 20 μs. Patent filtering procedures were obtained with minimal inflammation. There was no evidence of retinal detachment, cyclodialysis, or hyphema. Histologic examination showed a 20-μm to 150 μm zone of damage surrounded the fistula. A multi-center clinical trial is underway.

Application of Selective Photochemical Targeting of Cancer and Non-Cancer Cells

Deletion of selected cell populations by photochemical targeting is a powerful technique because of its inherent dual selectivity. Worldwide, the application has been focused largely on experimental treatment of a wide variety of cancers (photodynamic therapy of cancer, PDT). These include cutaneous and subcutaneous malignancies, cancers of the head and neck, central nervous system, ocular, endobronchial, esophageal, bladder, gynecological, and certain forms of blood cancers.

All of these treatments are based on the use of a poorly characterized mixture of chemicals, hematoporphyrin derivative (HPD) which has significant skin phototoxicity. The reported studies to date have been uncontrolled and often anecdotal. Randomized, phase III trials of PDT compared to standard therapy are now starting and it is expected to take at least two years or more before regulatory approval by U.S. FDA. Earlier approval is expected in Canada and Europe. These studies are being carried out by American Cyanamid Company, Lederle Laboratories Division, Pearl River, New York, for QLT Phototherapeutics Inc., of Vancouver, B.C.

The Wellman Laboratories made a limited commitment to photochemical targeting of cancer in 1985 and have in recent years expanded this to non-cancer applications. New photosensitizers (PS) belonging to the benzophenothiazinium, rhodamine, and porphyrin dyes have been synthesized and shown to be effective in vitro and in vivo in simple animal models. The use of molecular delivery systems to concentrate photosensitizers on target cells has been developed. Selective enhanced phototoxicity has been demonstrated with PS bound to monoclonal antibodies and microspheres. The possibility of modulating the efficacy of treatment by altering biochemical and physical parameters has also been demonstrated. The addition of biologic response modifiers, e.g. γ-Interferon and glutathione depletion gave enhanced phototoxicity. Similarly, increased efficacy can be obtained with pulsed high-intensity irradiation. These observations may ultimately have impact on the use of PDT clinically.

PRESENT AND FUTURE DIRECTIONS

There are several laser applications that have recently attracted significant theoretical, technical, research, and commercial interest:

1. Coronary angioplasty. Beginning with argon-ion laser exposure through optical fibers, laser-heated metal catheter probes, and now excimer pulsed-dye, and holmium lasers with or without fluorescence or imaging feedback, there have been many disparate approaches to this complex application. Several more years will be needed to place these modalities in proper perspective.

2. Photoacoustic/photomechanical interactions derived from laser-induced plasmas, including laser lithotripsy and ophthalmic photodisruption.

3. Selective photothermolysis in skin, of microvascular and now, pigmented lesions and tattoos.

4. Tissue "welding" to rapidly form leak-proof tissue functions, anastomosis with minimal scarring, or rejoining of tubular or axial structures such as blood vessels or nerves.

5. More controlled infrared laser surgery with variable cut-ablate-seal effects by combining new types of lasers and varying wavelengths and pulse durations. These are being developed, for example, in microsurgery of inner and middle ear and neurosurgical procedures.

6. Reshaping of the cornea using 193 nm ArF excimer laser ablation, to correct refractive abnormalities of the eye.

7. Development of new photosensitizers for photodynamic therapy of human cancer. Hematoporphyrin derivative, while sometimes effective and still not fully optimized, has major limitations for patient acceptability.

8. Biostimulation, acceleration of wound healing, or other advantageous metabolic effects dependent on photochemical reactions and, or electric field effects.

Many commercial and academic groups are exploring these applications. Efforts to develop these applications are fueled by theoretical possibility, scattered evidence of success, and potential high impact on the quality of cost-effectiveness of medical care. To date, none have achieved widespread acceptance based on documentation of safety, efficacy, and cost-effectiveness.

WHAT TECHNOLOGY DEVELOPMENTS DOES LASER MEDICINE NEED?

The technologic development of lasers and laser biomedicine, are intimately linked and mutually supportive. Beginning in 1960 with only one laser available, the ruby laser was used for eye surgery, melanoma, portwine stains, tattoo removal, and tissue cutting. Multiple complications occurred. All of these applications continued in the 1970's, but entirely without the ruby laser, as more controllable, continuous lasers such as CO_2, and argon-ion became available. The 1980's saw a re-emergence of pulsed lasers in medicine, with the development of ophthalmic photodisruption (Q-switched Nd:YAG), selective photothermolysis of portwine stains (pulsed dye), lithotripsy (pulsed dye and others), and various pulsed laser ablation techniques (see above). Interestingly, the Q-switched ruby laser has been "rediscovered", for treatment of tattoos and pigmented skin lesions by selective photothermolysis. More importantly, laser systems are being designed spectrally for biomedical problems, such that technology is now being driven in part by medicine. This is a complete reversal from the 1960's and 1970's.

During the coming decade, the process of medical problems driving the development of specific laser technology must and will continue. Fortunately, the convergence of laser and fiber optics technology for communications, military, and industry fits well with medical needs. Recent progress in diode laser technology may literally revolutionize laser medicine, given proper development. However, the days in which an off-the-shelf laser can be pointed at tissue in the hopes of significantly improving medical practice, are well over. An integrative, cost-conscious, competitive, and scientific approach to developing laser medical systems is necessary. In particular, the laser will more frequently be simply one component of an integrated system. In the next decade laser may cease to be the technological limiting element in such systems. Catheters, control systems, imaging systems, robotics, dosimetry, photochemical sensitizers and biologic delivery systems are and will become the limiting technologic factors. This ultimately motivates the close cooperation and integration of laser, electro-optic, medical, and biotechnology industries. The emerging area of laser-based medical diagnostics will also motivate this integration.

The specific technology needs certainly include efficient, reliable diode and diode-pumped small lasers. Single and phased-array diode lasers are likely to find instant application if desirable 800 nm photosensitizing dyes can be developed for PDT. Moreover, the ever-increasing need for reliable pulsed laser systems in medicine should be very well served if diode-pumped or other solid-state tunable pulsed lasers could eventually replace the larger, and maintenance-intensive pulsed dye lasers. Although titanium sapphire and alexandrite lasers appear promising, it remains to be seen whether these will become medically useful.

Without doubt, laser and fiber optics have brought photomedicine to organs never before accessible to light. The technologic limitations in laser angioplasty, thrombectomy, GI surgery, laparoscopic surgery, lithotripsy, invasive PDT, etc., are at present related to delivery systems, and ablation or treatment control systems.

Predictably, the intense development of such micro-invasive, well controlled laser procedures will ultimately fuel other, non-laser procedures. Simple, yet important technological developments in these related areas may have significant impact on laser medicine. One such example is the development of a flowing-fluid laser delivery catheter, in which angiography, laser delivery, and displacement of blood are provided by a simple catheter without the need for solid-core optical fibers.

In a manner similar to that of the computer industry, laser medicine needs to shift toward smaller, reliable, cheaper, and more available devices -- with a related increase in "software" and human interface development. It is theoretically possible to produce many pulsed-laser effects such as histologically localized thermal injury, selective ablation, and minimal thermal damage, using either pulsed or tightly-focussed continuous lasers in combination with dedicated diagnostic and/or laser scanning and modulation systems. The integration of microelectronics, optodiode, and feedback-driven control over exposure parameters should allow "smart lasers" and small, reliable lasers to gain increasing application. It should be possible, for example, to create laser-ablation devices which recognize and remove, coagulate, or expose only certain tissue structures with precision well beyond that of classical surgery.

Finally, it is imperative that the integration between academic and industrial efforts be facilitated as an essential part of the field of laser photomedicine. This challenge is not technologic per se, but political, managerial, and financial. Our ability to capture and sustain the immense opportunities now apparent in laser photomedicine will ultimately determine success in terms of improved medical care.

CONTRIBUTORS

R.R. Anderson
Department of Dermatology
Harvard Medical School
Massachusetts General Hospital
BOSTON, MA 02114, USA

G. von Bally
Laboratory of Biophysics
Institute of Experimental Audiology
University of Muenster
Kardinal-von-Galen-Ring 10
D-4400 MÜNSTER, Germany

J. Baraga
G.R.Harrison Spectroscopy Laboratory,
M.I.T.
77 Massachusets Avenue
CAMBRIDGE, MA 02139, USA

H. Barr
The Department of Surgery
John Radcliffe Hospital
OXFORD, England

J.L. Boulnois
Technomed International
North Woods Business Park
100 Rosenwood Drive, Suite 140
DANVERS, MA 01923, USA

S.G. Bown
National Medical Laser Center
The Rayne Institute
University College
5 University Street
LONDON WC1E 6JJ, England

R. Brancato
Clinica Oculistica
Ospedale S.Raffaele
Via Olgettina, 60
20132 SEGRATE (Milano), Italy

M. Dal Fante
Divisione di Endoscopia
Istituto Nazionale Tumori
Via Venezian, 1
20133 MILANO, Italy

B.L. Diffey
Regional Medical Physics Department
Dryburn Hospital
DURHAM DH1 5TW, England

J.F. Ennever
Department of Pediatrics
School of Medicine,Rainbow Babies and
Childrens Hospital,
Case Western Reserve University
CLEVELAND, OH 44106, USA

M.S. Feld
G.R.Harrison Spectroscopy Laboratory,
M.I.T.
77 Massachusets Avenue
CAMBRIDGE, MA 02139, USA

M. Fitzmaurice
Department of Pathology
Cleveland Clinic Foundation
CLEVELAND, OH 44106, USA

T.B. Fitzpatrick
Department of Dermatology
Massachusetts General Hospital
BOSTON, MA 02114, USA

R.W. Flower
Applied Physics Laboratory
J.Hopkins University
John Hopkins Rd.
LAWREL, MD 20707, USA

A. Fracasso
Istituto di Chirurgia
Cardiovascolare dell'Università,
Policlinico
Via Giustiniani, 2
35128 PADOVA, Italy

V. Gallucci
Istituto di Chirurgia
Cardiovascolare dell'Università,
Policlinico
Via Giustiniani, 2
35128 PADOVA, Italy

M.J.C. van Gemert
Academisch Medisch Centrum
Laser Centre
Meibergdreef 9
1105 AZ AMSTERDAM,
The Netherlands

H.J. Geschwind
Cardiac Catheterization Laboratory
University Hospital Henry Mondor
INSERUM U2
52, Av.du Marechal de Lattre de
Tassigny
94010 CRETEIL, France

P.C. Jackson
Department of Medical Physics
Bristol General Hospital
Guinea Street
BRISTOL BS1 6SY, England

A. Jakobsson
Division of Biomedical Instrumentation
Department of Biomedical Engineering
University Hospital
Lovsbergsvagen 13
S-581 85 LINKOPING, Sweden

G. Jori
Dipartimento di Biologia
Università di Padova
Via Trieste.75
35121 PADOVA, Italy

H. Key
Department of Medical Physics
Bristol General Hospital
Guinea Street
BRISTOL BS1 6SY, England

J. Kramer
Department of Cardiology
Cleveland Clinic Foundation
CLEVELAND, OH 44106, USA

G. Leoni
Clinica Oculistica
Ospedale S.Raffaele
Via Olgettina, 60
20132 SEGRATE (Milano), Italy

A. Mancini
Divisione di Endoscopia
Istituto Nazionale Tumori
Via Venezian, 1
20133 MILANO, Italy

G.E. Nilsson
Division of Biomedical Instrumentation
Department of Biomedical Engineering
University Hospital
Lovsbergsvagen 13
S-581 85 LINKOPING, Sweden

J.A. Parrish
Department of Dermatology
Harvard Medical School
Massachusetts General Hospital
BOSTON, MA 02114, USA

R. Pratesi
Istituto di Elettronica Quantistica
del CNR
Via Panciatichi 56/30
50127 FIRENZE, Italy

R. Rava
G.R.Harrison Spectroscopy Laboratory,
M.I.T.
77 Massachusets Avenue
CAMBRIDGE, MA 02139, USA

E. Reddi
Dipartimento di Biologia
Università di Padova
Via Trieste.75
35121 PADOVA, Italy

R. Richards-Kortum
G.R.Harrison Spectroscopy Laboratory,
M.I.T.
77 Massachusets Avenue
CAMBRIDGE, MA 02139, USA

P. Spinelli
Divisione di Endoscopia
Istituto Nazionale Tumori
Via Venezian, 1
20133 MILANO, Italy

G. Trabucchi
Clinica Oculistica
Ospedale S.Raffaele
Via Olgettina, 60
20132 SEGRATE (Milano), Italy

E. Unsöld
Gesellschaft für Strahlen
und Umweltforschung mbH
Zentrales Laserlaboratorium/ZLL
Inglostadter Landstrasse 1
8042 NEUHERBERG, Germany

K. Wårdell
Division of Biomedical Instrumentation
Department of Biomedical Engineering
University Hospital
Lovsbergsvagen 13
S-581 85 LINKOPING, Sweden

A.J. Welch
Biomedical Engineering Program
The University of Texas at Austin
AUSTIN, TX, USA

D.F. Welch
Spectra Diode Labs, Inc.
80 Rose Orchard Way
SAN JOSE', CA 95134-1356, USA

P.N.T. Wells
Department of Medical Physics
Bristol General Hospital
Guinea Street
BRISTOL BS1 6SY, England

B.C. Wilson
Ontario Cancer Treatment and
Research Foundation
Physics Department
711 Concession Street
HAMILTON, Ontario L8V 1C3, Canada

A.R. Young
Photobiology Unit
The Institute of Dermatology
Guy's and St.Thomas's Hospitals
Renfrew Road
LONDON, SE11 4TH, England

INDEX

Ablation, 34, 37, 39, 46, 213, 214, 252, 259, 287, 289
Ablative
    surgery, 36, 249, 259, 275
    interaction, 37
Absorption
    coefficient, 46
    spectroscopy, 81
        quantitative, 74
Action spectrum
    dermal bilirubin, 187
    melanosome, 219
    photothermal coagulation, 218
    photothermolysis, 217
    psoriasis, 155, 156
    tanning, 193
Angiography, 139, 266
    angiogram, 140
    ICG fluorescent angiogram, 141
    ocular
        NIR, 139, 283
    sodium fluorescein dye, 139
Angioplasty, 45, 249, 255, 256, 259, 264, 291
    atherosclerotic plaque, 45, 129, 261
    arterial spasm, 292
    balloon, 249, 261
    coronary, 255, 259
        stenosis, 262
        by-pass, 262
    catheter, 249, 253, 261, 290
    debridement 215, 291
    debris, 252
    vessel perforation, 249, 252, 253, 255, 256, 265, 292
Activation energy, 53
Adenine, 32
Albedo, 76
Albumin, 181
Alexandrite laser, see Laser
Anastomosis, 295

Angioscopy, 266
Anisotropy factor, 46
    see also Scattering
Aorta
    optical properties, 49, 50, 134
ArF laser, see Laser
Argon laser, see Laser
Arrhenius model, 55
Atherosclerotic plaque, see Angioplasty
Autofluorescence, 49, 119, 125, 223, 283

Balloon angioplasty, see Angioplasty
Bilirubin, 177
    configurational photoisomerization, 180
    conjugated, 177
    rotation of external pyrrole rings, 179
    photoequilibrium concentration, 180
    photooxidation, 179, 183
    quantum yield, 182
    structural photoisomerization, 181
        lumirubin, 182
Biliverdin, 177
Biostimulation, 276, 296
Blood
    oximetry, 45
    perfusion, 52
Blood flow, see also Doppler velocimetry
    Doppler ultrasound detection, 108
    gastric, 95
    microvascular, 96
    choroidal resistance, 145
Boltzmann equation, 75
Breakdown
    laser-induced, 40
Bronchology
    Nd:YAG laser application, 204
Burn
    ablation, 292

diagnostics, 293
    laser debridement, 215, 292
    wound depth, 96

Calculi
    biliary, 292
    fragmentation, 259, 290
Camera
    holographic, 65
Cancer
    breast, 104
        telediaphanography, 104
    carcinoma
        noncutaneous, 171
        squamous cell, 170
    colorectal, 203
    gastro-intestinal, 243
    melanoma, 171, 213
    tracheo-bronchial, 243
    urinary bladder, 243
Carbonization, 34
Carcinogenesis
    skin, 168
Catheter, see Angioplasty
Chemophotochemotherapy, 169
Choroid
    angiography, 139, 143, 147
    blood flow, 139
        resistance, 145
    circulation, 139
    vasculature, 139
Chromophore, 288
Cineholography, 69
Circulation, blood
    gut, 146
    microcirculation, 89
    peripheral, 89
CLED, 15, see also Diode Laser
CO_2 laser, see Laser
Coagulation, 34, 46
    necrosis, 34, 287
Coefficient
    absorption, 46, 82
    scattering, 46
    transport optical attenuation, 52
Coherence
    spatial, 197
    temporal, 197
Colon, 292
Colonscope, 146
Colostomy, 203
$CoMgF_2$ laser, see Laser
Configurational isomerization, 180
Configurational isomers, 184
Copper vapour laser, see Laser
Cornea

reshaping, 289, 290
Coronary angioplasty, 294
Coronary artery, 135, 261, 290
    fluorescence spectrum, 135
Coronary stenosis, 262
Cosmetic UVA lamp, 193
Cryptocyanine, 274
Cutaneous barrier function, 224

Damage
    biologic, 30
    threshold, 55
Denaturation, 34, 218
    molecular, 33
Depth
    penetration, 31, 33, 79, 90, 91, 289
    profiling, 80
Dermatology, 213, 276, 286
    laser Doppler flowmetry, 95, 223
Diabetic retinopathy, 209
Diagnostic techniques
    Doppler velocimetry, 89, 282
    fluorescence, 292
    holographic, 61
    laser-induced-fluorescence, 117, 129,
        282
        computer-assisted, 125
        real-time, 130
    NIR angiography, 139, 282
    reflectance spectroscopy, 73, 282
    transillumination imaging, 101, 283
Diode, light-emitting see Light-emitting
    diode
Diode laser, 3, 15, 83, 111, 146, 207,
    240
    active region, 6
    bandgap, 3
    bar, 8
    catastrophic failure, 9, 12
    efficiency, 5, 10
    conversion electrical to optical, 19
    gain guided, 3, 5
    index guided 5, 10
    linewidth, 11
    medical applications, 271
    mode hopping, 11
    mode stability, 10
    multi-lobe, 7
    noise, 11
    non-absprbing mirrors, 12
    power conversion, 10
    reliability, 12
    single mode operation, 19
    spectral distribution, 11
    thermal dissipation, 12
    two-dimensional array, 18

two-dimensional stack, 8
DNA repair, 192
Doppler flowmetry, 45, 89
   application to medical diagnostics,
      95
   blood flow, 96
   blood cell velocity, 90
   blood perfusion, 94
   concentration of red cells, 91
   Doppler shift, 90
   non-linear effects, 91
Doppler ultrasound detection
   blood flow, 108
Dose-response curve, 156
Dosimetry, 55, 79
Dye laser, see Laser
Dynamic light scattering, 89, 93
         see also Doppler flowmetry

Echography, 266
Electromechanical interaction, 39
Electron avalanche growth, 40
Endoscopy
   holographic, 67
   laser Doppler flowmetry, 95
   photocoagulation, 197
Energy transfer, 233
Enzyme activity, 123
Er:YAG, see Laser
Erythema, 191
   dose, 156
      minimal (MED), 156
   photoxic, 167
   UVR-induced, 155
Exchange transfusion, 178
Excimer laser, see Laser
Eye
   microsurgery
      cornea reshaping, 289
      glaucoma, 293
      sclerostomy, 293
      trabeculoplasty, 209
Eye-safe wavelength, 23

Fallopian tube
   elasticity, 68
Fluorescence, 45, 74, 117, 252
   emission, 254
   laser-induced, 283
   spectroscopy, 266
   time-resolved, 83
Fluorometry, 120
   remote, 124
Fluoroscopy, 252
Frequency
   non-linear conversion, 21

doubling, 5, 21
   self-doubling, 21
   tuning, 20
      microlaser, 20
Fundus camera, 139
Fundus holography, 62
Furocoumarins, 36

GaInAsP laser, 4
Gastroenterology, 202, 243
   endoscopic treatment, 202
General surgery, 286
Genetic disorders, 123
Glaucoma, see Eye
Glucuronic acid conjugate, 177
Graft, 95, 292
Gunn rat, 184
Gynaecology
   Nd:YAG surgery, 203

Hazard
   phototherapy, 162
He-Cd laser, see Laser
Healing, 252, 261
Heat
   adiabatic heating, 52
   characteristic time, 52
   conduction, 34, 38, 46, 51
   convection, 46
   diffusion, 38
      equation, 52
      length, 34
      theory, 34, 51, 75, 78-79
      time, 31
   diffusivity, 46
   dissipation, 259
   thermal relaxation time, 34, 216
   transfer, 46
      coefficients, 46
Hemangiomas, 213, 215, 218, 292
Hematoporphyrin, 36, 37, 227, 237, 277,
      278, 293
Hemoglobin, 32
Hemoglobin oxygenation, 82
Hemostasis, 31, 33, 287
Henyey-Greenstein phase function, 47,
      49
High performance liquid
   chromatography (HPLC), 182
Ho:YAG laser, see Laser
Holography, 61
   camera, 65
   double-exposure, 69, 70
   metrology, 65
   multiplex, 63
   subtraction, 67

Hyperbilirubinemia
    neonatal, phototherapy, 177, (*see*
        also Bilirubin, Phototherapy)
Hyperthermia, 79, 234, 290
    interstitial, 205
    thermal range, 34
    with diode lasers, 274

Image
    computer-aided analysis, 143
    fluorescence, 124
        difference imaging, 125
    of tissue perfusion, 97
    processing, 61, 144
    recording
        real-time, 143
Imaging technique, 96
Immunosuppression, 193
Indocyanine green, 140, 143, 208, 275,
        209
    absorption spectrum, 141
    angiography, 147
    fluorescence spectrum, 141
Infrared laser ablation, 290, 294
Injury
    photoacustic, 215
    photothermal, 213
Integrating sphere, 49
Interstitial laser treatment, 55
Ionization, 39

Jaundice of the newborn, *see* Hyperbili-
        rubinaemia,

Kernicterus, 178
Kidney stone, 292
Kryptocyanine, 234
KTP laser, *see* Laser

Lamp
    cosmetic UVA, 193
    emission spectrum, 153
    fluorescent tube, 152, 154, 277
    hot cathode
        indium iodide, 142
    high pressure, 154
    hot quartz, 154
    mercury arc, 152
    metal halides, 152
    phototherapy, 153
        spectrum, 159
    UV-A, 153
Laser
    gas
        argon, 252
        He-Cd, 253

    excimer, 38, 215, 219, 252, 259,
        290
    copper vapour, 217, 239
    liquid
        dye, 213, 291
    semiconductor, *see* Diode laser
    solid-state
        alexandrite, 296
        CLED-pumped, 15
        Co:MgF$_2$, 290
        Er:YAG, 290
        Ho:YAG, 256, 290, 294
        "KTP", 217
        ruby, 213, 293
            Q-switched, 219
        titanium-sapphire, 296
Laser Doppler flowmetry, *see* Doppler
        flowmetry
Laser-induced fluorescence, 283
LED *see* Light Emitting Diode
Light emitting diode (LED), 277
Light source, *see* Lamp, Laser
Light scattering, *see* Doppler flowmetry
Light-tissue interactions, 29, 287, 288
Lipoprotein, 234
Liposomal vesicles, 234
Lithotripsy, 42, 204, 293
Luminescence, 117
    decay time, 119
Lumirubin, 181

Macula, 139
    xanthophyll, 139
Matrices
    fluorescence excitation-emission,
        129
Mean free path
    of absorption or scattering, 46
Mechanical rupture, 41
Melanin, 32
Melanoma, *see* Cancer
Melanosome, 288
Methotrexate, 169
Microfluorometry
    laser-assisted, 124
Microlaser, 271
    medical applications, 271
    Cr;Tm:YAG, 24
    Er:YAG, 24
    Ho:YAG, 23
    microchip, 20, 282
    mode-locking, 21
    non-planar ring oscillator, 19
    narrow linewidth, 25
    rare earth, 22
    short-pulse operation, 21

tightly-folded resonator, 19
tunable, 24
Microlense, 272
Microplasma, 39
Microscope
    confocal scanning laser, 223
    slit-lamp microscope, 207
Mode-locking, 40
Modelling
    optical, 45
    thermal, 55
Molecular denaturation, 33
Monoadduct, 168
Monoclonal antibodies, 234
Monte Carlo modelling, 48, 49, 76, 79, 111
Multiple photon absorption, 39
Mutagenesis, 215, 228

Naphthalocyanines, 231, 274
Nd:YAG laser, *see* Laser
Necrosis, 34
Neuron activity, 121
Neurosurgery, 295
Non-linear effects, 289
Nonlinear materials, 31

Ophthalmology, 272, 286
Ophthalmoscope
    laser scanning, 147
Optical breakdown, 40, 289, 290
Optical fiber, 91, 276
    ball tip, 250
    bundle, 18
    coherent, 146
    delivery, 249
    sapphire, 276
Optical parametric oscillator, 22, 272
Optical trapping
    of cells, 283
Optical window, 80
Optics of turbid tissues, 287
Orthopedics, 286
Oxyhemoglobin, 32
Oxymetry, 45, 74

Pattern recognition
    ICG angiogram, 143
Penetration depth, *see* Depth
Perfusion
    blood, 94
    imager, 89, 97
    monitor, 89
    muscular, 91
    tissue, 89
Pharmacokinetics, 120

Phosphorescence, 117
Photoacoustic effects, 289, 290
Photobleaching 39
Photocarcinogenesis, 193
Photochemical interaction, 36
Photochemical reactions, 177, 287, 289
Photochemotherapy, 165
Photocoagulation, 29, 197,213, 286
    retinal, 207, 208
    with diode lasers, 274
Photoconjugation, 166
Photodilatation, 292
Photodisruption, 41, 259, 289, 294
Photodisruptive surgery, 287
Photodissociation, 38
Photodynamic therapy, 36, 37, 79, 218, 223, 227, 237, 294.
    with diode lasers, 274
    with LEDs, 278
Photofrin II, 121, 227
Photoluminescence, 117
Photomarking
    selective, 119
Photomechanical mechanisms, 287
Photooxidation
    bilirubin, 183, 184
Photosensitivity, 230
Photosensitization, 238
Photosensitizer, 37, 287, 289, 294
Phototherapeutic window, 274
Phototherapy
    of neonatal jaundice, 178 (*see also* Hyperbilirubinemia)
        action spectrum, 187
        blanket, 188, 278
        high-intensity, 188
        lamp, 153
        mutagenicity, 186
        optimal light, 186
        side effects, 186
        with LEDs, 277, 278
    of psoriasis, 165 (*see also* Psoriasis)
        index, 157
        risk factors, 170, 171
        units, 161
Photothermal process, 31
    histological change, 33
Photothermal radiometry, 223
    pulsed, 224
Photothermolysis, 35, 213, 216, 286
    selective, 216, 287, 288, 292
        action spectrum, 217
            melanosome, 219
Phototoxicity, 293
Photovasodilatation, 291
Phthalocyanines, 223, 228, 231, 239,

Pigment epithelium, 139
Pigmented lesions, 218
Plaque
    atherosclerotic, 129, 249
Plasma
    formation, 289
    shielding, 41
    shock wave, 41
Plume
    laser, 215
p-n junction, 3
Polarization, 73
Population inversion, 3
Port-wine stains, 36, 213, 215
Power spectral density, 93
Protein
    denaturation, 34
Psoralen, 165, 168, 192
Psoriasis, 165
    phototherapy
        action spectrum, 155
            for clearance, 156
            mechanism of action, 166
Pulse
    soliton Raman compression, 21

Quantum yield
    bilirubin
        isomerization, 181, 182
        intersystem, 184
Quantum well lasers, 6
Quenching
    fluorescence of ICG, 143

Radiation
    dosage, 170
    ionizing, 122
    pattern, 7
    solar
        long-term effects, 192
    transport, 31, 45, 48, 75
    ultraviolet, 151, 165, 191
Radiometry
    photothermal, 223
Radiophotoluminescence, 122
Recanalization, 253
Reflectance
    analysis, 223
    profiling
        depth, 80
        surface, 80
    specular, 73
    spectroscopy, 282
    spectrum
        spatially resolved, 74

temporaly resolved, 74
time-gated, 80
time-resolved, 80
Regioselectivity, 184
Repair process, 288
Retinoids, 169
Ring oscillator
    non-planar, 19
Robotic, 213, 292
    laser, 293
Ruby laser, 213, see also Laser

Safety, 294
    phototherapy unit
        patient, 161
        staff personell, 162
Sapphire control probes, 250
Scarring, 213, 215, 252
Scattering
    anisotropic scattering, 47
    anisotropy factor, 46, 74
    back-, 51
    coefficients, 46
    elastic, 75
    forward-, 47, 49, 76
    isotropic, 47, 76
    lifetime, 31
    multiple, 90
    single, 90
    transport coefficient, 77
Semiconductor lasers, 3
Senile lentigines, 218
Servicing
    phototherapy unit, 161
Shock wave, 39, 259 287, 290
Side effects, 168, 186
    phototherapy
        jaundice, 186
        psoriasis, 165
        UVA tanning, 191
Singlet states, 117
Skin
    ageing, 192
    cancer, 192, 292
    carcinogenesis, 168
    irritability, 95
    optics, 45, 73
    ulcers, 292
Slab microlaser, 17, 18
Smart lasers, 293, 297
Solar radiation, 192
Speckle
    noise, 65
    pattern, 89
Spectroscopy
    autofluorescence, 223

fluorescence, 129
reflectance
time-dependent, 83
Stenosis, 256
Stratum corneum, 215, 224
Structural isomerization, 181
Sunburn, 191
Superpulse mode, 35, 214
Surgery
ablative, 36, 275
contact, 275
robotically-controlled, 213

Tanning
action spectrum, 193
UVA, 191
Tattoo, 213, 219, 220, 276, 293, 294
Telangiectasias, 215, 292
Telediaphanography, 101
Therapeutic window, 36, 274, 289
Thermal chemistry, 287
Thermal damage, 31, 4141, 52, 214-216,
251, 252, 256, 266, 289
Thermal interaction, 31
Thermal mechanisms, 287
Thermalization, 31
Thermal modeling, 55
Thermal necrosis, 292
Thermionic emission, 39
Thrombolysis, 291
Time of flight gating, 111
Time-resolved micro-spectrofluorimetry,
120
Tin-protoporphyrin, 173
Tissue
density, 51
optical properties, 227
oxygenation, 89
perfusion, 89
phantoms, 49
reflectance, 73
removal, 31
welding, 13, 275, 295
Titanium sapphire, see Laser
Tomosynthesis, 67
Toxicity, 166, 234
photodynamic therapy, 234
phototherapy of jaundice, 177
PUVA, 171
long-term, 166
Transfusion
exchange, 178
Transillumination, 101
computed tomography, 108
imaging, 101
window, 101

Transport
equation, 48, 51
scattering coefficient, 77
Triplet state
deactivation, 117
Tumor
localization, 45
targeting, 234
Tunable dye laser, see Laser
Tunable solid-state, see Laser
Turbid materials, 46
Turbid tissue, 288
Two-photon absorption, 289
Two-wavelength laser fluorescence
excitation, 125

Ulcer
laser debridement, 215
Ultrasound, 252
Ultraviolet radiation, see Radiation
Urinary bladder, 243
Urology
Nd:YAG surgery, 203
UVA
lamps, 153
tanning, 191
UV-B, 165, 191
induced erythema, 155
phototherapy, 171

Vacuolation, 34
Vaporization, 34, 287
Vascular lesions, 213, 218, 292
colon, 294
Vascular malformations, 215, 292
laser surgery, 215, 293
Vasculature
choroidal, 140
internal organ , 146
intestinal, 146
microvasculature, 91
retinal, 141
Vessel perforation, 262
Vibronic state, 31

Welding
of tissue, 275, 294
Wound healing, 296

Xanthophyl, 139, 208

The manufacturer's authorised representative in the EU is Springer
Nature Customer Service Centre GmbH, Europaplatz 3, 69115 Heidelberg,
Germany. If you have any concerns regarding our products, please
contact ProductSafety@springernature.com

Printed and bound by CPI Group (UK) Ltd, Croydon, CR0 4YY
23/04/2026
02095629-0015